Quantum Gravity Phenomenology II

Quantum Gravity Phenomenology II

Editors

Arundhati Dasgupta
Alfredo Iorio

Basel • Beijing • Wuhan • Barcelona • Belgrade • Novi Sad • Cluj • Manchester

Editors
Arundhati Dasgupta
University of Lethbridge
Lethbridge
Canada

Alfredo Iorio
Charles University
Prague
Czech Republic

Editorial Office
MDPI AG
Grosspeteranlage 5
4052 Basel, Switzerland

This is a reprint of articles from the Special Issue published online in the open access journal *Universe* (ISSN 2218-1997) (available at: https://www.mdpi.com/journal/universe/special_issues/Quantum_Gravity_Phenomenology_II).

For citation purposes, cite each article independently as indicated on the article page online and as indicated below:

Lastname, A.A.; Lastname, B.B. Article Title. *Journal Name* **Year**, *Volume Number*, Page Range.

ISBN 978-3-7258-2371-0 (Hbk)
ISBN 978-3-7258-2372-7 (PDF)
doi.org/10.3390/books978-3-7258-2372-7

© 2024 by the authors. Articles in this book are Open Access and distributed under the Creative Commons Attribution (CC BY) license. The book as a whole is distributed by MDPI under the terms and conditions of the Creative Commons Attribution-NonCommercial-NoDerivs (CC BY-NC-ND) license.

Contents

About the Editors vii

Preface ix

Arundhati Dasgupta and José Fajardo-Montenegro
Aspects of Quantum Gravity Phenomenology and Astrophysics
Reprinted from: *Universe* **2023**, *9*, 128, doi:10.3390/universe9030128 1

Surajit Kalita and Banibrata Mukhopadhyay
Massive Neutron Stars and White Dwarfs as Noncommutative Fuzzy Spheres
Reprinted from: *Universe* **2022**, *8*, 388, doi:10.3390/universe8080388 25

Sergey Cherkas and Vladimir Kalashnikov
Vacuum Polarization Instead of "Dark Matter" in a Galaxy
Reprinted from: *Universe* **2022**, *8*, 456, doi:10.3390/universe8090456 37

Nikola Paunković and Marko Vojinović
Equivalence Principle in Classical and Quantum Gravity
Reprinted from: *Universe* **2022**, *8*, 598, doi:10.3390/universe8110598 57

J. M. Carmona, J. L. Cortés, J. J. Relancio and M. A. Reyes
A New Perspective on Doubly Special Relativity
Reprinted from: *Universe* **2023**, *9*, 150, doi:10.3390/universe9030150 73

Sergey Cherkas and Vladimir Kalashnikov
Æther as an Inevitable Consequence of Quantum Gravity
Reprinted from: *Universe* **2022**, *8*, 626, doi:10.3390/universe8120626 83

Soham Sen, Sukanta Bhattacharyya and Sunandan Gangopadhyay
Path Integral Action for a Resonant Detector of Gravitational Waves in the Generalized Uncertainty Principle Framework
Reprinted from: *Universe* **2022**, *8*, 450, doi:10.3390/universe8090450 101

Olamide Odutola and Arundhati Dasgupta
Gauss's Law and a Gravitational Wave
Reprinted from: *Universe* **2024**, *10*, 65, doi:10.3390/universe10020065 111

Martin Bojowald
Space–Time Physics in Background-Independent Theories of Quantum Gravity
Reprinted from: *Universe* **2021**, *7*, 251, doi:10.3390/universe7070251 122

Giulia Maniccia, Mariaveronica De Angelis and Giovanni Montani
WKB Approaches to Restore Time in Quantum Cosmology: Predictions and Shortcomings
Reprinted from: *Universe* **2022**, *8*, 556, doi:10.3390/universe8110556 143

Suddhasattwa Brahma
Constraints on the Duration of Inflation from Entanglement Entropy Bounds
Reprinted from: *Universe* **2022**, *8*, 438, doi:10.3390/universe8090438 182

Nephtalí E. Martínez-Pérez, Cupatitzio Ramírez-Romero and Víctor M. Vázquez-Báez
Phenomenological Inflationary Model in Supersymmetric Quantum Cosmology
Reprinted from: *Universe* **2022**, *8*, 414, doi:10.3390/universe8080414 188

About the Editors

Arundhati Dasgupta

Arundhati Dasgupta is an associate professor at the University of Lethbridge, Canada. Her research is in theoretical physics, gravitational physics, quantum gravity, and cosmology. She is currently working on studying interactions of gravitational waves with matter, which will also provide clues to the existence of the quanta of gravity, the graviton. She is very active in international efforts to change the under-representation of women in physics. She is currently an elected member of the International Union of Pure and Applied Physics Mathematical Commission and Regional Councilor for Alberta, Nunavut and Northwest Territories, Canadian Association of Physicists.

Alfredo Iorio

Alfredo Iorio is a full professor of theoretical physics at Charles University, Prague, Czech Republic. His research is in quantum field theories, both at high and low energy, gravitational physics, and mathematical modeling of various phenomena, even beyond high energy theory, such as virus structures and DNA condensation. In the last decade, he has focused on analog gravity, proposing a prominent role of graphene, and related materials, for the reproduction of black hole phenomena. He created an innovative elementary school in Prague and founded a scientific consulting company where theoretical physicists face the big challenges of climate change.

Preface

1. Introduction

In 1915, Albert Einstein presented his original work on General Relativity to the Prussian Academy of Sciences. It was a new theory of nature, changing previous paradigms about gravity. Spacetime curvature and gravitation became synonymous in this theory and the new paradigm has survived direct experimental tests. Yet, a hundred years later, General Relativity cannot be theoretically quantized using standard methods, and we are still searching for the quanta of gravity.

A decade earlier than the discovery of General Relativity, Max Planck postulated that light was a stream of photons, which led to the discovery of the quantum world, and Einstein's paper on photoelectric effect had used that result. The theoretical discovery of gravitational waves had followed the formulation of General Relativity as early as 1916, and they were observed in LIGO, a sensitive laser interferometer, in 2015. However, the quanta of the waves, the gravitons, the counterpart of the photons for gravity, have not been observed, or even theoretically formulated.

The reason that quantum gravity has not been discovered is connected to what is known as the hierarchy problem of nature. Gravitational force is 10^{-38} orders weaker than electrodynamics. One has to probe length scales of 10^{-35} m and obtain energies of 10^{19} GeV in colliders to see experimental evidence of quantum gravity. Theoretically too, the standard quantization schemes require new techniques to be invented due to the non-polynomial and non-linearity of the Einstein action. However, Hawking radiation, area quantization, and black hole thermodynamics were predicted theoretically, which suggested quantum origins, but these require experimental confirmation. The search for primordial black holes is an effort in that direction. The theoretical computation of graviton interactions is non-renormalizable, which makes the particles more mysterious. There are various studies including very recent ones that argue that gravity is perhaps classical all the way through. This places gravity on a pedestal, different from the other interactions of nature.

With the discovery of very weak gravitational waves, in 2016, the search for gravitons received new impetus, and 'Quantum Gravity Phenomenology' could have a future. Precision experiments have given us access to the gravitational waves emitted from distant events. Gravitational waves are measured up to an amplitude of 10^{-21} using interferometers. As the precision measurements of nature head towards the quantum gravity scale, there might be indirect verification of the quantum gravity physics of the microscopic spacetime. Quantum phenomena might have significant effects at macroscopic and large scales, as these are emergent from the microscopic scale. The very metric which measures the curvature of spacetime is the classical limit of a quantum operator, and therefore semi-classical fluctuations should be visible at some length scales larger than the Planck scale. At this time, there are several competing theoretical formulations of quantum gravity, some extensions of standard versions, and some new ones. These include loop quantum gravity, path-integral quantization, discrete models, causal dynamical triangulation, asymptotically safe gravity, and causal set theory. Various formulations have matured to discuss observational predictions. In this Special Issue of *Universe*, we discuss some of these, and provide new insights into the future direction of this research area.

Of the various ways to verify the theoretical predictions of quantum gravity, two broad categories exist. These are the following: (i) Using analog models which simulate gravitational systems and the quantum phenomena associated with these systems. (ii) Direct and indirect evidence of quantum gravity predictions in natural experiments.

There has been considerable research in both of the above avenues for exploring the existence of new quantum physics for gravity. A volume dedicated to papers in this field is the need of the hour, and *Universe*'s topical collections serve that purpose.

When we were asked to suggest topics for a Special Issue for the journal *Universe*, the topic of 'Quantum Gravity Phenomenology' was a natural choice. There have been a number of papers addressing both ways of obtaining experimental evidence on the nature of quantum gravitational physics in the past few decades. When we embarked on this venture of editing a Special Issue on quantum gravity phenomenology, we wanted a clear direction to emerge in this field. We have collected 22 paper contributions, now published in two volumes. We would like to thank profusely our Assistant Editor from *Universe*, Ms. Cici Xia, who has made the two volumes of paper contributions happen. Initially our aim was to seek papers only on the above two approaches (i and ii) for obtaining observational physics in quantum gravity. However, over the course of the time we took to finalize the volumes, the focus has diversified. We are very thankful to the authors for publishing their papers in *Universe*. Although we are still inconclusive about the true nature of spacetime at quantum length scales, we have some highlights on the current status of research in this field. In the following, we offer some general considerations and briefly discuss on the papers in the two volumes, with Volume I being primarily focused on analog models and Volume II on astrophysics.

2. Observational effects of Quantum Gravity in Astrophysics and Cosmology (Volume II)

Astrophysics and cosmology are two fields where the large-scale structure of the universe is emergent. What would the physics of these systems have to do with physics of the quantum? The very gravitational metric which describes these observed in nature is expected to be emergent from an underlying theory of quantum gravity; thus, everything has origins in the quantum. There would be direct and indirect evidence from early formation stages of the universe as well as in strong gravity events of this microscopic structure. The photon theory of light was invented in order to describe the black body radiation spectrum, a macroscopic effect of the microscopic. We expect similar evidence in macroscopic geometry, of Planck-scale physics. The phenomena of Hawking radiation from black holes, and particle creation in curved spacetime, are signatures of quantization in curved geometry. A detailed discussion of this including the search of Hawking radiation appears in the review by Dasgupta and Fajardo-Montenegro. The paper also discusses macroscopic effects such as magnification of quantum effects in the presence of strong gravitating regions like black hole horizons. Cosmology describes the beginning of a universe from a point, and indeed the physics of this Planck size universe was quantum. A discussion on quantum cosmology can be found in the paper of Bojowald in this volume.

Apart from this, we have valuable contributions to various angles of probing for evidence of quantum gravity. From the papers published in Volume II, we found that the field of quantum gravity research still has lot of theoretical puzzles. The aspects discussed include the following: (i) The choice of frames of reference in formulating the Planck scale or doubly special relativity (Carmona et al.). (ii) The choice of time in evolving the system (Maniccia et al) using WKB approach in cosmology. (iii) The validity of the equivalence principle in the formulation of a quantum theory (Paunkovic and Vojinovic). These papers also describe the observational consequences in nature and therefore are important contributions to the future of phenomenology. In addition, a very interesting paper reports on the existence of Aether, which might provide a way for a choice of the preferred frame for quantum mechanics of gravity (Cherkas and Kalashnikov).

For direct observational effects, we have two intriguing papers outlining (i) the consequence of a generalized uncertainty principle on gravitational wave detection (Sen et al.) and (ii) the effect of non-commutative geometry in the existence of massive compact objects such as neutron stars and white dwarfs (Kalita and Mukhopadhay). We also have two contributions to the discussion from super-symmetric and string motivated theories of cosmology and entropy of curved space-time (Martinez-Perez et al. and Brahma). Whereas supersymmetry and string theory have not been detected yet, in the current scenario, these dominate the narrative of the theoretical study of physics at Planck energies.

Another avenue for direct observation is the search for gravitons, the quanta for the gravitational wave. Dasgupta and Fajardo-Montenegro as well as the Odutola and Dasgupta papers discuss semi-classical coherent states for the linearized gravity system, and predict numbers for the observations. The papers use the formalism of Loop Quantum Gravity to probe semi-classical geometry, and provide an introduction to this approach to quantum gravity. Whereas at the current length scales the actual gravitational 'quanta' detection cannot be predicted, semi-classical corrections from these can be obtained. These corrections can be observable, e.g., in the presence of a charged particle, tiny magnetic fields will be generated from the spacetime fluctuations. Quantum devices such as SQUIDS which have an accuracy of 10^{-18} Tesla currently can aid in the detection of these. We require a slightly more accurate SQUID, but in principle the use of high Coulomb charges could bring the gravitational semi-classical corrections to current levels of sensitiveness.

We think the phenomenological field in the study of use quantum devices and nanodevices in detecting the quantum effects of gravity has lot of promise. Recent experimental work on testing gravitational fields using these nanodevices has shown progress.

Finally in Volume II, we have papers which discuss dark matter and dark energy, without which the discussion of gravitational physics is not complete. One very interesting paper discusses how vacuum polarization (Cherkas and Kalashnikov), a quantum effect, can provide an explanation for observations that require the existence of dark matter. There are other approaches to quantization, like Causal Set theory and asymptotically safe gravity, which have not been discussed in this Special Issue. Several quantization programs have advanced over the years and discuss observational effects. At the current time, with the recent discovery of gravitational waves, we think the most promising quantum search will be for the graviton, and the physics predicted with the gravitational wave as interferometers deal with the quantum regime.

3. Concluding Remarks

In conclusion, we ask the same question we had asked some years earlier when we began editing the Special Issue on Quantum Gravity Phenomenology. What might be the unique experiment which confirms to us that there is a quantum of gravity? The analog models of gravity tell us that Hawking radiation is a phenomenon for systems with horizons. The graphene experiment, which is in progress, will hopefully shed light on thermal effects in curved spacetime. At the current length scales, the semi-classical corrections to matter energy dispersion have been predicted theoretically. However, any anomaly at the quantum scale has to be systematically detected and verified. Currently, big science's search for dark matter, gravitational wave detectors, and cosmological telescopes on satellites can also be used for quantum gravity detection.

Arundhati Dasgupta and Alfredo Iorio
Editors

Review

Aspects of Quantum Gravity Phenomenology and Astrophysics

Arundhati Dasgupta [1,*,†] and José Fajardo-Montenegro [1,2,†]

[1] Physics and Astronomy, University of Lethbridge, Lethbridge, AB T1K 3M4, Canada
[2] Departamento de Física, Universidad del Valle, Cali 760032, Colombia; jose.luis.fajardo@correounivalle.edu.co
* Correspondence: arundhati.dasgupta@uleth.ca
† These authors contributed equally to this work.

Abstract: With the discovery of gravitational waves, the search for the quantum of gravity, the graviton, is imminent. We discuss the current status of the bounds on graviton mass from experiments as well as the theoretical understanding of these particles. We provide an overview of current experiments in astrophysics such as the search for Hawking radiation in gamma-ray observations and neutrino detectors, which will also shed light on the existence of primordial black holes. Finally, the semiclassical corrections to the image of the event horizon are discussed.

Keywords: quantum gravity; astrophysics; quantum gravity phenomenology; loop quantum gravity; primordial black holes; Hawking radiation

1. Introduction

The gravitational quantum is still elusive experimentally and somewhat "elusive" theoretically [1–3]. In electrodynamics, the quantum of the electromagnetic wave is known as the photon, and we work with the interactions of photons to derive quantum electrodynamics (QED) phenomena. In the case of gravity, gravitational waves have been discovered 100 years after their prediction. The question is, are there "gravitons" or quanta of these waves? Like QED, one can define the "Fock" space quantization for the linearized Einstein equations and study free gravitons. However, introducing interactions with gravitons to study scattering amplitudes leads to uncontrollable infinities [3]. This is known as the "non-renormalizability" of perturbative quantum gravity. General relativity might be nonperturbative in the quantum regime, and the story of the quanta could be present in the geometry measurements of area and volume [4]. These "nonperturbative" theoretical explorations cannot be verified, as they are still in the realm of the microscopic Planck length regime of 10^{-35} m. We investigate the semiclassical fluctuations of the flat geometry using loop quantum gravity (LQG) coherent states and discuss whether that can be interpreted as a graviton quantum.

Further in the 1970s, the discovery of black hole thermodynamics and Hawking radiation were studied as "semiclassical phenomena", where gravity remained classical and other fields were quantum. The isolated black hole was found to have a temperature proportional to its surface gravity and entropy equal to its horizon surface area. For a solar-mass black hole, which might have formed using stellar collapse, this temperature is of the order of 10^{-8} K. If we observe the current-day black holes, then they are immersed in the background cosmic radiation, which has a temperature of 2.783 K. As the heat flows from higher to lower temperatures, the black holes would not radiate into the surroundings, and as of now, there is no experimental evidence of Hawking radiation. The study of black hole mergers using gravitational waves has provided evidence for the area increase theorem [5]. How would one obtain a verification of the temperature and radiative properties of black holes? The existence of *primordial black holes* (PBH) of small mass, originating in density fluctuations of the early universe, would allow for high-temperature black holes and Hawking decays in the form of gamma-ray bursts. The search for PBH has been a subject of experimental study [6]. We discuss this in some detail, and the approximations which describe the theoretical derivation of Hawking radiation are also discussed. The current experiments provide stringent restrictions on the PBH contributions to photon and neutrino fluxes observed on earth, as well as as fractions of dark matter [7–11]. Strangely, new observations from gravitational wave data suggest that there are subsolar mass black holes. Recent work tries to find the origins of these, either as PBH or from other processes without the Chandrasekhar limit in the collapse process [12]. Whereas this is very interesting, this is not exactly the realm of quantum gravity, though the research in this area might shed light on semiclassical aspects.

However, astrophysical phenomena, such as the black hole merger event, the collapse of a supernova to form a black hole, and neutron star mergers, are strong gravitational events. The energies at which the events happen have strongly coupled gravitational interactions. The quantum dynamics near these events is interesting, and even though the effect is weak, one can try and find indirect evidence in the observational data. Using LQG coherent states, some of these can be studied semiclassically. We discuss these and also comment on other observational results from the semiclassical gravity program for astrophysical observations, including that for the image of the event horizon [13,14]. There are several collaborations in quantum gravity phenomenology which, in particular, discuss Lorentz violations and quantum anomalies. The appropriate discussions on these topics can be found in [15]. For a previous comprehensive review on quantum gravity phenomenology, see [16]. One of the aims of this current review is to also provide a pedagogical introduction to some aspects such as the search for primordial black holes which is a very active field currently.

This review has discussions on the (i) graviton, (ii) Hawking radiation, and (iii) semiclassical corrections to strong gravity systems such as the event horizon. The following section discusses the theory of the graviton and the experimental bounds. Section 3 describes the phenomena of Hawking radiation, as well as the experimental efforts to detect the emitted particles from PBH. Section 4 describes the physics of the event horizon and quantum correction predictions to the same. The final section concludes with the present status of the field of research in the above and future avenues of quantum gravity phenomenology.

2. Graviton

The electromagnetic (EM) wave is a solution to Maxwell's equation and is observed in nature. The visible spectrum is known as light, the infrared, which we interpret as heat, and radio waves. The ultraviolet radiation is also detectable and useful as are X-rays in many practical day-to-day events. These, when quantized, give us the photon description of the EM wave, and represent the source-free "free" EM fields. The actual production of EM radiation is from accelerated charges, but as the waves propagate out in space, they can be studied as "free" EM fields. In the case of gravity, Einstein's action is nonlinear, and the gravitational field has self-interactions. To find the "free" plane wave which propagates on its own, we take a linearized gravity, "weak fluctuations" over a flat background. Nonperturbative waves, produced using strong gravitational interactions have been studied in [17]. As the linearized gravitational waves represent classically "free" fields, one would expect that the Fock space quantization of these would be obtained similarly to the photon quantum electrodynamics description. However, herein lies the problem: the graviton theory is a nonrenormalizable theory [3]. Is it because the graviton vacuum, which represents the Minkowski spacetime is not a vacuum? Is flat space really a vacuum state in a true theory of quantum gravity? Can we have a perturbation over the flat-space system and describe a graviton as a quantum state in the flat-space background? In the case of the EM theory, the EM field propagates in a flat background that, however serves as a noninteractive arena for the EM fields to propagate. The photon is created and annihilated out of the QED vacuum, which is a state with the photon quantum number as zero. In the following, we discuss whether seeking a similar quantum field vacuum for the graviton is relevant. We also discuss the question of which physics of the systems we should experiment for the observation of the graviton.

2.1. The Linearized Theory of the Graviton

In the following, we discuss Einstein's theory of the linearized metric. The field equations for the Einstein action is "free" in its gauge-fixed form; however, if we try to write the full Einstein Lagrangian for the gravitational field, then there are interaction vertices to all orders for the graviton. The quantum amplitudes including these interactions do not converge, and neither can the theory be renormalized using standard techniques. To begin with, we write the metric of spacetime $g_{\mu\nu}$ as a flat space $\eta_{\mu\nu}$ and a weak fluctuation $h_{\mu\nu}$.

$$g_{\mu\nu} = \eta_{\mu\nu} + h_{\mu\nu}. \tag{1}$$

It is assumed that $|h_{\mu\nu}|_{max} \ll 1$ (μ, ν, α, β etc. $= 0, \ldots, 3$). Note that using standard convention, the metric is dimensionless and the amplitude of the fluctuations are defined using the absolute maximum value. From experiments [1], we are aware now that the amplitude of the "gravitational wave" is of the order of 10^{-22} as received on earth. One can write the Einstein Lagrangian density as a function of this metric, its determinant g, and scalar curvature R,

$$\mathcal{L} = \sqrt{g}\,R = -\frac{1}{2}\sqrt{-1+h}\left[(h^{\mu\nu})(\eta^{\alpha\beta}\partial_\alpha\partial_\mu h_{\nu\beta} - \Box\, h_{\mu\nu})\right]. \tag{2}$$

In the above, we have kept the terms in the Lagrangian which are quadratic in $h_{\mu\nu}$. The linear terms of the form $\eta^{\mu\nu}\eta^{\lambda\rho}\partial_\rho\partial_\mu h_{\lambda\nu}$ are total derivatives and contribute only at the boundaries, which we ignore. Further, $\Box \equiv \eta^{\alpha\beta}\partial_\alpha\partial_\beta$, and h is the trace of $h_{\mu\nu}$. The equation of motion from the above to a linear order in "$h_{\mu\nu}$" is

$$\eta^{\alpha\beta}\partial_\alpha\partial_\mu h_{\nu\beta} - \Box\, h_{\mu\nu} = 0. \tag{3}$$

This still has a gauge degree of freedom due to diffeomorphism invariance, which can be fixed by putting the $\partial^\alpha h_{\alpha\beta} = 0$ restriction on the linearized metric. The equation of motion reduces to a "wave equation"

$$\Box\, h_{\mu\nu} = 0. \tag{4}$$

The solution for this is a transverse wave (due to Lorentz's condition) and has two polarizations as additional restrictions to fix the residual gauge freedom keeping only two [18]. The two polarizations are taken as $h_+ = A_+ \cos(\omega z - \omega t)$ and $h_\times = A_\times \cos(\omega z - \omega t)$, if it is propagating in the z-direction [18], with angular frequency ω and amplitude $A_+, A_\times$. The question is: can these waves, when quantized, give us "quanta" as it is possible for photon quantization? In other words, can one define a Fock space representation for the perturbative Hilbert space of Einstein's gravity? The answer is surprisingly difficult, as the Einstein action introduces self-interactions of the gravitons to all orders, which cannot be renormalized using standard field theory techniques. The gravitational propagator can be calculated, but the quantum corrections cannot be made finite using regularization and renormalization techniques. One can see the origin of self-interactions even at this order in the Lagrangian in Equation (2) as the nonpolynomial "measure" $\sqrt{-1+h}$ can give rise to the interaction terms upon expanding the square root. A simple "degree of superficial divergence" counting of the gravitational perturbative Feynman diagram gives the number as $D = 2(k+1)$, where k is the number of independent momentum interactions [19]. This number therefore increases with the number of loops in the scattering calculations and cannot be absorbed by redefining the bare Lagrangian. For Yang–Mill's (YM) theory the same degree is given as $D = 4 - L_e$, where L_e is the number of external legs of the Feynman diagram. The YM theory is therefore renormalizable, as the number of terms in the Lagrangian which need to be renormalized is finite ($0 < L_e < 4$). One can use asymptotic techniques to obtain a renormalizable effective Lagrangian for gravity, but we do not discuss this in this review [20]. However, can there be a "free" graviton theory where we can ignore all the interactions? Up to a certain length scale, a "free graviton" quantization can be formulated, but the entire theory is also complicated by the definition of the "gravitational vacuum". In the theory of gravitational physics, the metric is the basic degree of freedom, and the graviton is a "perturbation" over the flat-space geometry. In a true quantization of the theory, the flat spacetime geometry is also an emergent "metric". If the metric is an operator, then causality and therefore quantization is not defined. The vacuum likely is the state with no metric or the state that is such that

$$\hat{g}_{\mu\nu} |0\rangle = 0. \tag{5}$$

There have been several attempts to obtain the perturbative quantum state using a polymer state in the nonperturbative quantization framework of loop quantum gravity. We report on those works briefly and then describe a semiclassical description of a "gravitational wave" using LQG. It remains though that the most complicated aspect of Einstein's gravity is the fact that the field which has to be quantized is the metric of the spacetime, the causality of the system is complicated by the quantization, and macroscopic configurations have to be emergent.

2.2. Gravitons in Loop Quantum Gravity

It was shown in [21] that the SU(2) generators of the loop quantum gravity (LQG) variables decouple into three independent gauge generators in the linearized approximation. In LQG, the basic variables are obtained from the ADM formulation of the canonical gravity. The spacetime is foliated by spatial slices Σ with a timelike normal vector along the fourth direction, specified using the coordinate t. The induced three-metric on Σ_t is given as q_{ab}, $(a, b = 1, 2, 3)$; the metric in the ADM formulation is given as

$$ds^2 = -(N^2 + N^a N_a)dt^2 + N^a dx_a dt + q_{ab}dx^a dx^b, \tag{6}$$

where N^2 is the lapse, N^a is the shift, and q_{ab} is the induced metric of the time slices Σ_t. The second fundamental form of this metric is $K_{ab} = \mathcal{L}_t q_{ab}$ and is the extrinsic curvature tensor which characterizes the embedding of the slice.

The LQG variables are defined using the soldering forms e_a^I which connect the tangent space ($I = 1, 2, 3$) of the three slices to the world volume. The canonical variables are defined as

$$e_a^I e_{bI} = q_{ab}, \quad E_I^a E^{bI} = q\, q^{ab}, \quad A_a^I = \Gamma_a^I - K_{ab} E^{bI}, \tag{7}$$

where e_a^I is the triad, E_I^a are densitized triads, and A_a^I have the properties of a connection due to their definition in terms of the spin connection Γ_a^I and the extrinsic curvature tensor K_{ab}. The details of the variables can be found in [22]. There is usually an Immirzi parameter in the definition of the gauge connection, and this reflects an ambiguity in the system. We chose to set it to one, for the purpose of this paper. The internal indices I transform in the SU(2) group, which is isomorphic to the group of rotations in the three-dimensional tangent space [22]. The generators of the transformations in the internal directions are the Gauss constraints

$$\mathcal{G}^I = \partial_a e^{aI} + \epsilon^{IJK} e_J^a A_{aK}. \tag{8}$$

In the linearized approximation, $q = 1$, $q^{ab} = \delta^{ab} + h^{ab}$ and $A_a^I = 0$, if one keeps the constraint up to a linear order in the fields, the constraint algebra commutes, i.e.,

$$\mathcal{G}^I_{\text{Lin}} = \partial_a(\delta e^{aI}) + \epsilon^{IJK} \delta_J^a \delta A_{aK}, \tag{9}$$

where due to the linearized metric, one has

$$e^{aI} = \delta^{aI} + \delta e^{aI}, \quad A_{aK} = 0 + \delta A_{aK}, \tag{10}$$

and
$$h^{ab} = \delta e^{aI} \delta^b_I, \tag{11}$$

$$\{\delta e^I_a(x), \delta A_{Kb}(y)\} = \kappa \delta^3(x-y) \delta^I_K \delta_{ab}, \tag{12}$$

where κ is related to Newton's constant G [22,23]. The δe^I_a and the δA_{Kb} are the linearized dynamical fields, which are quantized. In the limit $\kappa \to 0$,

$$\{\mathcal{G}^I_{\text{Lin}}, \mathcal{G}^J_{\text{Lin}}\} = 0. \tag{13}$$

Interestingly, if one keeps the next order in the constraint definition, the algebra is not zero to a linear order as the Poisson bracket gives a linear result in the fields.

$$\mathcal{G}^I_{\text{Lin}} = \partial_a(\delta e^{aI}) + \epsilon^{IJK}\left(\delta^a_J + \delta e^{aJ}\right)\delta A_{aK}, \tag{14}$$

and

$$\{\mathcal{G}^I_{\text{Lin}}, \mathcal{G}^J_{\text{Lin}}\} = \kappa\left(\delta A^{IJ} - \delta^{IJ}\delta A^b_b\right)\delta^3(x-y). \tag{15}$$

This term would go to zero in the limit $\kappa \to 0$. To avoid these confusions about the algebra and also questions about the Minkowski "quantum state" about which perturbation is being performed, we use the full $SU(2)$ degrees of freedom and imposed the linear metric only in the semiclassical approximation. The details of the calculations appear in [24].

For the polymer quantization of linearized gravity using the $U(1) \times U(1) \times U(1)$ Hilbert space, one can use the work of [25]. This approach is based on the linearized algebra of LQG variables, as given in Equation (13). The LQG phase space thus has a $U(1) \times U(1) \times U(1)$ symmetry in the linearized approximation, instead of the full $SU(2)$. The Hilbert space quantum states are of the form

$$|\vec{\alpha}, \{q\}\rangle = |\alpha_1, q_1\rangle|\alpha_2, q_2\rangle|\alpha_3, q_3\rangle, \tag{16}$$

where $|\alpha_i, q_i\rangle$ are elements of a $U(1)$ Hilbert space. q_i label integers and α labels the discrete network. The flux operator defined in terms of the triads is given as [25]

$$X^a_{\vec{\alpha},\{q\}(r)}(\vec{x}) = \sum_I q_I \int ds_I(\vec{e}_I(s^I), \vec{x}) \dot{e}^a_I, \tag{17}$$

where s_I is a surface in three dimensions, which the discrete edge e_I of the graph α intersects once.

The Fock space quantum vacuum for the graviton is a transform of the state in Equation (16). Whether this facilitates further study of the perturbation theory of the graviton is yet to be investigated. The transform is given as

$$\Phi_0 := \sum_{\alpha,q} c_{0\vec{\alpha},\{q\}}\langle\vec{\alpha}, q|, \tag{18}$$

where

$$c_{0\vec{\alpha},\{q\}} = \exp\left(-\frac{1}{4}\int d^3x\, G^{\vec{\alpha},\{q\}(r)}_{ab}(\vec{x}) * X^{ab}_{\vec{\alpha},\{q\}(r)}(\vec{x})\right), \tag{19}$$

where these are "smeared" operators in the LQG polymer space, and r is a measure of the Gaussian smearing ($X_r(\vec{x}) = \int d^3y X(\vec{y})\exp(-|\vec{x}-\vec{y}|^2/2r^2)/((2\pi r^2)^{3/2})$).

$$X^{ab}_{\vec{\alpha},\{q\}(r)} = \sum_i X^a_{\alpha_i,q_i}\delta^b_i. \tag{20}$$

The $G^{\vec{\alpha},\{q\}(r)}_{ab}(\vec{x})$ is related to the flux of the two "graviton" polarizations in the light cone. We refrain from getting into the details of the above, but the reader is urged to follow the details of the derivation in [25,26]. Whereas this approach to obtaining a "quantum" of linearized gravity is technically rather involved and involves an additional scale "r" apart from the usual discretization of quantum variables, it is believed to give a polymer representation of the "graviton".

The expectation values of the operators are preserved in the transform and therefore one loop corrections to the graviton propagator can be tested. A derivation of a one-loop correction using a perturbation of reduced loop quantum cosmology states exists in [27]. Another reference for the reduced phase-space quantization of linearized gravitational waves is [28]. Moreover, a more recent work uses the free graviton Lagrangian and obtains a "polymer state" for the same. This approach obtains some corrections to the gravitational wave propagator [29]. However, in none of the above papers the emergence of the background Minkowski metric is discussed. The self-interaction of gravitons is also not obtained to all orders, as predicted by the Einstein Lagrangian. In the next section, we try to find some phenomenological implications of the graviton's existence in observational data.

2.3. Gravitons in Semiclassical Gravity

In this subsection, we derive the semiclassical phase space of the gravitational wave metric and obtain a coherent state for the system using the techniques of [22,24]. To begin with, we find the triads for the metric and the LQG gauge connection, which are the classical variables for the system. The details can be found in [24]. The spatial metric for a standard gravitational wave metric in the tt-gauge is (the lapse is one and shift is zero in the ADM form of the four-metric)

$$q_{ab} = \begin{pmatrix} 1+h_+ & h_\times & 0 \\ h_\times & 1-h_+ & 0 \\ 0 & 0 & 1 \end{pmatrix}. \tag{21}$$

In the process of obtaining the coherent state for the above metric, we identify the classical phase space in terms of the LQG variables [30]. The triads $e_a^I e_{bI} = q_{ab}$ are obtained as

$$e_a^I = \begin{pmatrix} \sqrt{\frac{1-(h_+^2+h_\times^2)}{2(1-h_\times)}} & \sqrt{\frac{1-(h_+^2+h_\times^2)}{2(1-h_\times)}} & 0 \\ \frac{1-(h_\times-h_+)}{\sqrt{2(1-h_\times)}} & \frac{-1+(h_\times+h_+)}{\sqrt{2(1-h_\times)}} & 0 \\ 0 & 0 & 1 \end{pmatrix} = \begin{pmatrix} \frac{1}{\sqrt{2}}+\frac{h_\times}{2\sqrt{2}} & \frac{1}{\sqrt{2}}+\frac{h_\times}{2\sqrt{2}} & 0 \\ \frac{1}{\sqrt{2}}+\frac{1}{\sqrt{2}}(h_+ - \frac{h_\times}{2}) & -\frac{1}{\sqrt{2}}+\frac{1}{\sqrt{2}}(h_+ + \frac{h_\times}{2}) & 0 \\ 0 & 0 & 1 \end{pmatrix}. \tag{22}$$

Obviously, in our gauge choice, the triad is not diagonal at the zeroth order. The extrinsic curvature of the metric is obtained using the definition $K_{ab} = -\partial_t q_{ab}$, and the SU(2)-valued gauge connections defined in Equation (7) are:

$$\begin{aligned}
A_x^1 &= -\frac{1}{2\sqrt{2}}(\partial_z h_\times + \partial_z h_+) = A_y^2 \\
A_y^1 &= -\frac{1}{2\sqrt{2}}(\partial_z h_\times - \partial_z h_+) = -A_x^2 \\
A_z^1 &= A_z^2 = A_x^3 = A_y^3 = 0 \\
A_z^3 &= \frac{1}{2}\partial_z h_+.
\end{aligned}$$

We also computed the nonzero spin connections for this metric [30]. Next, we take a discretization of the background geometry. This smearing of variables is required to obtain smooth commutators of the quantum theory, instead of distributional delta functions. For details, see [22], and the smearing of the gauge connection on one-dimensional curves gives holonomies which involve path-ordering.

$$h_e(A) = \mathcal{P}\exp\left(\int A\right). \tag{23}$$

The discretization is not dictated by the theory but is motivated from the flat geometry of the classical three-metric. We take a planar graph, which form a cubic 3-d polyhedronal decomposition of the three-geometry, as shown in Figure 1. Therefore, there are six links and/or six faces meeting at a given vertex.

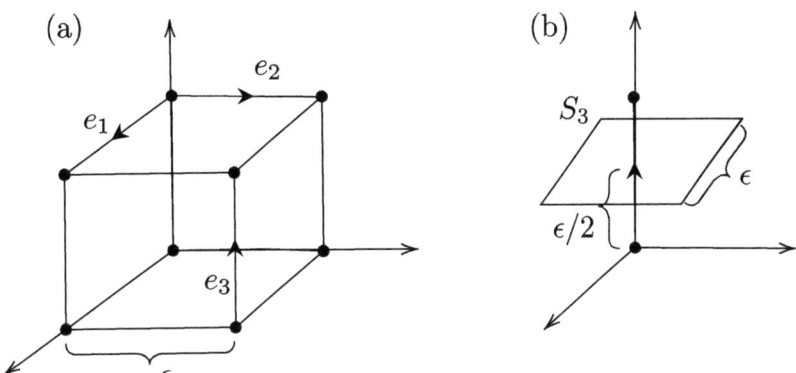

Figure 1. (a) Building block for the decomposition of the 3-geometry. (b) Example of one of the smearing surfaces to calculate the momenta.

The holonomies and the momentum are calculated as smeared along the one-dimensional edges of the graph, and the two-dimensional faces of the cube which the links intersect precisely at one point. These calculations are done using the techniques of [24]. The holonomies of the three independent links in the x, y, and z directions and the corresponding momenta are given up to a linear order in the amplitudes A_+, $A_\times$,

$$h_{e_x} = 1 - i\frac{\epsilon}{2} A_x^I \sigma_I \qquad (24)$$

$$h_{e_y} = 1 - i\frac{\epsilon}{2} A_y^I \sigma_I \qquad (25)$$

$$h_{e_z} = 1 + i\frac{A_+}{2} \sin\left(\omega\left(z_0 - t_0 + \frac{\epsilon}{2}\right)\right) \sin\left(\frac{\epsilon}{2}\right) \sigma_3, \qquad (26)$$

where one has taken a vertex at (x_0, y_0, z_0) and the links are of width ϵ. σ_I are the Pauli matrices. Next, one takes the faces centred at the middle of the links, i.e., at $x_0 + \epsilon/2$, $y_0 + \epsilon/2$, and $z_0 + \epsilon/2$, and of area ϵ^2. The momenta are labelled by the edges which intersect the faces. The momenta are defined as $P_e^I = \frac{1}{\kappa} \int_{S_e} *E^I$.

$$P_{e_x}^1 = \frac{1}{\sqrt{2}\kappa}\left(\epsilon^2 + \frac{\epsilon^2(A_\times)}{2}\cos(\omega(z_0 - t_0))\right) \qquad (27)$$

$$P_{e_x}^2 = \frac{1}{\sqrt{2}\kappa}\left(\epsilon^2 + \frac{\epsilon^2(2A_+ - A_\times)}{2}\cos(\omega(z_0 - t_0))\right) \qquad (28)$$

$$P_{e_y}^2 = \frac{1}{\sqrt{2}\kappa}\left(-\epsilon^2 + \frac{\epsilon^2(2A_+ + A_\times)}{2}\cos(\omega(z_0 - t_0))\right) \qquad (29)$$

$$P_{e_y}^1 = \frac{1}{\sqrt{2}\kappa}\left(\epsilon^2 + \frac{\epsilon^2(A_\times)}{2}\cos(\omega(z_0 - t_0))\right) \qquad (30)$$

$$P_{e_z}^3 = \frac{1}{\kappa}\epsilon^2. \qquad (31)$$

As the densitized triads are smeared over two-dimensional areas and acquire dimensions, the momenta are defined with the dimensional constant $1/\kappa$, $\kappa = 8\pi G/c^3$ to make the variables dimensionless. In the quantum version, this acquires the role of $1/\hbar\kappa = 1/l_p^2$, where l_p is the Planck length. The coherent states are defined as peaked at the classical values of a complexified SL(2,C) element as specified by Hall [31],

$$g_e = \exp(iT^I P_e^I) h_e,$$

and a detailed coherent state can be written for the above classical phase space, now described only using the discrete one-dimensional smeared holonomies and corresponding momenta. Note these "coherent states", as defined in [22] for LQG, are representative semiclassical states and are not exactly identifiable as "coherent states" as in completely solvable Hamiltonian systems. However, these states have minimal uncertainty in the time slice they are defined in. Next, we calculate the semiclassical corrections to the geometry by using the results of [13]. The coherent states are given for one such discrete element e and the LQG smeared variables as,

$$\psi^t(g_e, h_e) = \sum_j (2j+1) \exp(-\tilde{t} j(j+1)/2) \chi_j(g_e h_e^{-1}), \qquad (32)$$

where $\chi_j(h_e)$ is the character of the jth irreducible representation of SU(2). One can find the expectation value of the momentum operator $\hat{P}_e^I$ in this state, and one obtains it to the first order in the semiclassical parameter $\tilde{t}$ [13]

$$\langle \psi^t | \hat{P}_e^I | \psi^t \rangle = P_e^I \left(1 + \frac{\tilde{t}}{P_e} \left(\frac{1}{P_e} - \coth(P_e) \right) \right) = P_e^I (1 + \tilde{t} f(P_e)), \qquad (33)$$

where $P_e = \sqrt{P_e^I P_e^I}$ and $f(p) = (1/p)(1/p - \coth(p))$. Therefore, one can calculate the semiclassical corrections to the metric of the classical gravitational wave, if one writes a coherent state for each discrete element e which comprises the entire Minkowski three volume divided into cubic cells as in the figure. The vertices of the cube which are shared by three+three coherent states and these can have SU(2) intertwiners [32], but the nature of the corrections remain the same. Note these coherent states are not exactly similar to the coherent states for photons, which are Abelian. These coherent states are non-Abelian in nature.

In fact, if we take the pure Minkowski space and use the coherent state as a measure of the quantum fluctuation, what would we generate as the corrected metric? All the P_e^I's for the Minkowski metric can be obtained as given above and, in the limit, $A_{+,\times} = 0$ would represent the Minkowski metric. In this particular gauge, the corrections generate semiclassical fluctuations in the η_{xx}, η_{yy}, and η_{zz} components but not in the η_{xy} directions

Next, we discuss the fluctuations to the gravitational wave metric as generated from the coherent state which peaks at the gravitational wave metric. Obviously, the metric would fluctuate and generate semiclassical corrections to the geometry at order $\tilde{t}$. We set the semiclassical parameter (which has to be dimensionless) as a ratio of the Planck scale to the gravitational wave, wavelength, or $\tilde{t} = l_p^2/\lambda^2$. We take the wavelength as that is the length scale which characterizes the wave system. A relevant-frequency gravitational wave, which might generate detectable semiclassical fluctuations, has to be of very high frequency. Let us say a 10^{35} Hz gravitational wave will have the semiclassical parameter as $\tilde{t} \approx 10^{-16}$.

In the above, have we predicted a "quantum origin" of the gravitational wave that would comprise the "graviton"? Obviously, the story is not about particles in gravitational physics, or matter quanta, but the quantum of geometry. The tiny area measurements in each basis state of the operator $\hat{P}_e^I$ represent the "graviton", the condensate of which is represented by the coherent-state wave packet. It thus remains that from our perspective the Minkowski geometry is not the gravitational vacuum, but also emergent from a semi classical state. Therefore, one should not confuse the quantum gravity vacuum state with the "matter vacua". We suggest two ways to search for quantum gravity bounds/origins in a gravitational wave experiment:

(i) As the coherent states are non-Abelian in nature, the expectation values of operators have semiclassical corrections which originate due to self-interactions. These can be detected for high-frequency gravitational waves.

(ii) The search for individual "gravitons" or quanta of geometry would require much more precise instruments, able to resolve the coarse-graining of geometry itself.

The latter (ii) will require further investigations, in particular about what the dynamical fundamental "quanta" of LQG is. One also has to find if there is a gauge invariant observable which is measurable in experiments. Our questions seem to seek answers by quantizing matter and the gravitational degrees of freedom simultaneously. *However, due to the hierarchy problem, it is preferred that matter is quantized and the gravitational degrees of freedom are semiclassical in the current epoch.* In the combined Hilbert space of the matter and gravitational degrees of freedom $H_{\text{matter}} \otimes H_{\text{grav}}$, the combined matter–gravity state should be taken as

$$|\Psi\rangle = |\psi_{\text{matter}}\rangle \otimes |\psi^{\text{grav}}_{\text{semiclassical}}\rangle. \tag{34}$$

For previous work in adding matter interactions in LQG, refer to [23].

Using criterion (i) and the idea that matter quanta interact with gravitational degrees of freedom at semiclassical length scales, one finds that the semiclassical fluctuations of the metric are relevant. We therefore calculate the metric corrections as predicted from the coherent states for LQG constructed by Thiemann, Winkler, [22] and as observed in [13]. They emerge as

$$g_{xx} = (1+h_+)(1+2\tilde{t}\,f(P_{e_x})) \tag{35}$$
$$g_{yy} = (1-h_+)(1+2\tilde{t}\,f(P_{e_y})) \tag{36}$$
$$g_{xy} = h_\times(1+\tilde{t}\,f(P_{e_x})+\tilde{t}\,f(P_{e_y})) \tag{37}$$
$$g_{zz} = 1+t\,f(P_{e_z}). \tag{38}$$

The gauge invariant momenta are found to be:

$$P_{e_x} = \frac{\epsilon^2}{\kappa}\left(1+\frac{1}{2}h_+\right) \tag{39}$$

$$P_{e_y} = \frac{\epsilon^2}{\kappa}\left(1-\frac{1}{2}h_+\right) \tag{40}$$

$$P_{e_z} = \frac{\epsilon^2}{\kappa}. \tag{41}$$

The continuum limit is obtained using $\lim_{\epsilon \to 0} P_e/\epsilon^2$. This gives the metric fluctuations at a location (x_0, y_0, z_0) and one can solve the propagation of matter in this corrected metric. As evident in the continuum limit, the corrections are functions of the classical triads, and thus dependent only on the z coordinate. Moreover, the corrections are relevant only at one instant $t = t_0$ of the spacetime. For a 100 Hz frequency, the gravitational wave will have a semiclassical correction of the order of 10^{-84}, which is way smaller than the gravitational wave amplitude. If one probes higher-frequency gravitational waves, and therefore shorter wavelengths, the Planck scale coarse-graining will start manifesting itself and the effects might be evident in a gravitational wave detector. The Minkowski metric is also corrected semiclassically, and one can probe these using quantum fields in these geometries.

2.4. Summary

In this section, we gave a "semiclassical" state which could describe a gravitational wave at one instant. It predicted fluctuations which could be measurable for high-frequency waves $\geq 10^{30}$ Hz. These frequencies were way above the ones observed in the LIGO detectors. From the current observation of gravitational waves, there are bounds on the "graviton mass". From LIGO, the bound is 1.2×10^{-22} eV. This bound does not shed light on the origins of the mass from the methodology. Theoretically, the graviton mass can originate from quantum corrections to the Einstein theory, as well as from matter interactions which preserve diffeomorphism invariance. In this review, we do not discuss massive gravitons.

3. Search for Hawking Radiation and Primordial Black Holes

The discovery that quantum mechanics near black hole horizons results in particle creation originates in the paper by SW Hawking [33]. In that paper, a quantum field vacuum was time-evolved in the collapsing geometry of a star. The quantum state evolved into a thermal state, with a temperature inversely proportional to the mass of the black hole. In [33], it was shown that the exact temperature of a solar-mass black hole was 10^{-8} K. However, it would not radiate into the surrounding, which was at 2.78 K. This led to the search for black holes with mass $\sim 10^{14}$ g, and these could have formed in the early universe. Due to the Chandrasekhar limit, astrophysical black holes have a bounded mass if formed from stellar collapse. On the other hand, early universe density fluctuations can lead to the formation of tiny black holes, with horizon size fractions of a millimetre. These black holes have intrinsic temperatures higher than the current CMB temperature of 2.78 K. Even if the early universe had been hot, as the primordial universe cooled down, these black holes would start radiating and evaporate eventually or form Planck size remnants.

3.1. Formation of Primordial Black Holes (PBH)

The story of the collapse of matter to form black holes is well-studied in the work of Choptuik [34]. Scalar data in an initial slice undergo collapse, and the mass of the black hole formed has a scaling equation. This physics is true for early universe cosmology. It is noted that the matter undergoing collapse is taken as dust in most calculations and the Fermion/quark composition (required for the Chandrasekhar limit) of the cosmic soup is mostly ignored. For a comprehensive review of primordial black hole formation, one is referred to [6]. Here, we briefly outline the methods used to study matter collapse in the early universe. One of the main ingredient in the study of collapse in the early universe is Jean's instability. This instability characterizes density fluctuations in a fluid. The formula for Jean's instability is obtained by equating the time for free fall (or the time taken for an object of radius R to collapse under its own gravity) to the time taken by a sound wave to cross the radius. It is therefore a critical radius for which a pressure wave in the fluid gets trapped. Jean's critical length can also be obtained by solving for perturbations flowing in a fluid and the self-gravitational force generated by the perturbation. In the following, we discuss Jean's instability.

3.2. Jean's Instability

In this section, we discuss the collapse in a fluid of density ρ. This process also gives a rough description of the physics of a "density" collapsing under "perturbations" or under its own weight. The time for "free fall" of a mass in an elliptic orbit of eccentricity one according to Kepler's laws (of planetary motion) is

$$\tau^2 = \frac{\pi^2}{2} \frac{R^3}{GM}, \tag{42}$$

where M is the mass causing the orbit, and R is the distance from the focus of the ellipse. We use this to model self-collapse of a mass under its own gravity. If the mass collapses then only half of this time is taken. Given that the total mass in a radius R of a spherical distribution of constant density ρ is

$$M = \frac{4\pi}{3} R^3 \rho, \tag{43}$$

approximating the mass using this formula, the time for free fall is given as a function of density as

$$\tau = \sqrt{\frac{3\pi}{32G\rho}}. \tag{44}$$

If the speed of sound in the fluid is c_s, then the time for sound to flow through a distance R is

$$\frac{R}{c_s}. \tag{45}$$

This time would be the same as that a pressure wave flowing through the medium would take. If the gravitational collapse time is greater than the pressure wave time, the mass is unstable, and the critical length scale of the fluid region is given as

$$R_{JL} = \left(\frac{3\pi}{32}\right)^{1/2} \frac{c_s}{\sqrt{G\rho}}. \tag{46}$$

The same "collapse formula" can be derived using a spherical homogeneous mass M, whose radius increases by a perturbation $\Delta R = -\alpha R$, where α is a small perturbation. The change in pressure using the formula $\delta p/\delta \rho = c_s^2$ can be related to the change in density due to the compression, and this gives rise to a force and "acceleration" obtained as

$$a_p = \frac{\delta p}{\rho_0 R} = \frac{3\alpha c_s^2}{R}. \tag{47}$$

In the above, we took $\delta \rho = 3\alpha \rho_0$, where ρ_0 is the original density. Simultaneously the shrinking of the radius gives rise to an increase of the Newtonian acceleration

$$a_g = \frac{2GM\alpha}{R^2}. \tag{48}$$

If the gravitational acceleration exceeds the "pressure acceleration", the mass is expected to collapse, which gives a critical length

$$\frac{3\alpha c_s^2}{R_C} = \frac{2GM\alpha}{R_C^2} = \frac{4\pi}{3}\rho_0 R_C^3 \frac{2G}{R_C^2} \rightarrow R_c \propto \frac{c_s}{\sqrt{\rho_0 G}}. \tag{49}$$

Thus, the critical radius for the collapse in a fluid of density ρ is proportional to the speed of pressure waves c_s in the medium. Here, one of the important assumptions for the calculation of the speed of sound is the assumption that for the early universe fluid, entropy is conserved. We next discuss if a change in the description of the fluid of the early universe might change this Jean's length. The above discussion on Jean's instability can be found in many references, including [35,36].

3.3. A Quantum Entropy Production Fluid and Jean's Instability

In the above Newtonian derivation of gravitational collapse, the requirement that the fluid be isentropic may not be true in the early universe. In fact, entropy production causes the flow of the universe to be as in an "open system", where the big bang singularity is resolved [37]. We take a slight detour and discuss the situation where there is entropy production in the fluid as anticipated in [37]. In [37], it is conjectured that spacetime can generate particles which add to the fluid, the energy momentum tensor of the Einstein equation. This particle creation is a quantum process and might add insight to the origins of today's cosmological observations. In [37], it is shown that in such open systems, cosmological singularity is not formed. In this review, we briefly discuss whether the open system allows for PBH formation. The conservation law for open thermodynamic systems is given as

$$d(\rho V) + pdV - \frac{h}{n}d(nV) = 0, \tag{50}$$

where n is the particle number and $h = \rho + p$ is the "enthalpy" of the system. In most irreversible systems, as in systems with chemical reactions, enthalpy is a measure of the energy of the system, and is a path-independent quantity. The thermodynamics of these systems is controlled by the chemical potential μ, and the entropy per unit volume "s" is defined as

$$\mu n = h - Ts, \tag{51}$$

with T being the temperature of the system. The pressure for this fluid is given as

$$p = \frac{n\dot{\rho}}{\dot{n}} - \dot{\rho}. \tag{52}$$

If one assumes a fluid in the form of "radiation", i.e., $\rho = aT^4$ and $n = bT^3$, where a and b are dimensional constants [37], obviously, from Equation (52), the equation of state is $p = \rho/3$. In such an open system, if one obtains the propagation equation of a "pressure wave", then the conservation of mass and momentum equations are different. In previous work, the speed of sound in such a fluid was taken as $c_s = \sqrt{1/3}$, which was at constant entropy for the calculation of the Jean's instability. However, the speed of sound changes in a fluid with entropy production. We try to see the origin of the speed of a pressure wave in a gravitating fluid, and it is nonisentropic, with dynamics given by the equations above. To describe the propagation of pressure waves in a system, one uses the following equations. For the conservation of mass equation in the fluid, one has

$$\frac{\partial \rho}{\partial t} + \vec{\nabla} \cdot (\rho \vec{v}) = \dot{n}_i, \tag{53}$$

where we have the "convective" derivative of the density and any particle production on the other side of the equation. The conservation of momentum equation or Euler's equation gives (we assume that the fluid is not viscous)

$$\frac{\partial (\rho \vec{v})}{\partial t} + \vec{v} \cdot \vec{\nabla}(\rho \vec{v}) = -\vec{\nabla} p + \rho g. \tag{54}$$

In the above, the Navier–Stokes equations have been reduced by setting the viscosity to zero. On the right-hand side, there is a potential term which can be a gravitational potential term. In all discussions for the speed of sound, or the speed of pressure waves in the system, the velocity is taken to be small, and the density and pressure undergo perturbations. We assume no gravitational potential at this stage. If there is a linear perturbation in the velocity, density, and pressure of the fluid, with the $\dot{n}$ remaining the same, the perturbations lead to the following equations

$$\frac{\partial \delta \rho}{\partial t} + \rho_0 \vec{\nabla} \cdot \vec{\delta v} = 0, \tag{55}$$

and

$$\rho_0 \frac{\partial \vec{\delta v}}{\partial t} = -\vec{\nabla} \delta p. \tag{56}$$

If the system is isentropic, i.e., homogeneous, one can take a partial derivative of Equation (55) and obtain

$$\frac{\partial^2 \delta \rho}{\partial t^2} + \rho_0 \vec{\nabla} \cdot \frac{\partial \vec{\delta v}}{\partial t} = 0. \tag{57}$$

In the above, using Equation (56), one obtains

$$\frac{\partial^2 \delta \rho}{\partial t^2} - \nabla^2 \delta p = 0. \tag{58}$$

In the isentropic approximation

$$\delta\rho = \left(\frac{\partial\rho_0}{\partial p_0}\right)_s \delta p, \tag{59}$$

one plugs in the above and obtain

$$\frac{\partial^2 \delta\rho}{\partial t^2} - c_s^2 \nabla^2 \delta\rho = 0, \tag{60}$$

and one obtains the speed of propagation of the density perturbations as

$$\frac{1}{c_s} = \sqrt{\left(\frac{\partial\rho_0}{\partial p_0}\right)_s}. \tag{61}$$

In case the fluid has entropy changes, they induce a change in volume. One therefore can obtain for nonisentropic fluids

$$\delta\rho = \left(\frac{\partial\rho_0}{\partial p_0}\right)_s \delta p + \left(\frac{\partial\rho_0}{\partial s_0}\right)_p \delta s. \tag{62}$$

If we use the thermodynamic equation for entropy production as

$$\delta s = \left(\frac{\partial s_0}{\partial p_0}\right)_T \delta p, \tag{63}$$

then, in the formula for the "density perturbation" velocity, we have

$$c = \sqrt{\frac{c_s^2 c_p^2}{c_s^2 + c_p^2}}, \tag{64}$$

where

$$\frac{1}{c_p^2} = \left(\frac{\partial\rho_0}{\partial s_0}\right)_p \left(\frac{\partial s_0}{\partial p_0}\right)_T. \tag{65}$$

If we add the gravitational potential in Euler's equation, then the wave equation has an inhomogeneous term which has a "force driving term" obtained from the gradient of a gravitational potential. If we take the potential to originate from the density, we have $\nabla^2 \phi_1 = 4\pi G \rho_0$, then

$$\frac{\partial^2 \delta\rho}{\partial t^2} - c^2 \nabla^2 \delta\rho = -4\pi G \rho_0 \delta\rho. \tag{66}$$

We assume a plane wave solution for the density wave $\delta\rho \sim e^{i(\omega t + \vec{k}\cdot\vec{x})}$, and we find

$$\omega^2 - c^2 k^2 = 4\pi G \rho_0, \tag{67}$$

so a critical "pressure wave" is identified. For waves with wave numbers above that, the system will see instability. The critical wave number is given as

$$k^2 = \frac{4\pi G \rho_0}{c^2}. \tag{68}$$

Jean's instability is thus identified as perturbations having a wavelength greater than

$$\lambda_J > \sqrt{\frac{\pi}{G\rho_0}} c. \tag{69}$$

Unlike the previous estimate of the length scale where the gravitational instability sets in, here, the speed of sound is not a mere $\sqrt{1/3}$ as given in the formula for an isentropic radiation fluid but is obtained using Equation (64). In a turbulent early universe, therefore it is expected that the fluid would be nonisentropic. In addition, the open universe will ensure entropy production as spacetime generates particle species to add to the fluid. As the speed differs, so will the threshold for the formation of PBH. Note the origin of this change from an underlying quantum theory is implicit in the velocity change of the pressure wave. Note our results for a nonisentropic fluid is just one way to see how some of the formulas used for PBH might change; for other origins of change in Jean's instability formula in cosmic fluids, see [38].

3.4. PBH Formation

How does one obtain the dynamics of formation of PBH in the early universe? It is postulated that the FLRW universe metric could have perturbations induced by the density fluctuations of the fluid. These can be modelled using a spherical symmetry, and the conditions for the formation of "trapped surfaces" or apparent horizons derived using the "Misner–Sharp" equations. These PBH can then accrete and grow in size, and there can be PBH formed of masses which are bigger than the solar masses of $10 M_\circ$–$30 M_\circ$. A great deal of the current work on PBH discusses these and the fraction of PBH contributing to dark matter halos f_{PBH}. For further reading on the PBH production and the interest in them as contributors to dark matter and physical processes such as microlensing, etc. refer to [8]. As the black hole formation follows the same numerical flow as in the spherical collapse obtained by Choptuik, the PBH's mass has the following "scaling" formula

$$M_{PBH} = K\, M_H(t_H)(\delta_m - \delta_c)^\gamma, \tag{70}$$

where $\delta_m = (\rho - \rho_b)/\rho_b$ is the fluctuation in the fluid density over the Hubble density, at the radius where a compaction function is maximum. δ_c is the fluctuation at the critical radius related to the Jean's instability in the fluid found earlier. δ_c represents the threshold of black hole formation. This equation can only be trusted in the regime $\delta_m - \delta_c \sim 10^{-2}$. $M_H(t_H)$ is the Misner–Sharp mass of the horizon, K is a numerical constant. γ is a universal scaling exponent and varies depending on the fluctuation profile and the equation of state of the fluid. This equation provides the basis for PBH formation, though using classical equations. The compaction function $C(r,t)$ is defined as the excess of mass over the FLRW mass M_b defined as $M_b = 4\pi \rho_b R^3/3$,

$$C(r,t) = 2\frac{M(r,t) - M_b(r,t)}{R(r,t)}. \tag{71}$$

If one takes the perturbation of the FLRW metric to be modelled by a function $\zeta(r,t)$ in the FLRW metric three-slice as $a^2(t)e^{2\zeta(r,t)}r^2 d\Omega$, one gets a formula for the compaction function in terms of this parameterized fluctuation as

$$C(r) = \frac{2}{3}\left(1 - (1 - r\zeta'(r))^2\right). \tag{72}$$

This facilitates the study of this function in terms of the curvature fluctuations of the metric. The various calculations of the "peak" values of this compaction function use different ensembles for the fluctuations and accordingly, obtain different values. It is postulated that when the compaction function exceeds a critical value, a collapse occurs otherwise the fluctuation dissipates away. The density contrast parameter is related to the peak value of the compaction function as

$$\delta_m = C(r_m). \tag{73}$$

In this article, we refrain from discussing the various ways of finding PBH compaction function but only show a way the change in threshold value δ_c of PBH formation influences the collapse process. This critical value is related to Jean's instability in the cosmic fluid and as shown previously, vary according to the approximations used. A dependence on the formula for PBH on the nature of the fluid is discussed in [8]. As shown in Equations (69) and (64), the threshold of the onset of the instability of a fluid changes if quantum "particle creation" is allowed. In [37], the fluid exchanges particles with the gravitational "quantum field". In this open universe, there is no initial singularity [37], and as we anticipate, the formation of PBH would also differ. The masses would be different, and the nature of the cosmological fluctuations of the gravitational metric would also differ as per the "entropy production" of this open universe. A more detailed calculation using quantum cosmology is required for the exact changes required in the *theoretical* predictions of the PBH's mass, and the PBH formation from the cosmic soup.

The formation of PBH can vary from masses of the order of 10^5–10^{50} g, and therefore, they can range from small black holes to larger-than-solar-mass black holes. The lower limit is based on the Planck mass and the upper limit is based on the cosmological mass. How can we verify the existence of PBH? The existence of PBH can be verified using the observation of particles received on earth, which might have originated from the PBH using the Hawking radiation process. It is this process which we describe next. We discuss PBH whose evaporation time $\propto M^3$ is about the age of the universe. These PBH might have radiated away their mass in the form of photons and neutrinos and would provide evidence for the phenomena of Hawking radiation. The mass of these black holes is estimated as $<10^{14}$ g.

Curiously, there was an attempt to find quantum gravity effects on PBH production using loop quantum cosmology (LQC) corrections to the scale factor and the density [39]. The authors found that using the LQC-corrected early universe cosmology, the production of PBH was increased theoretically compared to estimates from other theoretical models as that of the Brans–Dicke gravity.

3.5. Evaporation of PBH

The mechanism of radiation from black holes can be studied using the power law for the emission of particles. In the 1970s [33,40,41], one typically calculated the power law using Hawking's formula for the particle flux from black holes. The total energy radiated per unit time from PBH of Hawking temperature T_H is given as

$$\frac{dE}{dt} = \int d\omega \int d\Omega \sum_{lm} \frac{\Gamma_{\omega slm}}{\exp(\omega/T_H) \pm 1} \tag{74}$$

where $\Gamma_{\omega slm}$ is the grey-body factor for the black hole geometry and represents matter waves scattering off the gravitational potential outside the black hole. s, l, m represent the spin and angular momentum quantum numbers of particles with frequency ω. The sign in the denominator is positive for bosons and negative for fermions. The Hawking temperature for a nonrotating black hole is inversely proportional to the mass. The grey-body factor is calculated using the solutions to the classical equation of motion of the particles in the black hole background and is a function of the spin, angular momentum, mass, and frequency of the emission. The fraction of power radiated in different species can be calculated. The total power radiated can be calculated numerically as

$$P = 2.011 \times 10^{-4} \hbar c^5 G^{-2} M^{-2}, \tag{75}$$

where M is the mass of the black hole. Most of the above is radiated out in the form of neutrinos (81.4%), 16.7% as photons and 1.9% as gravitons, as long as the black holes have mass $M > 10^{17}$g [40] After the black hole has shrunk further, the temperature being higher, and the mass being denser, the black hole radiates quarks in the form of muons and other particles such as electrons and positrons. For this range of black holes, 10^{14} g $< M < 10^{17}$ g the power radiated was found to be

$$P = 3.6 \times 10^{-4} \hbar c^5 G^{-2} M^{-2}, \tag{76}$$

90% is equally divided in electrons, positrons, and neutrinos, 9% in photons, and 1% in gravitons [40]. In this work, when computing the power of Hawking particles, the numerical calculations of the grey-body factors were used, and the above division into fractions were based on the spin of the particles. The emission of massive particles would have a different calculation, but for a detection on earth, the massless particles acquire relevance.

In a follow up work [41], the emission of gamma rays with energy of about 120 MeV was discussed, and a study of "gamma ray bursts" from evaporating PBH was introduced. In there, a mass distribution was assumed for PBH, and this is an ingredient in the current analysis of the data received on earth. The search for Hawking radiation phenomena in the universe is thus a search for primordial black holes and the particles emitted from them. There are several searches for primordial black holes using gamma-ray bursts which might be the evidence of these black holes evaporating. In the next, we describe some of these searches in detail and provide a bibliography.

3.6. Archived Data

The Imaging Compton Telescope (COMPTEL) [42] was decommissioned in 2007, but there remained the archived data to analyze gamma rays. The search from these data has shown bounds for the primordial black holes (PBH) $< 10^{17}$ g [43].

3.7. Gamma-Ray Bursts

There are several satellite-based experiments, which are functional or at the planning stage such as AMEGO and e-ASTROGRAM. AMEGO is an abbreviation for the All-sky Medium Energy Gamma-ray Observatory experiment and comprises a silicon tracker, a cesium iodide calorimeter, and a scintillator anticoincidence detector. All these will form the payload of a satellite. The detector will operate in the MeV range and provide a wider field of view than the Fermi-LAT detector. This detector is planned by NASA. e-Astrogram is a European Science Commission gamma-ray detector, based on similar instrumentation as AMEGO [44]. The e-Astrogram project aims to observe the frequency range of 0.3 MeV to 3 GeV. It is also aiming to be more sensitive at a particular frequency than previous instruments. These instruments will send data about the gamma-ray bursts and other sources which will give a clue on the existence of primordial black holes in the early universe.

3.8. HESS

The HESS is a gamma-ray observation experiment using an array of atmospheric imaging Cerenkov telescopes with energy in the TeV range. The telescopes are in Namibia. We report on the techniques of the HESS experiment in details here as an example, but it is one of several developments for PBH observations [45]. As the PBH which are smaller than 10^{17} g might have evaporated by now, one searches for gamma-ray burst signals. The PBHs are expected to have evaporated with an explosion of gamma rays, which have a high energy and last only for a few seconds. Using statistics and the methods of [46] Feldman and Cousins, one can estimate the "rate of" the PBH formation density $\dot{\rho}_{PBH}$, with 95% and 99% confidence levels. Further, we discuss this experiment's data analysis [45] in details to illustrate the methodology of the search of PBH. Let us say an unknown parameter μ is being assessed using the measurements of a variable x. Usually, one uses Bayesian statistics to estimate the "belief" in a system's parameter being μ_t. This is given using the formula

$$P(\mu_t|x_0) = \mathcal{L}(x_0|\mu)\frac{P(\mu_t)}{P(x_0)}, \qquad (77)$$

where $\mathcal{L}(x_0|\mu_t)$ is the "likelihood" of obtaining x_0 given μ_t. However, it is assumed that there is prior knowledge of the probability $P(\mu_t)$ of finding μ_t independent of what x_0 is, which might not be the case. The probability $P(x_0)$ can be absorbed in the normalization of the conditional probability. In Bayesian methods, the belief in finding μ_t given the measured values of x is expressed as a "confidence". This is mathematically

$$\int_{\mu_1}^{\mu_2} P(\mu_t|x_0)d\mu_t = \alpha, \qquad (78)$$

where α is the degree of confidence for μ_t to be in the confidence interval $[\mu_1, \mu_2]$. In [46], a variation of this is given, for estimating the value of a parameter μ given the measurements of the variable x. If one takes the ratio of two likelihoods, then the "prior knowledge" required in Bayesian statistics is not there.

$$R = \frac{\mathcal{L}(x|\mu)}{\mathcal{L}(x|\mu_{\text{best}})}, \qquad (79)$$

where μ_{best} is the value of the parameter which maximizes the conditional probability. This ratio determines the acceptance region in the x variable, for a given value of μ. A sum of the observation probabilities in decreasing order of R, until the required confidence limit is reached, provides a good estimate for the confidence intervals or upper limits for a parameter.

In the HESS observations, gamma rays were detected using the Cerenkov telescopes on earth. The number of photons detected could vary from one to infinity in a given time interval Δt. A time interval of $\Delta t = 10$ s was taken for the purpose. We assumed that the detection of photon "clusters" of size k followed a Poisson distribution

$$P(k, N) = e^{-N} \frac{N}{k!}, \qquad (80)$$

where $N(r, \alpha, \delta, \Delta t)$ is the number of γ rays emitted from PBH from a distance r in the angular interval in the sky specified by α, δ in unit time Δt. Integrating this over all space, i.e., r, α, δ, and over all runs of the experiment, the number of significant clusters of photons detected were estimated to be

$$n_{\text{sig}}(k, \Delta t) = \dot{\rho}_{PBH} V_{\text{eff}}(k, \Delta t), \qquad (81)$$

where

$$V_{\text{eff}}(k, \Delta t) = \sum_i T_i \int d\Omega_i \int dr\, r^2\, P(k, N) = \sum_i T_i \Omega_i \frac{(r_0\sqrt{N_0})^3}{2} \frac{\Gamma(k-3/2)}{\Gamma(k+1)}, \qquad (82)$$

where N_0 is the number of photons emitted from PBH at a distance of r_0. T_i is the run's live time of the experiment, and Ω_i is the solid angle of the observations. Based on the observed data, the statistical analysis using the techniques of Feldman and Cousins was implemented. The parameter being sought was n_{sig} given n as the observed variable. Note that these photon clusters, which might be from evaporating PBH, were received along with the background photons, whose number was taken as $\bar{n}$, or off photons.

$$R = \prod_n \frac{\mathcal{L}(n|\bar{n} + n_{\text{sig}})}{\mathcal{L}(n|\bar{n})}. \qquad (83)$$

Here, the maximal value of the likelihood function was taken as that of the background $\bar{n}$. The χ^2 estimate of the above can be found as [46]:

$$\text{LNR} = -2\ln(R) = 2\sum_n n_{\text{sig}} + n(\ln(n) - \ln(\bar{n} - n_{\text{sig}})), \qquad (84)$$

where n is the number of observed photon signals in the on position of the telescopes and $\bar{n}$ is the number of mean observed signals in the off data. This is an estimate of the background photons, obtained by averaging over "scrambled" time intervals. In deriving the above, we used the Poisson distribution.

This LNR had a maximum of 0.006 in the preliminary data for $\Delta t = 10$ s and 6240 runs of four of the five telescopes [45]. This showed that there was not much of the PBH excess data. However, if one sets LNR = 4, 9, one can obtain an upper-limit estimate for $\dot{\rho}_{PBH}$ with 95% and 99% confidence levels. The upper limit was found to be

$$\dot{\rho}_{PBH} < 2.5 \times 10^4/pc^3yr \quad (95\%), \tag{85}$$
$$\dot{\rho}_{PBH} < 5 \times 10^4/pc^3yr \quad (99\%), \tag{86}$$

These data points were further updated with other experiments such as VERITAS, MILAGRO, FERMI-LAT, and SWGO [47]. A comparative plot of the experimental predictions of evaporating PBH or final bursts at the 99% confidence limit is given in Figure 2. The data for this are quoted from [48] (2021). For some recent updates in the field of constraints on PBH see [49].

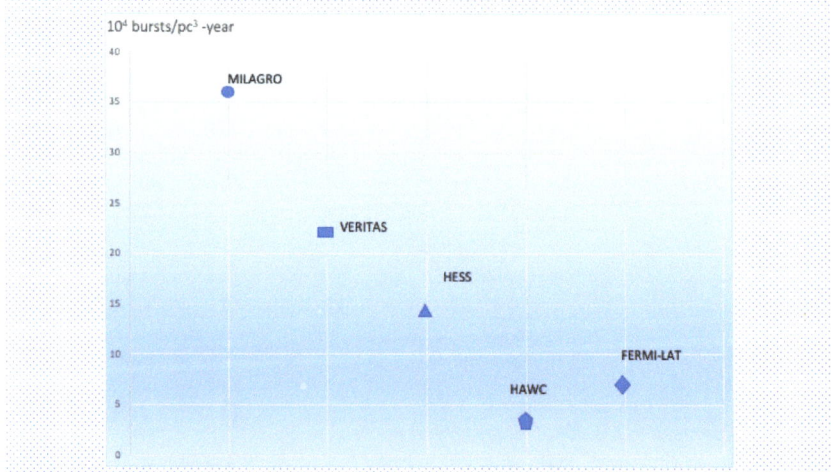

Figure 2. The upper estimates of the number of final bursts at the 99% confidence limit from some experiments [48].

For recent data on HESS, one can refer to the experiment's website [50].

3.9. Neutrino Experiments

The Hawking radiation from PBH releases neutrinos. The flux of these as a function of the PBH production and then a further analysis for "secondary effects" producing neutrinos were analyzed. The data from several experiments were taken and showed almost no or a very small estimation of the PBHs. Using a recent work [51], we comment on the results. A neutrino spectrum rate was defined using the Hawking emission spectrum as in Equation (74). Further, there can be secondary neutrino production due to the decay of hadrons produced initially:

$$\frac{d^2N_\nu}{d\omega_\nu dt} = \int_0^\infty dM \frac{dN}{dM} \left(\frac{d^2N_\nu}{d\omega_\nu dt}\bigg|_{prim} + \frac{d^2N_\nu}{d\omega_\nu dt}\bigg|_{sec} \right). \tag{87}$$

where the black hole's mass distribution could be taken as a Gaussian log-normal profile,

$$\frac{dN}{dM} = \frac{1}{\sqrt{2\pi}\sigma M} \exp\left(-\frac{\ln^2(M/M_{PBH})}{2\sigma^2}\right), \tag{88}$$

or simply a delta function profile centred at $M = M_{PBH}$. In the above, M_{PBH} is an average mass, and σ is the standard deviation, as the mass of the black hole is allowed to vary. A plot of the differential neutrino flux from extragalactic sources and the milky way can be calculated using publicly available software [51] and plotted. The differential flux of the neutrino plotted as a function of the energy ω_ν varied between 10^2 and 10^{-5}, as the energy varied from 1 to 100 MeV for PBH of mass 10^{13} g. The evaporated PBH were taken as a fraction of the cosmic background which is 10^{-18} to obtain this result.

The experimental bounds obtained from the Super-Kamiokande data showed that for PBH which were already evaporated, the abundance ratio was about 10^{-17} for 10^{13} g black holes and a confidence limit of 90%. The question is of course what the above bounds imply for quantum gravity phenomenology? Whereas the PBH production cannot be ruled out completely, using the above estimation methods, it remains that the mechanism of PBH formation could be different, and the emission flux calculations greatly modified by intervening cosmic flows and quantum effects. In this aspect, one has to wait for future experiments such as JUNO, DARWIN, ARGO, and DUNE, and perhaps quantum cosmology predictions of the PBH formation from a more fundamental theory such as loop quantum gravity.

It is obvious from the above discussions that the detection of bursts of photons and neutrinos on earth gives a very small window for the PBH to exist which would be evaporating now, i.e., those having masses 10^5–10^{14} g. However, as we know, there can still be the option that there are PBH which have not evaporated away but have formed remnants. These will still be candidate dark matter contributors. The fraction of PBH which contribute to dark matter and have not been evaporated yet is also estimated as $\sim 10^{-3}$ for masses of the order of 10^{16} g as in [52]. There are other papers investigating this using various data sources such as microlensing, accretion disk luminosity, radio signals, anisotropies of the CMB, etc. We refer the reader to reviews in this field [8]; there are also discussions of the PBH formation and evaporation using LQG corrected metrics, though in reduced phase-space formulations [53]. In our opinion, whereas the search is now much focused than earlier on what a gamma-ray burst or a neutrino flux from PBH may be, the research is still nascent.

4. Event Horizon

In the initial days of the discovery of the black hole metric solution to Einstein's equation, the existence of the horizon was one of the most bizarre predictions. The existence of trapped surfaces in general relativity was later firmly established with the Ray–Chowdhury equations and Hawking–Penrose singularity theorems. However, the debate continued on whether the event horizon existed, as it was unobservable. With the discovery of compact objects and the observation of X-rays from them, various models were tested for the existence of the event horizon. As the conclusions were model-dependent, the search continued, until the event horizon telescope project produced an "assembled image" of the photon sphere surrounding a black hole [14,54]. This confirmed some of the predictions about the behaviour of geodesics near a black hole's horizon, but did it confirm the presence of an event horizon? Perhaps not, but this is as "good as it gets". The snapshot of the photon sphere assimilated from eight infrared telescopes captured the electromagnetic waves circulating a compact object. The question we are asking in this article is: can we use the observations of geodesics around a black hole to measure semiclassical physics? In a work using semiclassical states in loop quantum gravity [13], it was shown that quantum fluctuations could cause instabilities in black holes, and these could produce tangible detectable effects for astrophysical black holes [13]. The main results of the paper were the calculation of a nonpolynomial correction to the metric of the Schwarzschild black hole. The semiclassically corrected metric was shown to be of the following form

$$ds^2 = -\left(1 - \frac{r_g}{r} - \tilde{t}\, h_{tt}\right)dt^2 + \tilde{t}\, h_{rt}\, dtdr + \left\{\frac{1}{(1 - r_g/r)} + \tilde{t}\, h_{rr}\right\}dr^2 +$$
$$+ \left(r^2 + \tilde{t}\, h_{\theta\theta}\right)d\theta^2 + \left(r^2 \sin^2\theta + \tilde{t}\, h_{\phi\phi}\right)d\phi^2. \tag{89}$$

where r_g is the Schwarzschild radius, and the location of the horizon is at $r_g = 2GM$, where M is the mass of the black hole. $h_{tt}, h_{rt}, h_{rr}, h_{\theta\theta}$, and $h_{\phi\phi}$ are the perturbations motivated from the corrections to the metric [13]. The perturbations of the metric could be attributed to other quantum models of gravity, but we used the one motivated from [13], and a shift was generated, h_{rt}, breaking the "static" nature of the metric. The $\tilde{t}$ which appears in this coherent state was obtained using the length scales of the system and was thus a ratio of Planck's area to the area of the horizon $\tilde{t} = l_p^2/r_g^2$. Using this, we solved for the geodesics of the black hole. The geodesics were taken as circular orbits and the radial coordinate r was solved as a function of the coordinate ϕ. These orbits described the trajectory of light rays which were incident on the black hole geometry from a distance, and the impact parameter measured the perpendicular distance of the light ray from the horizon. Using the invariant distance on the Schwarzschild geometry, one can write the equation of motion for the geodesic of a photon as a differential equation in the azimuth ϕ, which was taken as the affine parameter along the geodesic. The deviations in geodesic computations for the rotating black hole from the nonrotating black holes were small [55] but detectable. For rotating black holes, the cross section of the photon scattering might not be circular [55], but the difference was about 4%. However, quantum corrections might be different, and one needs to formulate coherent states for rotating black holes separately. The effect of the presence of "echoes" might still be true. The results stated in this paper thus apply to nonrotating black holes strictly but pave the way for realistic ones.

If we arrange the terms in a way they can be grouped into terms which are zeroth order in $\tilde{t}$ and then first order in $\tilde{t}$ (in the equatorial plane), one gets [14]:

$$\frac{1}{r^4}\left(\frac{dr}{d\phi}\right)^2 + \frac{1}{r^2}\left(1 - \frac{r_g}{r}\right)\left(1 + \tilde{t}\,\frac{h_{\phi\phi}}{r^2} - \tilde{t}\, h_{rr}\right) = \frac{1}{b^2}\left(1 + 2\tilde{t}\,\frac{h_{\phi\phi}}{r^2} - \tilde{t}\, h_{rr} + \tilde{t}\,\frac{h_{tt}}{1 - r_g/r}\right). \tag{90}$$

As one traces the trajectory through the entire path, the asymptotic angle of "scattering" from the black hole geometry emerges as a function of the impact parameter of the photon. The solution is obtained using a set of elliptic integrals and one finds

$$\exp(-\phi_\infty) = \delta^{1 + 0.0203\,\tilde{t}} \exp\left(+\frac{0.47\,\tilde{t}^{1/2}}{(0.67\delta + 0.225\,\tilde{t})^{1/2}} + 0.23\,\tilde{t} + 1.712\,\frac{\tilde{t}}{\delta}\right), \tag{91}$$

where $\delta = b - b_c$, and ϕ_∞ is the asymptotic angle the geodesic makes as it re-emerges to the asymptotic region. The difference of the photon geodesic impact parameter with the impact parameter of the critical orbit $b_c = 3\sqrt{3}M$ is expected to be zero as the photon can orbit an infinite number of times round the horizon. One can see that in Equation (91), the $\tilde{t} \to 0$ reduces to a linear term in δ. Most importantly, $\delta \to 0$ as $\phi_\infty = \mu + 2n\pi \to \infty$. n counts the number of times the geodesic encircles the black hole, and this goes to infinity for the critical geodesic with the critical impact parameter. The photon circles the black hole an infinite number of times, when the critical impact parameter is reached. If we take the semiclassical corrections, then the plot of $w(\delta)$ (the RHS of Equation (91) as a function of δ shows that the function does not reach zero but bounces off (see Figures 3 and 4), and this we can associate with the presence of a quantization.

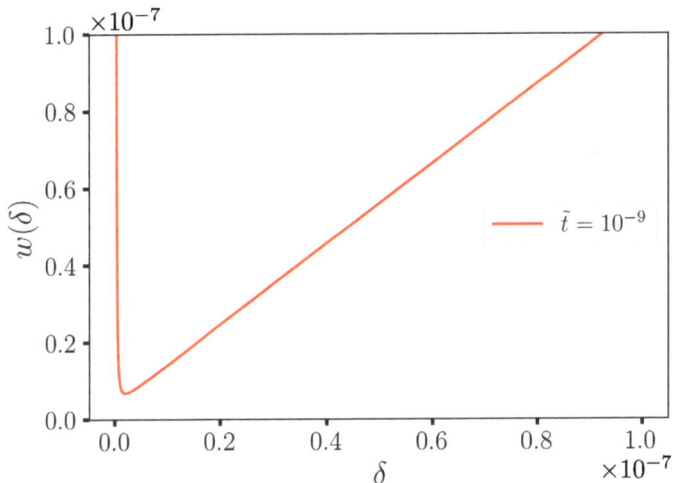

Figure 3. Plot of the semiclassically corrected photon geodesic impact parameter relation. The plot shows a bounce as the distance from the critical radius approaches the semiclassical length scale of $\tilde{t} \sim 10^{-8}$ units.

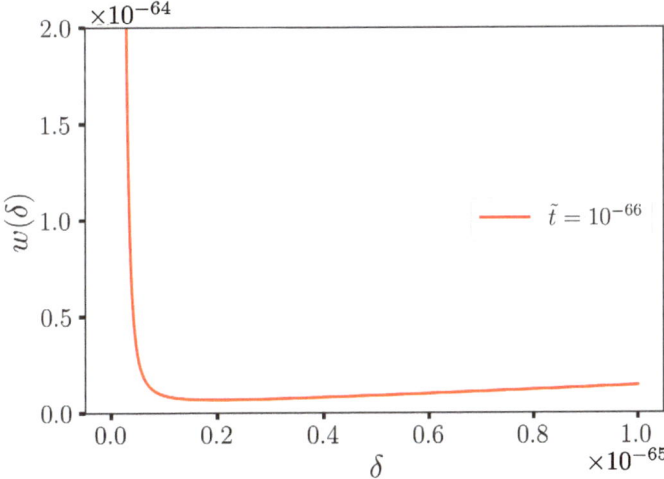

Figure 4. Plot of the semiclassically corrected photon geodesic impact parameter relation. The plot shows a bounce as the distance from the critical radius approaches the semiclassical length scale of $\tilde{t} \sim 10^{-66}$ units.

This observation is commensurate with the work in fuzzballs and ECHOS [56]. In these models, the horizon is replaced by a "wall" at a particular distance from the black hole. In our calculations with the LQG coherent states [13], we found the explicit location of the "wall" as a function of the semiclassical parameter $\tilde{t}$. We expect that our results can be eventually verified from observational data from astrophysical black holes [56].

5. Conclusions

As it happens, the search for quantum gravity in experiments is still nascent. However, we expect that in the early universe, the length scales were quantum, and therefore the search for relics of quantum gravity is ongoing. There are a number of papers in this

Universe special issue in quantum gravity phenomenology which discuss cosmology and the effect of quantum cosmology in observational physics. In this review, the experiments we discussed only provided bounds on the mass of the graviton, the PBH production. We discussed the quantum effects which could be "directly" observable in recent experiments including in gravitational wave detectors and event horizon telescope images. We also reported on the numerous experiments which observe particles from distant celestial events on earth. We showed theoretical calculations and reported on bounds from experiments on Hawking emission from PBH. The experimental bounds did not violate any theoretical predictions. The observations provide directions for the experimental community to seek for more precise measurements. The plot of the electric and magnetic polarizations from the EHT [57] and the launching of LISA [58] are ongoing efforts in that direction. The study of fast radio bursts (FRB) provided an effort towards finding the quantum origins of astrophysical phenomena. The most promising experiments on earth for the quantum effects of gravity remain the GW detectors and the possibility that one would detect a "graviton" or its semiclassical version in the near future.

Funding: This research is partially supported by MITACS internship of José Fajardo-Montenegro.

Acknowledgments: This article is written for universe special issue in Quantum Gravity Phenomenology. AD would like to thank the universe editorial team, particularly Cici Xia for making this two volumes possible. AD is also thankful to co-editor Alfredo Iorio for collaboration.

Conflicts of Interest: The authors declare no conflict of interest.

References

1. Abbott, B.P.; Abbott, R.; Abbott, T.D.; Abernathy, M.R.; Acernese, F.; Ackley, K.; Adams, C.; Adams, T.; Addesso, P.; Adhikari, R.X.; et al. LIGO Scientific collaboration and Virgo Collaboration, Observation of Gravitational Waves from a Binary Black Hole Merger. *Phys. Rev. Lett.* **2016**, *116*, 061102. [CrossRef] [PubMed]
2. Shao, L.; Wex, N.; Zhou, S. New graviton mass bound from binary pulsars. *Phys. Rev.* **2020**, *D102*, 024069. [CrossRef]
3. Bern, Z. Perturbative Quantum Gravity and its Relation to Gauge Theory, Living Reviews in Relativity. *Living Rev. Relativ.* **2002**, *5*, 5. [CrossRef] [PubMed]
4. Rovelli, C. *Loop Quantum Gravity*; Cambridge University Press: Cambridge, UK, 2007.
5. Isi, M.; Farr, W.M.; Giesler, M.; Scheel, M.A.; Teukolsky, S. Testing the Black-Hole Area Law with GW150914. *Phys. Rev. Lett.* **2021**, *121*, 011103. [CrossRef]
6. Escriva, A. PBH formation from spherically symmetric hydrodynamical perturbations: A review. *Universe* **2022**, *8*, 66. [CrossRef]
7. Carr, B.; Kohri, K.; Sendouda, Y.; Yokoyama, J. Constraints on Primordial Black Holes. *Rep. Prog. Phys.* **2021**, *84*, 116902. [CrossRef]
8. Escriva, A.; Kuehnel, F.; Tada, Y. Primordial Black Holes. *arXiv* **2022**, arXiv:2211.05767v3.
9. Laha, R. Primordial Black Holes as a Dark Matter Candidate Are Severely Constrained by the Galactic Center 511 keV γ-Ray Line *Phys. Rev. Lett.* **2019**, *123*, 251101. [CrossRef]
10. Dasgupta, B.; Laha, R.; Ray, A. Neutrino and Positron Constraints on Spinning Primordial Black Hole Dark Matter. *Phys. Rev. Lett.* **2020**, *125*, 101101. [CrossRef]
11. Cappelutti, N.; Hasinger, G.; Natarajan, P. Exploring the High-redshift PBH-ΛCDM Universe: Early Black Hole Seeding, the First Stars and Cosmic Radiation Backgrounds. *Astro. J.* **2022**, *926*, 205. [CrossRef]
12. Dasgupta, B.; Laha, R.; Ray, A. Low Mass Black Holes from Dark Core Collapse. *Phys. Rev. Lett.* **2021**, *126*, 141105. [CrossRef] [PubMed]
13. Dasgupta, A. Quantum Gravity Effects on Unstable Orbits in Schwarzschild Space-time. *Journ. Cosm. Astro. Phys.* **2010**, *5*, 011 [CrossRef]
14. Maharana, S.; Dasgupta, A. Semiclassical corrections to the photon orbits of a non-rotating black hole. *arXiv* **2011**, arXiv:2011.00676
15. Addazi, A.; Alvarez-Muniz, J.; Alves Batista, R.; Amelino-Camelia, G.; Antonelli, V.; Arzano, M.; Asorey, M.; Atteia, J.L. Bahamonde, S.; Bajardi, F.; et al. Quantum gravity phenomenology at the dawn of the multi-messenger era—A review. *Prog. Part. Nucl. Phys.* **2022**, *125*, 103948. [CrossRef]
16. Amelino-Camelina, G. Are We at the Dawn of Quantum Gravity Phenomenology. *Lect. Notes. Phys.* **2000**, *549*, 1–49.
17. Alcubierre, M.; Allen, G.; Brügmann, B.; Lanfermann, G.; Seidel, E.; Suen, W.-M.; Tobias, M. Gravitational Collapse of Gravitational Waves in 3D Numerical Relativity. *Phys. Rev.* **2000**, *D61*, 041501. [CrossRef]
18. Bishop, N.T.; Rezzolla, L. Extraction of Gravitational Waves in Numerical Relativity. *Liv. Rev. Relat.* **2016**, *19*, 2. [CrossRef]
19. Dewitt, B.S. Quantum Theory of Gravity. III. Applications of the Covariant Theory. *Phys. Rev.* **1967**, *162*, 1239. [CrossRef]
20. Lauscher, O.; Reuter, M. Is Quantum Einstein Gravity Nonperturbatively Renormalizable? *Class. Quant. Grav.* **2002**, *19*, 483 [CrossRef]
21. Ashtekar, A.; Rovelli, C.; Smolin, L. Gravitons and loops. *Phys. Rev.* **1991**, *D44*, 1740. [CrossRef] [PubMed]

22. Thiemann, T.; Winkler, O. Gauge field theory coherent states (GCS): II. Peakedness properties. *Class. Quant. Grav.* **2001**, *18*, 2561–2636. [CrossRef]
23. Sahlmann, H. Coupling Matter to Loop Quantum Gravity. Ph.D. Thesis, University of Potsdam, Potsdam, Germany, 2002.
24. Dasgupta, A. Coherent states for black holes. *J. Cosmol. Astropart. Phys.* **2003**, *8*, 004. [CrossRef]
25. Varadarajan, M. Gravitons from a loop representation of linearized gravity. *Phys. Rev.* **2002**, *66* 024017. [CrossRef]
26. Varadarajan, M. The graviton vacuum as a distributional state in kinematic loop quantum gravity. *Class. Quant. Grav.* **2005**, *22*, 1207. [CrossRef]
27. Mortuza-Hossain, G. Large volume quantum correction in loop quantum cosmology: Graviton illusion? *arXiv* **2005**, arXiv:gr-qc/0504125.
28. Hinterleitner, F.; Major, S. Toward loop quantization of plane gravitational waves. *Class. Quant. Grav.* **2012**, *29*, 065019. [CrossRef]
29. Garcia-Chung, A.; Mertens, J.B.; Rastgoo, S.; Tavakoli, Y.; Moniz, P.V. Propagation of quantum gravity-modified gravitational waves on a classical FLRW spacetime. *Phys. Rev. D* **2021**, *103*, 084053. [CrossRef]
30. Fajardo-Montenegro, J.L.; Dasgupta, A. *Semiclassical States for Gravitons*. work in progress.
31. Hall, B. The Segal-Bargmann "Coherent State" Transform for Compact Lie Groups. *J. Funct. Anal.* **1994**, *122*, 103. [CrossRef]
32. Dasgupta, A. Semiclassical Loop Quantum Gravity and Black Hole Thermodynamics. *SIGMA* **2013**, *19*, 013. SIGMA.2013.013. [CrossRef]
33. Hawking, S.W. Particle creation by black holes. *Comm. Math. Phys.* **1975**, *43*, 199. [CrossRef]
34. Choptuik, M.W. Universality and scaling in gravitational collapse of a massless scalar field. *Phys. Rev. Lett.* **1993**, *70*, 12. [CrossRef] [PubMed]
35. Mcmillan, S. Available online: http://www.physics.drexel.edu/~\steve/Courses/Physics-431/jeans$\_$instability.pdf (accessed on 31 January 2023).
36. Tomisaka, K. Lecture Notes. Available online: http://th.nao.ac.jp/MEMBER/tomisaka/Lecture$\_$Notes/StarFormation/6/node36.html (accessed on 31 January 2023).
37. Prigogine, I.; Geheniau, J.; Gunzig, E.; Nardone, P. Thermodynamics of cosmological matter creation. *Proc. Natl. Acad. Sci. USA* **1988**, *85*, 7428. [CrossRef] [PubMed]
38. Kremer, G.M.; Richarte, M. G.; Teston, F. Jeans instability in a universe with dissipation. *Phys. Rev.* **2018**, *D97*, 023515. [CrossRef]
39. Dwivedee, D.; Nayak, B.; Jamil, M.; Singh, L.P.; Myrzakulov, R. Evolution of Primordial Black Holes in Loop Quantum Cosmology. *J. Astrophys. Astron.* **2014**, *35*, 97. [CrossRef]
40. Page, D. Particle emission rates from a black hole: Massless particles from an uncharged, non-rotating hole. *Phys. Rev.* **1976**, *D13*, 198. [CrossRef]
41. Page, D.; Hawking, S.W. Gamma rays from primordial black holes. *Astrophys. J.* **1976**, *206*, 1–7. [CrossRef]
42. COMPTEL Data Source. Available online: http://cta.irap.omp.eu/ctools/users/tutorials/howto/comptel/index.html (accessed on 31 January 2023).
43. Coogan, A.; Morrison, L. Profumo, Direct Detection of Hawking Radiation from Asteroid-Mass Primordial Black Holes. *Phys. Rev. Lett.* **2021**, *126*, 171101. [CrossRef]
44. AMEGO Collaboration. Available online: https://asd.gsfc.nasa.gov/amego/ (accessed on 31 January 2023).
45. Tavernier, T.; Glicenstein, J.-F.; Brun, F. Search for Primordial Black Hole evaporations with H.E.S.S. *arXiv* **2019**, arXiv:1909.01620.
46. Feldman, G.; Cousins, R.J. A Unified Approach to the Classical Statistical Analysis of Small Signals. *Phys. Rev.* **1998**, *D57*, 3873. [CrossRef]
47. Lopez-Coto, R.; Doro, M.; de Angelis, A.; Mariotti, M.; Harding, J.P. Prospects for the observation of Primordial Black Hole evaporation with the Southern Wide field of view Gamma-ray Observatory. *J. Cosmol. Astropart. Phys.* **2021**, *8*, 040. [CrossRef]
48. Engel, K.; Peisker, A.; Harding, P.; Wood, J.; Martinez-Castellanos, I.; Albert, A.; Tollefson, K.; on behalf of HAWC Collaboration. Setting Upper Limits on the Local Burst Rate Density of Primordial Black Holes Using HAWC, PoS ICRC2019 516 (2021). Available online: https://inspirehep.net/literature/1820186 (accessed on 31 January 2023).
49. Korwar, M.; Profumo, S. Updated Constraints on Primordial Black Hole Evaporation. *arXiv* **2023**, arXiv:2302.04408.
50. HESS Experiment. Available online: https://www.mpi-hd.mpg.de/hfm/HESS/ (accessed on 31 January 2023).
51. Bernal, N.; Chu, X.; Garcia-Cely, C.; Hambye, T.; Zaldivar, B. Production Regimes for Self-Interacting Dark Matter. *J. Cosmol. Astropart. Phys.* **2022**, *10*, 68. [CrossRef]
52. Zhang, Z.H.; Yang, L.T.; Yue, Q.; Kang, K.J.; Li, Y.J.; An, H.P.; Greeshma, C.; Chang, J.P.; Chen, Y.H.; Cheng, J.P.; et al, Search for keV–MeV Light Dark Matter from Evaporating Primordial Black Holes at the CDEX-10 Experiment. *arXiv* **2022**, arXiv:2211.07477v1.
53. Barrau, A.; Martineau, K.; Moulin, F. A Status Report on the Phenomenology of Black Holes in Loop Quantum Gravity: Evaporation, Tunneling to White Holes, Dark Matter and Gravitational Waves. *Universe* **2018**, *4*, 102. [CrossRef]
54. Event Horizon Telescope Collaboration. Available online: http://eventhorizontelescope.org/ (accessed on 31 January 2023).
55. Event Horizon Telescope Collaboration. First M87 Event Horizon Telescope Results. I. The Shadow of the Supermassive Black Hole. *Astrophys. Phys. J. Lett.* **2019**, *875*. [CrossRef]
56. Cardoso, V.; Pani, P. Testing the nature of dark compact objects: A status report. *Living Rev. Relativ.* **2019**, *22*, 4. [CrossRef]

57. The Event Horizon Telescope collaboration, First M87 Event Horizon Telescope Results. VIII. Magnetic Field Structure near The Event Horizon. *Astrophys. J. Lett.* **2021**, *910*, L12. [CrossRef]
58. The LISA Space Telescope. Available online: https://lisa.nasa.gov/ (accessed on 31 January 2023).

Disclaimer/Publisher's Note: The statements, opinions and data contained in all publications are solely those of the individual author(s) and contributor(s) and not of MDPI and/or the editor(s). MDPI and/or the editor(s) disclaim responsibility for any injury to people or property resulting from any ideas, methods, instructions or products referred to in the content.

Article

Massive Neutron Stars and White Dwarfs as Noncommutative Fuzzy Spheres

Surajit Kalita [1,2] and Banibrata Mukhopadhyay [2,*]

[1] High Energy Physics, Cosmology & Astrophysics Theory (HEPCAT) Group, Department of Mathematics & Applied Mathematics, University of Cape Town, Cape Town 7700, South Africa; surajit.kalita@uct.ac.za
[2] Department of Physics, Indian Institute of Science, Bangalore 560012, India
* Correspondence: bm@iisc.ac.in

Abstract: Over the last couple of decades, there have been direct and indirect evidences for massive compact objects than their conventional counterparts. A couple of such examples are super-Chandrasekhar white dwarfs and massive neutron stars. The observations of more than a dozen peculiar over-luminous type Ia supernovae predict their origins from super-Chandrasekhar white dwarf progenitors. On the other hand, recent gravitational wave detection and some pulsar observations provide arguments for massive neutron stars, lying in the famous mass-gap between lowest astrophysical black hole and conventional highest neutron star masses. We show that the idea of a squashed fuzzy sphere, which brings in noncommutative geometry, can self-consistently explain either of the massive objects as if they are actually fuzzy or squashed fuzzy spheres. Noncommutative geometry is a branch of quantum gravity. If the above proposal is correct, it will provide observational evidences for noncommutativity.

Keywords: noncommutative geometry; white dwarf; neutron star; equation of state; Chandrasekhar limit

1. Introduction

Quantum mechanics (QM) and general theory of relativity (GR) are widely regarded as the two most promising discoveries of the twentieth century. QM is used to describe different microscopic phenomena, whereas GR is used to explain phenomena in which gravity plays a significant role. QM is primarily based on the Heisenberg algebra, which relates the position operator ($\hat{x}_i$) and the momentum operator ($\hat{p}_i$) as $[\hat{x}_i, \hat{p}_j] = i\hbar \delta_{ij}$, where $\hbar = h/2\pi$, with h being the Planck constant. Note that in QM, position and momentum operators commute among themselves, i.e., $[\hat{x}_i, \hat{x}_j] = [\hat{p}_i, \hat{p}_j] = 0$. GR, on the other hand, is based on the equivalence principle, which can account for the perihelion precision of Mercury, the generation of gravitational waves (GWs), gravitational lensing, and a variety of other fascinating phenomena. Both QM and GR are required to understand the structure of compact objects, such as white dwarfs (WDs) and neutron stars (NSs). GR primarily governs the hydrostatic balance of a star, which is a macroscopic property; whereas, QM determines the equation of state (EoS), i.e., the relation between pressure and density of the constituent particles.

If a progenitor star has mass approximately in between 10 and $20\,M_\odot$, it becomes a NS at the end of its lifetime. A NS typically possesses central density, ρ_c of about 10^{14} to a few factors of $10^{15}\,\mathrm{g\,cm^{-3}}$ [1]. Although NSs predominantly consist of neutrons, various other particles, including hyperons, may also be present at such a high density. This uncertainty arises from the fact that such a high density has yet to be achieved in the laboratory, and hence, the specific nuclear reactions, as well as their rates, are unknown. Researchers have so far provided various NS EoSs, each comprising different particle contributions and strong nuclear forces. Most of these EoSs are based on the relativistic energy dispersion relation $E^2 = p^2 c^2 + m^2 c^4$, where c is the speed of light and E denotes the energy of the particle with mass m with p being its momentum. Although most NSs have masses of approximately 1 to 2 $M_\odot$, recent pulsar observations PSR J2215+5135 and PSR B1957+20 show that they have masses of about 2.3 and 2.4 $M_\odot$, respectively [2,3]. Similarly, the LIGO/Virgo collaboration detected a GW merger event, GW 190814, where one of the merged objects has a mass of about 2.6 $M_\odot$ [4], which is mostly thought to be a NS [5–8]. Nevertheless, there was no detection of electromagnetic counterpart for this GW event, and hence various other proposals for this object, such as black hole [9,10], quark star [11], etc. have been put forward. In this article, however, we only talk about NSs while referring to this GW event. Based on these observations, various simulations have been performed and it has been suggested that those EoSs, which give the maximum mass of a non-rotating and non-magnetized NS less than 2 $M_\odot$, should be ruled out [5,12,13]. Hence, considering GR formalism, various EoSs, such as FPS [14], ALF1 [15], etc., seem to be inappropriate for NSs. Modified gravity, on the other hand, has emerged as a popular alternative to replace GR in the high-density regime over the last decades. It can be shown that modified gravity alters the hydrostatic balance of the star and thereby increases the mass of a NS [16–18]. As a result, some of these EoSs may still be valid in the modified gravity formalism.

On the other hand, WDs are the end-state of stars with mass $\lesssim (10 \pm 2)\,M_\odot$ [19]. They possess ρ_c typically ranging approximately from 10^5 g cm^{-3} to a few factor of 10^{10} g cm^{-3}. A WD achieves its stable equilibrium configuration by balancing the outward force of the degenerate electron gas with the inward force of gravity. If the WD has a binary companion, it pulls out matter from the companion, resulting in the increase of WD mass. Once the WD hits the Chandrasekhar mass-limit, which is about 1.4 $M_\odot$ for a carbon-oxygen non-rotating, nonmagnetised WD [20], this pressure balance is lost, and it bursts out to create a type Ia supernova (SN Ia). However, recent observations of more than a dozen of peculiar over-luminous SNe Ia [21–29] reveal that they had to be produced from super-Chandrasekhar limiting mass WDs, i.e., the WDs burst significantly above the Chandrasekhar mass-limit [30,31]. Various theories incorporating magnetic fields [32,33], modified gravity [34–36], etc., can explain this violation of the Chandrasekhar mass-limit, although each has its own set of limitations.

The goal of this work is to introduce noncommutativity (NC) among position and momentum variables and examine how it affects WDs and NSs. A popular way of proposing NC is by defining $[x_i, x_j] = i\eta$ and $[p_i, p_j] = i\theta$ with η and θ being the NC parameters. It was shown that in the presence of NC, the spacetime metric alters [37], causing the event horizon to shift and the singularity at the centre of a black hole to vanish, which is replaced by a regular de-Sitter core [38–40]. It further alters some other properties associated with black holes, such as the stability of Cauchy horizon [41], mini black hole formation with the central singularity replaced by a self-gravitating droplet [42], the Hawking temperature [43]. Various researchers also utilised this NC to describe a variety of other phenomena including Berry curvature, fundamental length-scale, Landau levels, gamma-ray bursts, and many more [44–49]. Note that the basic assumption in the structure of this NC is quite ad-hoc. In 1992, Madore introduced the idea of a three-dimensional fuzzy sphere NC [50], which has been used to better understand the thermodynamical features of non-interacting degenerate electron gas [51,52]. This formalism was later refined by projecting all the points of the fuzzy sphere onto an equatorial plane and named this configuration a squashed fuzzy sphere [53]. This NC model was also proven to imitate the magnetic field by producing distinct energy levels, which are similar to the Landau levels created in the presence of a magnetic field [54].

Apart from a few black hole applications, the implication of NC on compact objects is a relatively novel concept. We earlier showed its applications on the structure of WDs. We considered both the formalism of NC separately and showed that they modify the energy dispersion relation of electrons [55,56]. We further used this relation to obtain a new EoS of the degenerate electrons present in WDs and showed that it can explain the super-Chandrasekhar limiting mass WDs, which are believed to be the progenitors of the observed over-luminous type Ia supernovae. We obtained the maximum mass of a WD to be about 2.6 $M_\odot$ in the presence of NC, and this mass-limit decreases as the strength of NC reduces. We further showed that the NC is prominent if the separation of electrons is less than the Compton wavelength of electrons, and it turns out to be an emergent phenomenon.

The EoS obtained for WD is valid only up to neutron drip density, above which neutron starts contributing to the degenerate pressure. In this article, we obtain a new EoS above the neutron drip density taking into account of NC and derive a new mass–radius relation for NSs. With the advancement of technology, different proposed electromagnetic and GW detectors are likely to detect numerous WDs and NSs. If their observed masses and radii follow the mass–radius relations predicted based on NC, it would be a direct proof of NC's existence.

The following is a breakdown of how this article is structured. In Section 2, we briefly review the squashed fuzzy sphere formalism and the modified energy dispersion relation, which we utilize in Section 3 to derive the EoS for degenerate particles reside inside WD and NS in the presence of NC. We further use this EoS to obtain the new mass–radius relation of the NS in Section 4. Finally, we present our concluding remarks in Section 5.

2. Squashed Fuzzy Sphere Formalism and Modified Energy Dispersion Relation

In this section, we recapitulate the basic formalism of a squashed fuzzy sphere. In $\mathbb{R}^3$, the equation of a sphere with radius r is given by

$$x_1^2 + x_2^2 + x_3^2 = r^2, \tag{1}$$

where (x_1, x_2, x_3) are the Cartesian coordinates of the points on the sphere. A fuzzy sphere is similar to a regular sphere, except that its coordinates x_i ($i = 1, 2, 3$) follow the regular QM angular momentum algebra [50]. Hence, if J_i are the generators of $SU(2)$ group in an N-dimensional irreducible representation, we have

$$x_i = \kappa J_i, \tag{2}$$

with

$$J_1^2 + J_2^2 + J_3^2 = \frac{\hbar^2}{4}\left(N^2 - 1\right)\mathbb{I} = C_N \mathbb{I}, \tag{3}$$

where κ is the proportionality (scaling) constant, $C_N = \hbar^2(N^2 - 1)/4$, and $\mathbb{I}$ is the N-dimensional identity matrix. Substituting J_i in terms of x_i and defining $k = \kappa r$, we obtain

$$\kappa = \frac{r}{\sqrt{C_N}} \quad \text{and} \quad k = \frac{r^2}{\sqrt{C_N}}. \tag{4}$$

Since the angular momentum algebra follows the commutation relation $[J_j, J_k] = i\hbar \epsilon_{jkl} J_l$, the coordinates of the fuzzy sphere follow [50]

$$[x_j, x_k] = i\frac{k\hbar}{r}\epsilon_{jkl} x_l. \tag{5}$$

When all the points of a fuzzy sphere are projected on any of its equatorial planes, the result is a squashed fuzzy sphere. It should be noted that this is not a stereographic projection. The projection of all the points of a fuzzy sphere on the $x_1 - x_2$ equatorial plane is shown in Figure 1. The points of the upper hemisphere are projected on the equatorial plane's top side, while the points of the lower hemisphere are projected on the plane's lower side, and then they are glued together. Writing x_3 in terms of x_1 and x_2, using Equation (1) and replacing it in Equation (5), we obtain the squashed fuzzy sphere's commutation relation, given by [53]

$$[x_1, x_2] = \pm i\frac{k\hbar}{r}\sqrt{r^2 - x_1^2 - x_2^2}. \tag{6}$$

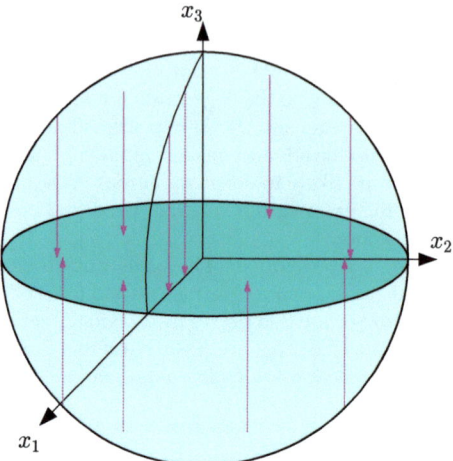

Figure 1. Schematic diagram of a squashed fuzzy sphere where all the points of the fuzzy sphere are projected on x_1-x_2 plane.

The Laplacian for the squashed fuzzy sphere is given by [53]

$$\Box_s = \frac{1}{k^2} \sum_{i=1}^{2} [X_i, [X_i, \cdot]], \qquad (7)$$

which satisfies the following eigenvalue equation

$$\Box_s \hat{Y}_{\tilde{m}}^{\tilde{l}} = \frac{\hbar^2}{r^2} \{\tilde{l}(\tilde{l}+1) - \tilde{m}^2\} \hat{Y}_{\tilde{m}}^{\tilde{l}}, \qquad (8)$$

where $\tilde{l}(\tilde{l}+1) - \tilde{m}^2$ are eigenvalues of the squashed fuzzy Laplacian with $\tilde{l}$ taking all the integer values from 0 to $N-1$ and $\tilde{m}$ taking all the integer values from $-\tilde{l}$ to $\tilde{l}$. Using this Laplacian, one can obtain the energy dispersion relation in the squashed fuzzy sphere given by [53,56]

$$E_{\tilde{l},\tilde{m}}^2 = \frac{2\hbar c^2}{k\sqrt{N^2-1}} [\tilde{l}(\tilde{l}+1) - \tilde{m}(\tilde{m}\pm 1)]. \qquad (9)$$

Moreover, Equation (6) in spherical polar coordinates (r, θ, ϕ) can be recast as

$$[\sin\theta\cos\phi, \sin\theta\sin\phi] = \pm i \frac{k\hbar}{r^2} \cos\theta. \qquad (10)$$

This shows the NC is between θ and ϕ alone, while they are commutative with r-coordinate. In other words, the formalism of a squashed fuzzy sphere is such that it actually provides a NC between the azimuthal and polar coordinates. This is because the squashed plane in a fuzzy sphere can be any of its equatorial planes, which means that the squashed fuzzy sphere possesses rotational symmetry about the equatorial plane. Regardless of the squashed plane, the above energy dispersion remains unchanged. As a result, a particle travelling along the r-coordinate in a squashed fuzzy sphere is not affected by NC and the exact energy dispersion relation is given by

$$E^2 = p_r^2 c^2 + m^2 c^4 \left[1 + \{\tilde{l}(\tilde{l}+1) - \tilde{m}(\tilde{m}\pm 1)\} \frac{2\hbar}{m^2 c^2 k \sqrt{N^2-1}} \right], \qquad (11)$$

where p_r is the momentum of the particle in the radial direction. In the limit $N \gg 1$, the above expression reduces to [56]

$$E^2 = p_r^2 c^2 + m^2 c^4 (1 + 2\nu \theta_D), \quad \nu \in \mathbb{Z}^{0+}, \qquad (12)$$

where $\theta_D = 2\hbar/m^2 c^2 k$. It is noticeable that this expression is very similar to the dispersion relation of Landau levels in the presence of a magnetic field. If the magnetic field is present along z-direction with strength B, the energy dispersion relation for an electron with mass m_e is given by [54]

$$E^2 = p_z^2 c^2 + m_e^2 c^4 \left(1 + 2\nu \frac{B}{B_c}\right), \quad \nu \in \mathbb{Z}^{0+}, \qquad (13)$$

where p_z is the momentum of the electron along the z-direction and $B_c = m_e^2 c^3/\hbar e$ is the critical magnetic field (Schwinger limit) with e being the charge of an electron. Comparing Equations (12) and (13), we obtain

$$B \equiv \frac{2c}{ek}. \qquad (14)$$

Hence, in a squashed fuzzy sphere, k^{-1} behaves as the strength of NC. A detailed discussion on the equivalence of magnetic field and NC was given by Kalita et al. [56]. Equation (12) provides the energy dispersion relation of one squashed fuzzy sphere, inside which k is constant. If we consider a sequence of concentric squashed fuzzy spheres with same N, from Equation (4), we have $k \propto r^2$, i.e., k increases and thus the strength of NC reduces from centre to the surface. As a result, all concentric spheres with a radius greater than r contribute to the effective NC at a point with radius r. From Equation (6), it is evident that NC vanishes at the surface.

3. Noncommutative Equation of State for Degenerate Particles

In this section, we first discuss the commutative cases. In 1935, Chandrasekhar provided EoS for the degenerate electrons [20]. This EoS is valid for a system whose density is less than the neutron drip density (approximately 3.18×10^{11} g cm^{-3}), above which neutron also starts contributing to the degenerate pressure. Harrison and Wheeler (hereinafter HW), in 1958, provided an EoS considering a semi-empirical mass formula, which is valid even at higher densities than neutron drip density. Denoting ρ to be the matter density and $\mathcal{P}$ the total pressure, HW EoS is given by [1]

$$\rho = \frac{n_{\text{ion}} M(A,Z) + \epsilon_e(n_e) - n_e m_e c^2 + \epsilon_n(n_n)}{c^2}, \qquad (15)$$

$$\mathcal{P} = \mathcal{P}_e + \mathcal{P}_n,$$

where ϵ_n is the energy density of neutrons and ϵ_e is the same for electrons. Similarly, $\mathcal{P}_e$ and $\mathcal{P}_n$ are, respectively, the pressures due to electrons and neutrons. Here, n_e, n_n, and n_{ion} are the number densities of electron, neutron, and ion, respectively, while $M(A,Z)$ is the energy of nucleus with mass number A and atomic number Z.

In commutative physics, where $E^2 = p^2 c^2 + m^2 c^4$ holds true, the pressures and energy densities are given by

$$\mathcal{P}_e = \frac{m_e c^2}{\lambda_e^3} \phi(x_{F,e}), \quad \mathcal{P}_n = \frac{m_n c^2}{\lambda_n^3} \phi(x_{F,n}), \quad \mathcal{P}_p = \frac{m_p c^2}{\lambda_p^3} \phi(x_{F,p}), \qquad (16)$$

$$\epsilon_e = \frac{m_e c^2}{\lambda_e^3} \chi(x_{F,e}), \quad \epsilon_n = \frac{m_n c^2}{\lambda_n^3} \chi(x_{F,n}), \quad \epsilon_p = \frac{m_p c^2}{\lambda_p^3} \chi(x_{F,p}), \qquad (17)$$

where $\lambda_e = \hbar/m_e c$, $\lambda_n = \hbar/m_n c$, and $\lambda_p = \hbar/m_p c$ are the reduced Compton wavelengths of electron, neutron, and proton, respectively, with m_n being the mass of a neutron and m_p the mass of a proton. Moreover, $x_{F,e} = p_{F,e}/m_e c$, $x_{F,n} = p_{F,n}/m_n c$, and $x_{F,p} = p_{F,p}/m_p c$ with $p_{F,e}$, $p_{F,n}$, and $p_{F,p}$ being the Fermi momentum of electron, neutron, and proton respectively, and

$$\phi(x_F) = \frac{1}{8\pi^2}\left[x_F\sqrt{1+x_F^2}\left(\frac{2x_F^2}{3}-1\right) + \ln\left\{x_F + \sqrt{1+x_F^2}\right\}\right],$$

$$\chi(x_F) = \frac{1}{8\pi^2}\left[x_F\sqrt{1+x_F^2}\left(2x_F^2+1\right) - \ln\left\{x_F + \sqrt{1+x_F^2}\right\}\right].$$

This EoS can explain physics beyond the neutron drip density regime. However, above 4.54×10^{12} g cm^{-3}, the neutrons contribute most in the pressure and density. Hence, beyond this density, HW used the idea n-p-e EoS where neutrons, protons, and electrons are considered to be degenerate and non-interacting. In the commutative picture, the n-p-e EoS is given by [1]

$$\mathcal{P} = \mathcal{P}_e + \mathcal{P}_n + \mathcal{P}_p,$$
$$\rho = \frac{\epsilon_e + \epsilon_n + \epsilon_p}{c^2}. \tag{18}$$

HW and n-p-e EoSs together provide the pressure–density relation of the non-interacting degenerate particles.

In NC, these EoSs are expected to be modified. Vishal and Mukhopadhyay earlier derived a modified HW EoS of degenerate particles in the presence of a constant magnetic field [57]. Later, to study the effect of varying NC on degenerate electron gas, we obtained the following relation [56]

$$\theta_D = \frac{1}{\xi}\frac{h^2 n_e^{2/3}}{\pi m_e^2 c^2}, \tag{19}$$

where ξ is a dimensionless proportionality constant. The dependency $\theta_D \propto n_e^{2/3}$ is required to match the modified EoS with the Chandrasekhar EoS at a low density where NC does not have any significant influence. Thus, we obtained the modified EoS for degenerate electrons when all the electrons reside in the ground level, given by [55,56]

$$\mathcal{P}_e = \frac{2\rho_e^{2/3}}{\xi h \mu_e^{2/3} m_p^{2/3}}\left\{p_{F,e} E_{F,e} - m_e^2 c^3 \ln\left(\frac{E_{F,e} + p_{F,e} c}{m_e c^2}\right)\right\}, \tag{20}$$

$$\rho_e = \frac{64 \mu_e m_p p_{F,e}^3}{\xi^3 h^3}, \tag{21}$$

where μ_e is the mean molecular weight per electron and $E_{F,e}$ is the Fermi energy of electrons which is related to $p_{F,e}$ as

$$E_{F,e}^2 = p_{F,e}^2 c^2 + m_e^2 c^4 (1 + 2\nu\theta_D). \tag{22}$$

Since, for the present purpose, we require the modified HW and n-p-e EoSs in the presence of NC, we also assume a similar form of pressure–density relation except that the various properties of the electron are now replaced by the same for the corresponding particle. After performing some simplifications using Equations (19) and (21), we obtain

$$\theta_D = \frac{16}{\xi^3}\frac{x_F^2}{\pi}. \tag{23}$$

Note that we do not put any subscript for the electron in this equation, which means that it is valid for electrons, protons, and neutrons. We further denote the NC parameters of neutron, proton, and electron as $\theta_{D,n}$, $\theta_{D,p}$, and $\theta_{D,e}$ respectively. Thus, the modified HW and n-p-e EoSs are given by the same expressions of Equations (15) and (18), except the pressures and energy densities of the respective particles are modified as follows:

$$\mathcal{P}_e = \frac{m_e c^2 \theta_{D,e}}{2\pi^2 \lambda_e^3} \eta(x_{F,e}), \quad \mathcal{P}_n = \frac{m_n c^2 \theta_{D,n}}{2\pi^2 \lambda_n^3} \eta(x_{F,n}), \quad \mathcal{P}_p = \frac{m_p c^2 \theta_{D,p}}{2\pi^2 \lambda_p^3} \eta(x_{F,p}), \quad (24)$$

$$\epsilon_e = \frac{m_e c^2 \theta_{D,e}}{2\pi^2 \lambda_e^3} \psi(x_{F,e}), \quad \epsilon_n = \frac{m_n c^2 \theta_{D,n}}{2\pi^2 \lambda_n^3} \psi(x_{F,n}), \quad \epsilon_p = \frac{m_p c^2 \theta_{D,p}}{2\pi^2 \lambda_p^3} \psi(x_{F,p}), \quad (25)$$

where

$$\eta(x_F) = \frac{1}{2} x_F \sqrt{1 + x_F^2} - \frac{1}{2} \ln\left(x_F + \sqrt{1 + x_F^2}\right),$$

$$\psi(x_F) = \frac{1}{2} x_F \sqrt{1 + x_F^2} + \frac{1}{2} \ln\left(x_F + \sqrt{1 + x_F^2}\right).$$

We already showed that if all the electrons reside only in the ground energy level, we require $\zeta_e \approx 1.5$ to match the noncommutative EoS with the Chandrasekhar EoS at the low density [55]. However, the corresponding parameters for neutron and proton (ζ_n and ζ_p) remain arbitrary. We choose ζ_n and ζ_p in such a way that the maximum mass of NS in the mass–radius curve is above $2\,M_\odot$, which we discuss in the next section. Thereby, we calculate both the noncommutative HW and n-p-e EoSs when all the particles are in their respective ground levels (see Figure 2). Note that the neutron drip density changes in the presence of NC, which was also shown earlier in the presence of strong magnetic fields forming Landau levels [57].

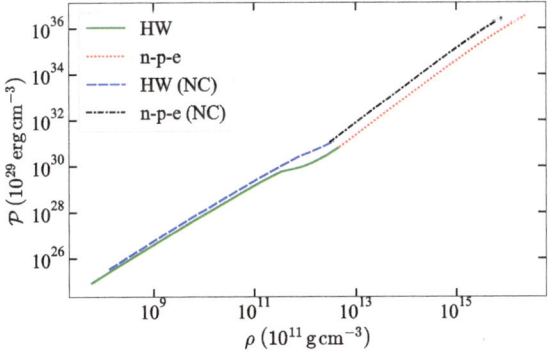

Figure 2. HW and n-p-e EoSs in the commutative and noncommutative formalisms.

4. Mass—Radius Relation of Noncommutativity Inspired White Dwarfs and Neutron Stars

We assume a semi-classical approach to obtain the mass–radius relations for WDs and NSs. In other words, we use classical pressure balance and mass estimate equations (also known as the Tolman-Oppenheimer-Volkoff or TOV equations) while the EoS is governed by the NC. The TOV equations are given by [58]

$$\frac{dM}{dr} = 4\pi r^2 \rho,$$

$$\frac{d\mathcal{P}}{dr} = -\frac{G}{r^2}\left(\rho + \frac{\mathcal{P}}{c^2}\right)\left(M + \frac{4\pi r^3 \mathcal{P}}{c^2}\right)\left(1 - \frac{2GM}{c^2 r}\right)^{-1}, \quad (26)$$

where M is the mass of the star inside a volume of radius r and G is the Newton gravitational constant. We earlier showed that NC is prominent when the inter-particle separation is less than the Compton wavelength of the respective particles [55,56]. When we consider the hydrostatic balance equations for the entire star having a macroscopic size, the length-scale of the stellar fluid is much larger than the Compton wavelength of the constituent particles. Thus, the TOV equations remain commutative in the semi-classical limit. Furthermore, when all the electrons reside in the ground energy level, we already found the mass–radius curve earlier [55,56], and for recapitulation, we display it again in Figure 3. It is evident that NC inspired WDs can possess more mass than the conventional WDs following the Heisenberg algebra. The maximum mass of such a non-rotating WD is estimated to be around 2.6 $M_\odot$, explaining the origins of many over-luminous SNe Ia.

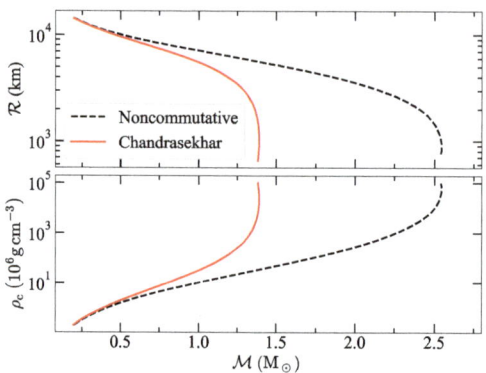

Figure 3. **Upper** figure: Mass–radius relation; **Lower** figure: variation of central density with the mass of WDs. Here $\mathcal{M}$ and $\mathcal{R}$ are the mass and radius of the star, respectively.

In the case of a NS, ρ_c is high, and we employ a combination of HW and n-p-e EoSs to derive its mass–radius relation, as illustrated in Figure 4. In the commutative picture, the maximum mass turns out to be just 0.7 $M_\odot$, while it is increased to about 2 $M_\odot$ in the case of NC, which is supported by the observations of massive pulsars. However, the radius increases to 20 km in this situation, which is almost ruled out by existing GW observations [59–61]. Note that the relation $\theta_D \propto x_F^2$ in Equation (23), is valid for electrons and we extrapolate it for neutrons and protons too. If we choose a different dependency of θ_D on x_F, the EoS alters and so does the mass–radius relation for NS. Figure 5 depicts several mass–radius relations for various powers of x_F. It is evident that as the power decreases, the radius for maximum mass falls as well and when $\theta_D \propto \sqrt{x_F}$, the maximum mass is about 2.08 $M_\odot$, with radius being 12 km. These masses and radii obey the observational bounds of NSs, and hence, such an EoS is a realistic one.

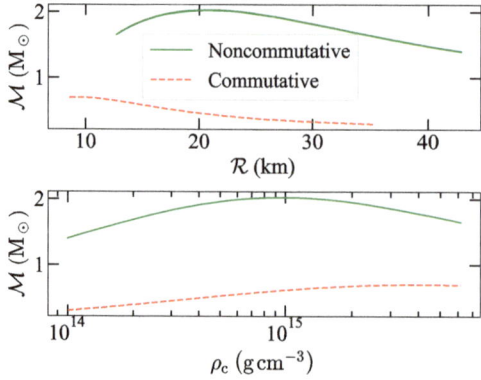

Figure 4. **Upper** figure: Mass–radius relation; **Lower** figure: variation of central density with the mass of NSs.

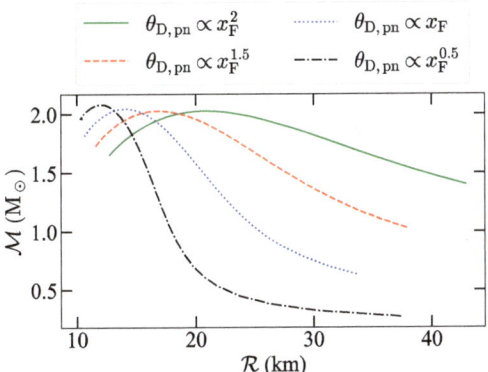

Figure 5. Mass–radius relations of NC induced NSs for various $\theta_D - x_F$ relations. $\theta_{D,pn}$ means θ_D for proton and neutron.

5. Conclusions

For a long time, scientists have been fascinated by the possibility of massive WDs and NSs from several direct or indirect observations. Various ideas, such as magnetic fields and rotation, modified gravity, etc., have been thoroughly investigated in recent years. Rotation can explain massive NSs, which, however, alone fails to elucidate the massive WDs with a mass of about 2.8 $M_\odot$. High magnetic fields can, in principle, explain both these massive objects. Nonetheless, the maximum field that a compact object can possess is always a source of contention. Similarly, despite the fact that modified gravity can explain such high masses, it has so far been impossible to identify the most appropriate one from the hundreds of such modified gravity models. In this regard, each of these theories suffers its own limitations.

In the context of astronomical objects, the concept of NC is relatively new. With the exception of a few applications on black holes and wormholes, it has received little attention in astrophysics. We earlier self-consistently used NC for the first time to explain the super-Chandrasekhar WDs [55,56]. We first employed a basic planar NC model and later used a squashed fuzzy sphere model to modify the EoS of the degenerate electrons present in a WD. This modification leads to increasing the mass of a WD. If the electrons solely occupy the ground energy level, i.e., NC is the strongest, the new mass-limit of WD turns out to be about 2.6 $M_\odot$. As NC weakens and electrons occupy higher energy levels, this mass-limit decreases. It is to be noted that the effect of NC is only prominent at sufficiently high densities and negligible at low densities. Hence, our model supports the observed bigger WDs, which generally have very low densities, and it does not violate any observable at such low densities. We have already established that the strength of NC depends on the length scale of the system. If the inter-particle separation distance is smaller than the Compton wavelength of the corresponding particle, NC starts becoming prominent [55]. Furthermore, NC does not have any classical effect, unlike magnetic fields (i.e., field pressure, tension, etc.), and hence, the problem of instabilities that occurred in magnetic fields does not arise in the case of NC, making the NC model preferable over that of magnetic fields.

In this article, we have extrapolated NC to higher densities and investigated for its effect on the structure of NSs. For simplicity, we have only considered the effects of neutrons, protons, and electrons, and assumed they are non-interacting. In commutative physics, it is well known that such an EoS gives the maximum mass of a NS to be about 0.7 $M_\odot$ [1]. However, current observations demand the maximum mass of a non-rotating NS has to be at least 2 $M_\odot$ [5,12,13]. Once we introduce NC, we have found that even such non-interacting particles can constitute an EoS which generates NS with a mass of about 2.1 $M_\odot$ and radius 12 km. Such an EoS is perfectly legitimate with the current observation constraints. Note that we have only considered the case where all the particles are in their respective ground energy levels, which is the scenario for the strongest NC. One can, in principle, consider higher occupancy in the energy levels. However, it only reduces the mass of the NS, as we have seen in the case of WDs [56], and the maximum mass could fall below 2 $M_\odot$ and those cases would be unrealistic. In such instances, one must account for the interactions that may occur between the various particles at these high densities, which, however, is beyond the scope of this paper. In such cases, even the EoSs, which are considered non-physical, might not be ruled out if they are affected by NC. In the future, GW observations may detect numerous massive WDs and NSs [33,62,63], allowing us to constrain more EoSs and examine the NC effect on these compact objects more closely. If observed masses and radii of WDs and NSs follow the respective predicted mass–radius relations based on NC, it would be a direct confirmation for the existence of NC at scales far away from the Planck scale.

Author Contributions: S.K. and B.M. conceptualized the project; S.K. carried out the formal analysis and original writing; B.M. supervised the project and edited the draft. All authors have read and agreed to the published version of the manuscript.

Funding: This research received no external funding.

Institutional Review Board Statement: Not applicable.

Informed Consent Statement: Not applicable.

Data Availability Statement: The data presented in this study are available on request from the corresponding author.

Acknowledgments: S.K. would like to acknowledge support from the South African Research Chairs Initiative of the Department of Science and Technology and the National Research Foundation.

Conflicts of Interest: The authors declare no conflict of interest.

References

1. Shapiro, S.L.; Teukolsky, S.A. *Black Holes, White Dwarfs, and Neutron Stars: The Physics of Compact Objects*; John Wiley and Sons Hoboken, NJ, USA, 1983.
2. Linares, M.; Shahbaz, T.; Casares, J. Peering into the Dark Side: Magnesium Lines Establish a Massive Neutron Star in PSR J2215 + 5135. *Astrophys. J.* **2018**, *859*, 54. [CrossRef]
3. van Kerkwijk, M.H.; Breton, R.P.; Kulkarni, S.R. Evidence for a Massive Neutron Star from a Radial-velocity Study of the Companion to the Black-widow Pulsar PSR B1957+20. *Astrophys. J.* **2011**, *728*, 95. [CrossRef]
4. Abbott, R.; Abbott, T.D.; Abraham, S.; Acernese, F.; Ackley, K.; Adams, C.; Adhikari, R.X.; Adya, V.B.; Affeldt, C.; Agathos M.; et al. GW190814: Gravitational Waves from the Coalescence of a 23 Solar Mass Black Hole with a 2.6 Solar Mass Compact Object. *Astrophys. J. Lett.* **2020**, *896*, L44. [CrossRef]
5. Most, E.R.; Papenfort, L.J.; Weih, L.R.; Rezzolla, L. A lower bound on the maximum mass if the secondary in GW190814 was once a rapidly spinning neutron star. *Mon. Not. R. Astron. Soc. Lett.* **2020**, *499*, L82–L86. [CrossRef]
6. Huang, K.; Hu, J.; Zhang, Y.; Shen, H. The Possibility of the Secondary Object in GW190814 as a Neutron Star. *Astrophys. J.* **2020**, *904*, 39. [CrossRef]
7. Tsokaros, A.; Ruiz, M.; Shapiro, S.L. GW190814: Spin and Equation of State of a Neutron Star Companion. *Astrophys. J.* **2020**, *905*, 48. [CrossRef]
8. Dexheimer, V.; Gomes, R.O.; Klähn, T.; Han, S.; Salinas, M. GW190814 as a massive rapidly rotating neutron star with exotic degrees of freedom. *Astrophys. J.* **2021**, *103*, 025808. [CrossRef]
9. Yang, Y.; Gayathri, V.; Bartos, I.; Haiman, Z.; Safarzadeh, M.; Tagawa, H. Black Hole Formation in the Lower Mass Gap through Mergers and Accretion in AGN Disks. *Astrophys. J.* **2020**, *901*, L34. [CrossRef]
10. Vattis, K.; Goldstein, I.S.; Koushiappas, S.M. Could the 2.6 $M_\odot$ object in GW190814 be a primordial black hole? *Phys. Rev.* **2020** *102*, 061301. [CrossRef]
11. Bombaci, I.; Drago, A.; Logoteta, D.; Pagliara, G.; Vidaña, I. Was GW190814 a Black Hole-Strange Quark Star System? *Phys. Rev. Lett.* **2021**, *126*, 162702. [CrossRef]
12. Rezzolla, L.; Most, E.R.; Weih, L.R. Using Gravitational-wave Observations and Quasi-universal Relations to Constrain the Maximum Mass of Neutron Stars. *Astrophys. J.* **2018**, *852*, L25. [CrossRef]
13. Malik, T.; Agrawal, B.K.; De, J.N.; Samaddar, S.K.; Providência, C.; Mondal, C.; Jha, T.K. Tides in merging neutron stars Consistency of the GW170817 event with experimental data on finite nuclei. *Phys. Rev. C* **2019**, *99*, 052801. [CrossRef]
14. Pandharipande, V.R.; Ravenhall, D.G. Hot Nuclear Matter. In *Proceedings of the Nuclear Matter and Heavy Ion Collisions*, Les Houches, France, 7–16 February 1989; NATO Advanced Study Institute (ASI) Series B; Volume 205, p. 103.

15. Alford, M.; Braby, M.; Paris, M.; Reddy, S. Hybrid Stars that Masquerade as Neutron Stars. *Astrophys. J.* **2005**, *629*, 969–978. [CrossRef]
16. Astashenok, A.V.; Capozziello, S.; Odintsov, S.D. Maximal neutron star mass and the resolution of the hyperon puzzle in modified gravity. *Phys. Rev. D* **2014**, *89*, 103509. [CrossRef]
17. Astashenok, A.V.; Odintsov, S.D.; de la Cruz-Dombriz, Á. The realistic models of relativistic stars in $f(R) = R + \alpha R^2$ gravity. *Class. Quantum Gravity* **2017**, *34*, 205008. [CrossRef]
18. Astashenok, A.V. Neutron and quark stars in $f(R)$ gravity. *Int. Mod. Phys. Conf. Ser.* **2016**, *41*, 1660130. [CrossRef]
19. Lauffer, G.R.; Romero, A.D.; Kepler, S.O. New full evolutionary sequences of H- and He-atmosphere massive white dwarf stars using MESA. *Mon. Not. R. Astron. Soc.* **2018**, *480*, 1547–1562. [CrossRef]
20. Chandrasekhar, S. The highly collapsed configurations of a stellar mass (Second paper). *Mon. Not. R. Astron. Soc.* **1935**, *95*, 207–225. [CrossRef]
21. Howell, D.A.; Sullivan, M.; Nugent, P.E.; Ellis, R.S.; Conley, A.J.; Le Borgne, D.; Carlberg, R.G.; Guy, J.; Balam, D.; Basa, S.; et al. The type Ia supernova SNLS-03D3bb from a super-Chandrasekhar-mass white dwarf star. *Nature* **2006**, *443*, 308–311. [CrossRef]
22. Hicken, M.; Garnavich, P.M.; Prieto, J.L.; Blondin, S.; DePoy, D.L.; Kirshner, R.P.; Parrent, J. The Luminous and Carbon-rich Supernova 2006gz: A Double Degenerate Merger? *Astrophys. J.* **2007**, *669*, L17–L20. [CrossRef]
23. Yamanaka, M.; Kawabata, K.S.; Kinugasa, K.; Tanaka, M.; Imada, A.; Maeda, K.; Nomoto, K.; Arai, A.; Chiyonobu, S.; Fukazawa, Y.; et al. Early Phase Observations of Extremely Luminous Type Ia Supernova 2009dc. *Astrophys. J.* **2009**, *707*, L118–L122. [CrossRef]
24. Tanaka, M.; Kawabata, K.S.; Yamanaka, M.; Maeda, K.; Hattori, T.; Aoki, K.; Nomoto, K.; Iye, M.; Sasaki, T.; Mazzali, P.A.; et al. Spectropolarimetry of Extremely Luminous Type Ia Supernova 2009dc: Nearly Spherical Explosion of Super-Chandrasekhar Mass White Dwarf. *Astrophys. J.* **2010**, *714*, 1209–1216. [CrossRef]
25. Silverman, J.M.; Ganeshalingam, M.; Li, W.; Filippenko, A.V.; Miller, A.A.; Poznanski, D. Fourteen months of observations of the possible super-Chandrasekhar mass Type Ia Supernova 2009dc. *Mon. Not. R. Astron. Soc.* **2011**, *410*, 585–611. [CrossRef]
26. Taubenberger, S.; Benetti, S.; Childress, M.; Pakmor, R.; Hachinger, S.; Mazzali, P.A.; Stanishev, V.; Elias-Rosa, N.; Agnoletto, I.; Bufano, F.; et al. High luminosity, slow ejecta and persistent carbon lines: SN 2009dc challenges thermonuclear explosion scenarios. *Mon. Not. R. Astron. Soc.* **2011**, *412*, 2735–2762. [CrossRef]
27. Yuan, F.; Quimby, R.M.; Wheeler, J.C.; Vinkó, J.; Chatzopoulos, E.; Akerlof, C.W.; Kulkarni, S.; Miller, J.M.; McKay, T.A.; Aharonian, F. The Exceptionally Luminous Type Ia Supernova 2007if. *Astrophys. J.* **2010**, *715*, 1338–1343. [CrossRef]
28. Scalzo, R.; Aldering, G.; Antilogus, P.; Aragon, C.; Bailey, S.; Baltay, C.; Bongard, S.; Buton, C.; Canto, A., Cellier-Holzem, F.; et al. A Search for New Candidate Super-Chandrasekhar-mass Type Ia Supernovae in the Nearby Supernova Factory Data Set. *Astrophys. J.* **2012**, *757*, 12. [CrossRef]
29. Cao, Y.; Johansson, J.; Nugent, P.E.; Goobar, A.; Nordin, J.; Kulkarni, S.R.; Cenko, S.B.; Fox, O.D.; Kasliwal, M.M.; Fremling, C.; et al. Absence of Fast-moving Iron in an Intermediate Type Ia Supernova between Normal and Super-Chandrasekhar. *Astrophys. J.* **2016**, *823*, 147. [CrossRef]
30. Scalzo, R.A.; Aldering, G.; Antilogus, P.; Aragon, C.; Bailey, S.; Baltay, C.; Bongard, S.; Buton, C.; Childress, M.; Chotard, N.; et al. Nearby Supernova Factory Observations of SN 2007if: First Total Mass Measurement of a Super-Chandrasekhar-Mass Progenitor. *Astrophys. J.* **2010**, *713*, 1073–1094. [CrossRef]
31. Kamiya, Y.; Tanaka, M.; Nomoto, K.; Blinnikov, S.I.; Sorokina, E.I.; Suzuki, T. Super-Chandrasekhar-mass Light Curve Models for the Highly Luminous Type Ia Supernova 2009dc. *Astrophys. J.* **2012**, *756*, 191. [CrossRef]
32. Das, U.; Mukhopadhyay, B. New Mass Limit for White Dwarfs: Super-Chandrasekhar Type Ia Supernova as a New Standard Candle. *Phys. Rev. Lett.* **2013**, *110*, 071102. [CrossRef]
33. Kalita, S.; Mukhopadhyay, B. Continuous gravitational wave from magnetized white dwarfs and neutron stars: Possible missions for LISA, DECIGO, BBO, ET detectors. *Mon. Not. R. Astron. Soc.* **2019**, *490*, 2692–2705. [CrossRef]
34. Kalita, S.; Mukhopadhyay, B. Modified Einstein's gravity to probe the sub- and super-Chandrasekhar limiting mass white dwarfs: A new perspective to unify under- and over-luminous type Ia supernovae. *J. Cosmol. Astropart. Phys.* **2018**, *9*, 007. [CrossRef]
35. Sarmah, L.; Kalita, S.; Wojnar, A. Stability criterion for white dwarfs in Palatini $f(R)$ gravity. *Phys. Rev. D* **2022**, *105*, 024028. [CrossRef]
36. Kalita, S.; Sarmah, L. Weak-field limit of $f(R)$ gravity to unify peculiar white dwarfs. *Phys. Lett. B* **2022**, *827*, 136942. [CrossRef]
37. Nair, V.P. Noncommutative Mechanics, Landau Levels, Twistors, and Yang-Mills Amplitudes. *Lect. Notes Phys.* **2006**, *1*, 97.
38. Nicolini, P.; Smailagic, A.; Spallucci, E. Noncommutative geometry inspired Schwarzschild black hole. *Phys. Lett. B* **2006**, *632*, 547–551. [CrossRef]
39. Nicolini, P. Noncommutative Black Holes, the Final Appeal to Quantum Gravity: A Review. *Int. J. Mod. Phys. A* **2009**, *24*, 1229–1308. [CrossRef]
40. Kumar, R.; Ghosh, S.G. Accretion onto a noncommutative geometry inspired black hole. *Eur. Phys. J. C* **2017**, *77*, 577. [CrossRef]
41. Batic, D.; Nicolini, P. Fuzziness at the horizon. *Phys. Lett. B* **2010**, *692*, 32–35. [CrossRef]
42. Arraut, I.; Batic, D.; Nowakowski, M. A noncommutative model for a mini black hole. *Class. Quantum Gravity* **2009**, *26*, 245006. [CrossRef]
43. Franchino-Viñas, S.A.; Pisani, P. Thermodynamics in the NC disc. *Eur. Phys. J. Plus* **2018**, *133*, 421. [CrossRef]
44. Seiberg, N.; Witten, E. String theory and noncommutative geometry. *J. High Energy Phys.* **1999**, *1999*, 032. [CrossRef]

45. Amelino-Camelia, G.; Majid, S. Waves on Noncommutative Space-Time and Gamma-Ray Bursts. *Int. J. Mod. Phys. A* **2000**, *15*, 4301–4323. [CrossRef]
46. Amelino-Camelia, G. Testable scenario for relativity with minimum length. *Phys. Lett. B* **2001**, *510*, 255–263. [CrossRef]
47. Magueijo, J.; Smolin, L. Lorentz Invariance with an Invariant Energy Scale. *Phys. Rev. Lett.* **2002**, *88*, 190403. [CrossRef]
48. Amelino-Camelia, G. Relativity: Special treatment. *Nature* **2002**, *418*, 34–35. [CrossRef]
49. Son, D.T.; Yamamoto, N. Berry Curvature, Triangle Anomalies, and the Chiral Magnetic Effect in Fermi Liquids. *Phys. Rev. Lett.* **2012**, *109*, 181602. [CrossRef]
50. Madore, J. The fuzzy sphere. *Class. Quantum Gravity* **1992**, *9*, 69–87. [CrossRef]
51. Chandra, N.; Groenewald, H.W.; Kriel, J.N.; Scholtz, F.G.; Vaidya, S. Spectrum of the three-dimensional fuzzy well. *J. Phys. Math. Gen.* **2014**, *47*, 445203. [CrossRef]
52. Scholtz, F.G.; Kriel, J.N.; Groenewald, H.W. Thermodynamics of Fermi gases in three dimensional fuzzy space. *Phys. Rev. D* **2015**, *92*, 125013. [CrossRef]
53. Andronache, S.; Steinacker, H.C. The squashed fuzzy sphere, fuzzy strings and the Landau problem. *J. Phys. Math. Gen.* **2015**, *48*, 295401. [CrossRef]
54. Lai, D.; Shapiro, S.L. Cold Equation of State in a Strong Magnetic Field: Effects of Inverse beta -Decay. *Astrophys. J.* **1991**, *383*, 745 [CrossRef]
55. Kalita, S.; Mukhopadhyay, B.; Govindarajan, T.R. Significantly super-Chandrasekhar mass-limit of white dwarfs in noncommutative geometry. *Int. J. Mod. Phys. D* **2021**, *30*, 2150034. [CrossRef]
56. Kalita, S.; Govindarajan, T.R.; Mukhopadhyay, B. Super-Chandrasekhar limiting mass white dwarfs as emergent phenomena of noncommutative squashed fuzzy spheres. *Int. J. Mod. Phys. D* **2021**, *30*, 2150101. [CrossRef]
57. Vishal, M.V.; Mukhopadhyay, B. Revised density of magnetized nuclear matter at the neutron drip line. *Phys. Rev. C* **2014**, *89*, 065804. [CrossRef]
58. Ryder, L. *Introduction to General Relativity*; Cambridge University Press: Cambridge, UK, 2009. [CrossRef]
59. Annala, E.; Gorda, T.; Kurkela, A.; Vuorinen, A. Gravitational-Wave Constraints on the Neutron-Star-Matter Equation of State *Phys. Rev. Lett.* **2018**, *120*, 172703. [CrossRef]
60. Capano, C.D.; Tews, I.; Brown, S.M.; Margalit, B.; De, S.; Kumar, S.; Brown, D.A.; Krishnan, B.; Reddy, S. Stringent constraints on neutron-star radii from multimessenger observations and nuclear theory. *Nat. Astron.* **2020**, *4*, 625–632. [CrossRef]
61. Annala, E.; Gorda, T.; Katerini, E.; Kurkela, A.; Nättilä, J.; Paschalidis, V.; Vuorinen, A. Multimessenger Constraints for Ultradense Matter. *Phys. Rev. X* **2022**, *12*, 011058. [CrossRef]
62. Kalita, S.; Mukhopadhyay, B.; Mondal, T.; Bulik, T. Timescales for Detection of Super-Chandrasekhar White Dwarfs by Gravitational-wave Astronomy. *Astrophys. J.* **2020**, *896*, 69. [CrossRef]
63. Kalita, S.; Mondal, T.; Tout, C.A.; Bulik, T.; Mukhopadhyay, B. Resolving dichotomy in compact objects through continuous gravitational waves observation. *Mon. Not. R. Astron. Soc.* **2021**, *508*, 842–851. [CrossRef]

Article

Vacuum Polarization Instead of "Dark Matter" in a Galaxy

Sergey L. Cherkas [1,*,†] and Vladimir L. Kalashnikov [2,†]

1. Institute for Nuclear Problems, Belarusian State University, Bobruiskaya 11, 220006 Minsk, Belarus
2. Department of Physics, Norwegian University of Science and Technology, Høgskoleringen 5, Realfagbygget, 7491 Trondheim, Norway
* Correspondence: cherkas@inp.bsu.by
† These authors contributed equally to this work.

Abstract: We considered a vacuum polarization inside a galaxy in the eikonal approximation and found that two possible types of polarization exist. The first type is described by the equation of state $p = \rho/3$, similar to radiation. Using the conformally unimodular metric allows us to construct a non-singular solution for this vacuum "substance" if a compact astrophysical object exists in the galaxy's center. As a result, a "dark" galactical halo appears that increases the rotation velocity of a test particle as a function of the distance from a galactic center. The second type of vacuum polarization has a more complicated equation of state. As a static physical effect, it produces the renormalization of the gravitational constant, thus, causing no static halo. However, a non-stationary polarization of the second type, resulting from an exponential increase (or decrease) of the galactic nuclei mass with time in some hypothetical time-dependent process, produces a gravitational potential, appearing similar to a dark matter halo.

Keywords: vacuum energy; dark matter; vacuum polarization; active galaxy nuclei

1. Introduction

Among the various issues of combining general relativity (GR) and quantum mechanics, one encounters the problems of vacuum energy and black holes.

The first problem is to explain why enormous zero-point vacuum energy density $\rho_v \sim k_{max}^4$ (here k_{max} is the UV energy scale of quantum field theory associated with a hard 3-momentum cutoff of the order of the Planck mass M_p) does not influence a universe expansion (e.g., see [1–3] and references herein). The second problem is associated with the loss of unitarity and information inside of the black hole horizon (e.g., see [4,5] and references therein), that prevents the definition of a pure quantum state.

On the other hand, the basis of GR is a notion of manifold [6], i.e., a metric space, which could be covered by coordinate maps. When a concrete space–time possessing some symmetry is considered, one aims to introduce a system of coordinates allowing maximal covering of this particular manifold. For instance, the Schwarzschild solution only describes the region outside the horizon, and one has to introduce the Kruskal coordinates to cover the complete domain [7]. Nevertheless, one could admit an opposite view: restricting the manifold by sewing all the black hole horizons by some coordinate transformation. This approach is similar to a case when a man finds a hole in their trousers at the knee. In such a case, he steps back a little from the hole border and then subtends it into a node with the help of sewing.

It is allowed using the conformally unimodular (CUM) metric [8], where an ultra-compact black hole-like astrophysical object appears as a non-singular ball named "eicheons" [9]. Besides, the vacuum energy problem could be partially solved in the CUM metric if one builds a gravity theory admitting an arbitrary choice of the energy density level [8]. That is possible because the equations for evolution of the Hamiltonian $\mathcal{H}$ and the momentum constraints $\mathcal{P}$ admit not only the trivial solution $\mathcal{P} = 0$, $\mathcal{H} = 0$, but also $\mathcal{P} = 0$, $\mathcal{H} = const$.

The constant compensates for the main part of the vacuum energy density proportional to the Planck mass in the fourth degree [8,10]. Residual energy density, remaining after omitting the main part of the vacuum energy density, is some kind of dark energy and results in a cosmological picture containing a period of linear evolution in cosmic time [10,11] followed by late accelerated expansion.

Both dark energy and dark matter (DM) are unknown "substances" appearing in modern cosmology and astrophysics [12–14]. DM appears not only on cosmological scales but also on galaxy scales. The lowest scale at which there is evidence for DM is of $\approx$kpc [15,16]. Dark energy is associated with vacuum energy, whereas DM is expected to be some kind of a non-baryonic matter weakly interacting with the known particles of the standard model [17–19]. Nevertheless, there are attempts to explain the DM by a DM-like behavior of vacuum energy [20], or a vacuum polarization induced by the gravitational field. Heuristic models of vacuum polarization such as [21–25], which would demand dipolar fluid [26], anti-gravitation [27] or hydrodynamical phenomena in a vacuum treated as hypothetical (super-)fluid [28–30], are of interest.

The conventional renormalization procedure of the quantum field theory applied to vacuum energy near a massive object [31–36] leads to the modification of the gravitational potential only at small distances of the order of gravitational radius that are unobservable with current technologies. That is, the renormalization excludes the manifestation of micro-scale phenomena on the macro-scales (nevertheless, see [20]). This conclusion assumes the general covariance of the mean vacuum value of stress–energy tensor $<0|T_{\mu\nu}|0>$ on a curved background. However, regarding the vacuum state $|0>$, the invariant relatively general transformation of coordinates does not exist [37]. That raises a question: is it reasonable to demand the covariance of $<0|T_{\mu\nu}|0>$ in the absence of invariant $|0>$? If invariance violation, which implies the existence of "æther", takes place, then, similar to condensed-matter physics, DM still could be treated as an emergent phenomenon produced by vacuum polarization.

The outline of this paper is as follows: In Section 2, we argue the necessity of considering a vacuum polarization from a cosmological point of view and explain that the CUM metric is needed to omit the main part of vacuum energy. Section 3 contains a perturbation formalism in the CUM metric, which is required to introduce a vacuum polarization as some media, i.e., "æther". The eikonal approximation is used in Section 4 to obtain the vacuum energy density and pressure of a quantum scalar field by summating the contributions from the distorted virtual plane waves. The expression for a vacuum equation of state is obtained. In Section 5, the F-type vacuum polarization, possessing a radiation equation of state, is used in the Tolman–Volkov–Oppenheimer (TOV) equations for two substances. This type of vacuum polarization results in a dark halo if eicheon is situated in the galactic center. In Section 6, the Φ-type of vacuum polarization is considered. This type of polarization leads to the renormalization of the gravitational constant in the stationary case. However, it can contribute to the DM halo for the non-stationary processes. In the Conclusion, we summarize the consequences of two types of vacuum polarization for galaxies. In the Appendix A, we consider the static and empty universe to demonstrate an example of an exact solution for the system of perturbations, taking into account the F-type vacuum polarization.

2. A Spatially Uniform Universe in the CUM Metric

2.1. CUM Metric in the Five Vectors Theory of Gravity

We based our analysis on using the CUM metric, which is the foundation of the so-called five vectors theory (FVT) [8]. In the course of this analysis, we will use the particular cases of the CUM metric appropriate to the physics considered.

A general class of CUM metrics is defined as [8]

$$ds^2 \equiv g_{\mu\nu}dx^\mu dx^\nu = a^2(1 - \partial_m P^m)^2 d\eta^2 - \gamma_{ij}(dx^i + N^i d\eta)(dx^j + N^j d\eta), \quad (1)$$

where $x^\mu = \{\eta, \mathbf{x}\}$, γ_{ij} is a spatial metric, $a = \gamma^{1/6}$ is a locally defined scale factor, and $\gamma = \det \gamma_{ij}$, η is a conformal time connected with a cosmic time t through $dt = a(\eta, \mathbf{x})d\eta$. The spatial part of the interval (1) looks as

$$dl^2 \equiv \gamma_{ij}dx^i dx^j = a^2(\eta, \mathbf{x})\tilde{\gamma}_{ij}dx^i dx^j, \qquad (2)$$

where $\tilde{\gamma}_{ij} = \gamma_{ij}/a^2$ is a matrix with the unit determinant. The interval (1) is similar formally to the ADM one [38], but the lapse function is taken in the form of $a(1 - \partial_m P^m)$, where P^m is a three-dimensional vector, and ∂_m is a conventional partial derivative. In the gravity theory admitting arbitrary choice of the energy density level [8], there are the Lagrange multipliers P, the shift function N, and three triads e^a to parameterize the spatial metric $\gamma_{ij} = e_i^a e_j^a$. Such model was named the FVT of gravity [8]. In contrast to GR, where the lapse and shift functions are arbitrary, the restrictions $\partial_n(\partial_m N^m) = 0$ and $\partial_n(\partial_m P^m) = 0$ arise in FVT. The Hamiltonian $\mathcal{H}$ and momentum $\mathcal{P}_i$ constraints in the particular gauge $P^i = 0$, $N^i = 0$ obey the constraint evolution Equations [8]:

$$\partial_\eta \mathcal{H} = \partial_i \left(\tilde{\gamma}^{ij} \mathcal{P}_j \right), \qquad (3)$$

$$\partial_\eta \mathcal{P}_i = \frac{1}{3} \partial_i \mathcal{H}, \qquad (4)$$

which admits adding some constant to $\mathcal{H}$. In the FVT frame, it is not necessary that $\mathcal{H} = 0$, but $\mathcal{H} = \mathrm{const}$ is also allowed. The particular cases of the CUM metric corresponding to the Bianchi I model and the spherically symmetric space–time were analyzed in [39,40].

2.2. Uniform, Isotropic and Flat Universe

Let us consider a particular case of (1)

$$ds^2 = a(\eta)^2(d\eta^2 - d\mathbf{x}^2) \qquad (5)$$

corresponding to a spatially uniform, isotropic and flat universe, where the Friedmann equations take the form [11,41,42]:

$$M_p^{-2} e^{4\alpha} \rho - \frac{1}{2} e^{2\alpha} \alpha'^2 = \mathrm{const}, \qquad (6)$$

$$\alpha'' + \alpha'^2 = M_p^{-2} e^{2\alpha} (\rho - 3p). \qquad (7)$$

Here $\alpha(\eta) = \log a(\eta)$, the prime denotes the derivative with respect to the conformal time. We use the system of units $\hbar = c = 1$ and the reduced Planck mass $M_p = \sqrt{\frac{3}{4\pi G}}$ (in physical units $M_p = \sqrt{\frac{3\hbar c}{4\pi G}}$). According to FVT [8], the first Friedmann Equation (6) is satisfied up to some constant, and the main parts of the vacuum energy density and pressure

$$\rho_v \approx (N_{boson} - N_{ferm}) \frac{k_{max}^4}{16\pi^2 a^4}, \qquad (8)$$

$$p_v = \frac{1}{3}\rho_v \qquad (9)$$

do not contribute to the universe expansion because the constant in (6) compensates the vacuum energy density, whereas there is no vacuum contribution in Equation (7) by virtue of the equation of state (9).

Bosons and fermions contribute with opposite signs into a vacuum energy density (8) [43,44]. Here, we do not consider the supersymmetry hypotheses $N_{boson} - N_{ferm}$ due to the absence of evidence for supersymmetric partners to date [45].

For the contributions of massive particles and condensates, we imply the Pauli sum rules [44,46]. These rules are not fulfilled at this moment due to the incompleteness of the standard model of particle physics. Nevertheless, one may hope that they will be satisfied after possible discoveries beyond the standard model.

Other contributors to the vacuum energy density are the terms depending on the derivatives of the universe expansion rate [10,41,42,46]. They have the correct order of magnitude $\rho_v \sim M_p^2 H^2$, where H is the Hubble constant, and explain the accelerated expansion of the universe driven by the residual energy density and pressure [10,41,42,46]:

$$\rho_v = \frac{a'^2}{2a^6} M_p^2 (2 + N_{sc}) \mathcal{S}_0, \quad p_v = \frac{M_p^2 (2 + N_{sc}) \mathcal{S}_0}{a^6} \left(\frac{1}{2} a'^2 - \frac{1}{3} a'' a \right), \tag{10}$$

where $\mathcal{S}_0 = \frac{k_{max}^2}{8\pi^2 M_p^2}$. Equation (10) includes the number of minimally coupled scalar fields N_{sc} plus two degrees of freedom of the gravitational waves [41]. The massless fermions and photons do not contribute to (10) [41].

According to (10), the accelerated expansion of universe allows finding a value of the momentum UV cut-off

$$k_{max} \approx \frac{12 M_p}{\sqrt{2 + N_{sc}}} \tag{11}$$

from the measured value of the universe deceleration parameter and other cosmological observations [10,41]. It should be noted that the UV cut-off of the 3-momentums k_{max} in (8) and hereafter also reflects the diffeomorphism symmetry violation[1] (e.g., see [47–52] and references herein).

3. Perturbations of a Uniform Background in the CUM Metric

In this section, the scalar perturbations[2] of the CUM metric (1) are considered [53]:

$$ds^2 = a(\eta, x)^2 \left(d\eta^2 - \left(\left(1 + \frac{1}{3} \sum_{m=1}^{3} \partial_m^2 F(\eta, x) \right) \delta_{ij} - \partial_i \partial_j F(\eta, x) \right) dx^i dx^j \right). \tag{12}$$

Here the perturbations of the locally defined scale factor are expressed through a gravitational potential Φ:

$$a(\eta, x) = e^{\alpha(\eta, x)} \approx e^{\alpha(\eta)} (1 + \Phi(\eta, x)). \tag{13}$$

A stress–energy tensor can be written in the hydrodynamic approximation

$$T_{\mu\nu} = (p + \rho) u_\mu u_\nu - p g_{\mu\nu}. \tag{14}$$

The perturbations of the energy density $\rho(\eta, x) = \rho(\eta) + \delta\rho(\eta, x)$ and pressure $p(\eta, x) = p(\eta) + \delta p(\eta, x)$ are considered around spatially uniform values.

Let us introduce new variables

$$\wp(\eta, x) = a^4(\eta, x) \rho(\eta, x), \tag{15}$$
$$\Pi(\eta, x) = a^4(\eta, x) p(\eta, x) \tag{16}$$

for reasons which will be explained below. The perturbations of (15), (16) around the uniform values can be written now as $\wp(\eta, x) = e^{4\alpha(\eta)} \rho(\eta) + \delta\wp(\eta, x)$, $\Pi(\eta, x) = e^{4\alpha(\eta)} p(\eta) + \delta\Pi(\eta, x)$. The 4-velocity u is represented in the form of

$$u^\mu = e^{-\alpha(\eta)} \{ 1, \nabla \frac{v(\eta, x)}{\rho(\eta) + p(\eta)} \} \approx \{ e^{-\alpha(\eta)} (1 - \Phi(\eta, x)), e^{3\alpha(\eta)} \nabla \frac{v(\eta, x)}{\wp(\eta) + \Pi(\eta)} \}, \tag{17}$$

where $v(\eta, x)$ is a scalar function. Expanding all perturbations into the Fourier series $\delta\wp(\eta, x) = \sum_k \delta\wp_k(\eta) e^{ikx}$... etc. results in the equations for the perturbations:

$$-6\Phi'_k + 6\alpha'\Phi_k + k^2 F'_k + \frac{18}{M_p^2} e^{-2\alpha} \sum_i v_{ki} = 0, \quad (18)$$

$$-18\alpha'\Phi'_k - 6(k^2 + 3\alpha'^2)\Phi_k + k^4 F_k + \frac{18}{M_p^2} e^{-2\alpha} \sum_i \delta\wp_{ki} = 0, \quad (19)$$

$$-12\Phi_k - 3(F''_k + 2\alpha' F'_k) + k^2 F_k = 0, \quad (20)$$

$$-9(\Phi''_k + 2\alpha'\Phi'_k) - 9(2\alpha'' + 2\alpha'^2 + k^2)\Phi_k + k^4 F_k - \frac{9}{M_p^2} e^{-2\alpha} \left(\sum_i 3\delta\Pi_{ki} - \delta\wp_{ki} \right) = 0, \quad (21)$$

where the index i denoting the kind of substance has been introduced. It is remarkable that, as a result of the choice of the variables (15)–(17), the unperturbed values ρ and p do not appear in the system (18)–(21). This allows us to avoid the influence of the large uniform energy density and pressure (8) and (9) on the evolution of perturbation. Equations (18) and (19) are consequences of the Hamiltonian and momentum constraints, while Equations (20) and (21) are equations of motion. For consistency of the constraints with the equations of motion, every kind of fluid has to satisfy the continuity and Euler equations:

$$\alpha'(\delta\wp_{ki} - 3\delta\Pi_{ki}) - (3\Pi_i - \wp_i)(\Phi'_k - 4\Phi_k \alpha') + 4\wp'_i \Phi_k - \delta\wp'_{ki} + k^2 v_{ki} = 0, \quad (22)$$

$$\Phi_k(\wp_i - 3\Pi_i) + \delta\Pi_{ki} + v'_{ki} = 0. \quad (23)$$

Equations (18)–(23) have the same form as in GR, but for the consistency of Hamiltonian and momentum constraints (18) and (19) with the equations of motions (20)–(23), it is sufficient for the first Friedmann Equation (6) to be valid up to some constant. Namely, for such consistency, it is necessary that the differentiation of constraints with the subsequent substitution of the second-time derivatives from the equations of motion (7), (20)–(23) leads to identical equalities. This consistency is a feature of using the CUM metric, in particular, and the FVT theory, in general. In any other metrics different from CUM (that is, in a frame of GR), the first Friedmann Equation (6) with the $const = 0$ in the right hand side is needed for consistency of the constraints and the equations of motion.

4. Vacuum as a Medium: The Eikonal Approximation for Quantum Fields

Generally, a vacuum could also be considered as some fluid (e.g., see [28–30]), i.e., "æther" [54], but with some stochastic properties along with its elastic ones [42,46,55]. Here we are interested in its elastic properties only. In Refs. [42,46], the speed of sound for the scalar waves of vacuum polarization $c_s^2 = \frac{p'_v(\eta)}{\rho'_v(\eta)}$ was introduced, where p_v and ρ_v are given by (10). That is the only heuristic conjecture.

Here, the actual calculations of the vacuum density and pressure on the curved background are performed in the eikonal approximation. The last one has a very transparent background. In the Minkowski's space–time, the virtual plane waves penetrate space–time and, to obtain a vacuum energy density, we must summarize the contributions of every wave. In the curved space–time, it is necessary to summarize the contributions of the distorted waves to obtain the spatially non-uniform energy density and pressure. It should be mentioned that the eikonal approximation was successfully used in high energy physics [56] and even in gravity [57], where the small-angle scattering amplitude of two massive particles were calculated in all orders on gravitational constant G.

A massless scalar field in the external gravitational field obeys the equation

$$\frac{1}{\sqrt{-g}} \partial_\mu (\sqrt{-g} g^{\mu\nu} \partial_\nu) \phi = 0. \quad (24)$$

Using the gauge $N = 0$, $P = 0$ in (1) reduces the CUM metric to

$$ds^2 = a^2(d\eta^2 - \tilde{\gamma}_{ij}dx^i dx^j), \tag{25}$$

so that Equation (24) takes the form

$$\phi'' + 2\frac{a'}{a}\phi' - \frac{1}{a^2}\partial_i\left(a^2\tilde{\gamma}^{ij}\partial_j\right)\phi = 0. \tag{26}$$

This leads to

$$\chi'' - \chi\frac{a''}{a} - \tilde{\gamma}^{ij}\partial_i\partial_j\chi - \partial_i\tilde{\gamma}^{ij}\partial_j\chi + \frac{\chi}{a}\left(\tilde{\gamma}^{ij}\partial_i\partial_j a + \partial_j a \partial_i \tilde{\gamma}^{ij}\right) = 0 \tag{27}$$

after the change of variables $\phi = \chi/a$. Further, in the terms of the metric perturbations Φ and F, we come to

$$\chi'' - \Delta\chi + \hat{V}\chi = 0, \tag{28}$$

where a "potential" operator $\hat{V}$ has the form

$$\hat{V} = -\alpha'' - \alpha'^2 - 2\alpha'\Phi' - \Phi'' + \Delta\Phi + \frac{1}{3}\Delta F \Delta - \frac{\partial^2 F}{\partial x^j \partial x^i}\frac{\partial^2}{\partial x^j \partial x^i} - \frac{2}{3}(\nabla(\Delta F))\cdot \nabla. \tag{29}$$

A quantization of the scalar field in terms of creation and annihilation operators implies [37]

$$\hat{\chi}(\eta,x) = \sum_k u_k(\eta,x)\hat{a}_k + u_k^*(\eta,x)\hat{a}_k^+, \tag{30}$$

where the function u_k satisfies Equation (27), and the orthogonality condition is [37]

$$\int (u_k \partial_\eta u_q^* - u_k^* \partial_\eta u_q)d^3x = i\delta_{kq}. \tag{31}$$

A solution of Equations (27) and (29) for the functions u_k can be written in the eikonal approximation

$$u_k(\eta,x) = \frac{1}{\sqrt{2k}}\exp(-i\eta k + ikx - i\Theta_k(\eta,x)), \tag{32}$$

which leads to the equation for the eikonal function

$$2k\Theta'_k + \left(2k_m\tilde{\gamma}^{mj} - i\partial_m \tilde{\gamma}^{mj}\right)\partial_j\Theta_k + \frac{1}{a}(a'' - \tilde{\gamma}^{ij}\partial_i\partial_j a - \partial_j a \partial_i \tilde{\gamma}^{ij}) + ik_j\partial_m \tilde{h}^{mj} - k_m k_j \tilde{h}^{mj} = 0, \tag{33}$$

and, according to Equations (12) and (13), is written in the terms of the metric perturbations $\Phi(\eta,x)$, $F(\eta,x)$:

$$k\Theta'_k + k\nabla\Theta_k(\eta,x) = \frac{1}{2}V_k, \tag{34}$$

where

$$V_k(\eta,x) = -2\alpha'\Phi' - \Phi'' + \Delta\Phi + k_i k_j \partial_i \partial_j F - \frac{k^2}{3}\Delta F. \tag{35}$$

A solution of (34) can be obtained in the form

$$\Theta_k(\eta,x) = \frac{1}{2k}\int_{\eta_0}^{\eta} V_k\left(\tau, x + \frac{k}{k}(\tau - \eta)\right)d\tau, \tag{36}$$

where the lower integration limit η_0 depends on the cosmological model. In particular, it could be 0 or $-\infty$. The mean value of the stress–energy tensor of a massless scalar field

$$\hat{T}_{\mu\nu} = \partial_\mu \hat{\phi} \partial_\nu \hat{\phi} - \frac{1}{2} g_{\mu\nu} g^{\alpha\beta} \partial_\alpha \hat{\phi} \partial_\beta \hat{\phi} \tag{37}$$

can be averaged over the vacuum state and compared with the hydrodynamic expression (14). This gives

$$\delta\wp(\eta, x) = e^{2\alpha(\eta)} < 0 \left| \frac{\hat{\phi}'^2}{2} + \frac{(\nabla\hat{\phi})^2}{2} \right| 0 > \approx \frac{1}{2} \sum_k \frac{\alpha' \Phi'}{k} + \Theta'_k - \frac{k \nabla \Theta_k}{k}, \tag{38}$$

$$\delta\Pi(\eta, x) = e^{2\alpha(\eta)} < 0 \left| \frac{\hat{\phi}'^2}{2} - \frac{(\nabla\hat{\phi})^2}{6} \right| 0 > \approx \frac{1}{2} \sum_k \frac{\alpha' \Phi'}{k} + \Theta'_k + \frac{k \nabla \Theta_k}{3k}, \tag{39}$$

$$\nabla v = -e^{2\alpha(\eta)} < 0 |\hat{\phi}' \nabla \hat{\phi}| 0 > \approx \sum_k \frac{k \Theta'_k}{k} - \nabla \Theta_k - \frac{\alpha' \nabla \Phi}{k}, \tag{40}$$

where only spatially non-uniform parts of the vacuum averages are implied in the second equalities on the right-hand side of (38)–(40). The last depends on the metric perturbations $F(\eta, x)$ and $\Phi(\eta, x)$ contained in Equations (12) and (13). The final equalities in (38)–(40) result from calculations in the eikonal approximation (32).

Considering the quantity $\delta\wp(\eta, x) - 3\delta\Pi(\eta, x)$ and using Equations (34) and (35), result in

$$\delta\wp(\eta, x) - 3\delta\Pi(\eta, x) = -\sum_k \frac{k \nabla \Theta_k}{k} + \Theta'_k + \frac{\alpha' \Phi'}{k} =$$

$$-\sum_k \frac{1}{2k} V_k + \frac{\alpha' \Phi'}{k} = \sum_k \frac{1}{2k} \left(\Phi'' - \Delta\Phi - k_i k_j \partial_i \partial_j F + \frac{k^2}{3} \Delta F \right) = \frac{N_{sc}}{8\pi^2} k_{max}^2 (\Phi'' - \Delta\Phi), \tag{41}$$

where summation has been changed by integration $\sum_k \to \int d^3k/(2\pi)^3$ and it is taken into account that $\int_{k<k_{max}} \frac{1}{2k} \left(k_i k_j - \frac{k^2}{3} \delta_{ij} \right) d^3k = 0$. The number N_{sc} of the scalar fields minimally coupled with gravity has been introduced as in (10).

In consequence of Equation (41), two types of spatially non-uniform vacuum polarization exist. Namely, the F-polarization has a radiation-type equation of state[3]

$$\delta\Pi_{vF}(\eta, x) = \frac{1}{3} \delta\wp_{vF}(\eta, x), \tag{42}$$

whereas the Φ-polarization has an equation of state

$$\delta\Pi_{v\Phi}(\eta, x) = \frac{1}{3} \delta\wp_{v\Phi}(\eta, x) - \frac{N_{sc}}{24\pi^2} k_{max}^2 (\Phi'' - \Delta\Phi). \tag{43}$$

Both types of spatially non-uniform vacuum polarizations correspond to the uniform component of (8), (9), whereas the uniform polarization given by (10) has no non-uniform counterpart with an accuracy of our consideration, i.e., in the second order on derivatives. It must be emphasized that it is easy to obtain the equation of state (9) for a spatially uniform main part of the vacuum energy density, but it is not so trivial to do that for a spatially non-uniform vacuum energy density.

In principle, the system (18)–(23), (42) and (43) is a fundamental system allowing to consider a broad range of cosmological and astrophysical phenomena including CMB and BAO. However, below, we restrict ourselves to a galactic DM, which scales from kpc to Mpc.

5. Galactic DM as a F-Vacuum Polarization

As it was shown in Section 4, the F-component of vacuum polarization has the equation of state analogous to radiation (see Equation (9)). In this sense, it is similar to the uniform part of vacuum energy density in Equation (8).

At the same time, it is difficult to determine the concrete value of the non-uniform vacuum energy density because, according to (38), it contains an eikonal function Θ_k, which is determined by the integral (36). For instance, one has $\Theta_k(\eta, r) = \sum_q \tilde{\Theta}_{k,q}(\eta) e^{iqr}$ and from (35), (36) finds

$$\tilde{\Theta}_{k,q}(\eta) = \frac{1}{k}\left(\frac{1}{3}k^2 q^2 - (qk)^2\right) \int_{\eta_0}^{\eta} F_q(\tau) e^{ikq(\tau-\eta)/k} d\tau. \tag{44}$$

Calculation of the integral (44) requires one to know the full evolution history of $F_q(\tau)$. It is simpler to use only the fact that the F-contribution to the vacuum polarization has the equation of state

$$p_{vF} = \rho_{vF}/3. \tag{45}$$

The distributions of matter–energy density and potential are not determined for the static case in the first order on perturbations (see Appendix A). However, it is possible to consider a nonlinear heuristic model treating the F-vacuum as an abstract substance with the above equation of state. The model consists of a core of some incompressible substance modeling a baryonic-like matter placed on the radiation background, i.e., the F-polarized vacuum or "dark radiation", which interacts with this core only gravitationally. Below, we find a spherically symmetric solution for an incompressible substance with the constant energy density ρ_1 on the background of radiation density ρ_2.

5.1. Equations in the CUM Metric

The CUM metric in the case of spherical symmetry acquires the form [9]

$$ds^2 = a^2(d\eta^2 - \tilde{\gamma}_{ij} dx^i dx^j) = e^{2\alpha}\left(d\eta^2 - e^{-2\lambda}(dx)^2 - (e^{4\lambda} - e^{-2\lambda})(xdx)^2/r^2\right), \tag{46}$$

where $r = |x|$, $a = \exp\alpha$, and λ are functions of η, r. The matrix $\tilde{\gamma}_{ij}$ with the unit determinant is expressed through $\lambda(\eta, r)$. The interval (46) could be also rewritten in the spherical coordinates:

$$x = r\sin\theta\cos\phi, \quad y = r\sin\theta\sin\phi, \quad z = r\cos\theta \tag{47}$$

to give

$$ds^2 = e^{2\alpha}\left(d\eta^2 - dr^2 e^{4\lambda} - e^{-2\lambda} r^2 \left(d\theta^2 + \sin^2\theta d\phi^2\right)\right). \tag{48}$$

Restricting ourselves to static solutions, the equations for the functions $\alpha(r)$ and $\lambda(r)$ are written as [9]

$$\mathcal{H} = e^{2\alpha}\left(-\frac{e^{2\lambda}}{6r^2} + e^{-4\lambda}\left(\frac{1}{6r^2} - \frac{4}{3}\frac{d\alpha}{dr}\frac{d\lambda}{dr} + \frac{1}{6}\left(\frac{d\alpha}{dr}\right)^2 + \frac{2}{3r}\frac{d\alpha}{dr} + \frac{1}{3}\frac{d^2\alpha}{dr^2} + \right.\right.$$
$$\left.\left.\frac{7}{6}\left(\frac{d\lambda}{dr}\right)^2 - \frac{5}{3r}\frac{d\lambda}{dr} - \frac{1}{3}\frac{d^2\lambda}{dr^2}\right) + \frac{e^{2\alpha}}{M_p^2}\sum_j \rho_j(r)\right) = \text{const}, \tag{49}$$

$$\frac{d^2\alpha}{dr^2} = -\frac{3e^{6\lambda}}{r^2} + \frac{3}{r^2} - 8\frac{d\alpha}{dr}\frac{d\lambda}{dr} + 7\left(\frac{d\alpha}{dr}\right)^2 + \frac{10}{r}\frac{d\alpha}{dr} + 3\left(\frac{d\lambda}{dr}\right)^2 - \frac{6}{r}\frac{d\lambda}{dr} +$$
$$3\frac{e^{2\alpha+4\lambda}}{M_p^2}\sum_j \rho_j - 3p_j, \tag{50}$$

$$\frac{d^2\lambda}{dr^2} = -\frac{5e^{6\lambda}}{r^2} + \frac{5}{r^2} - 18\frac{d\alpha}{dr}\frac{d\lambda}{dr} + 12\left(\frac{d\alpha}{dr}\right)^2 + \frac{18}{r}\frac{d\alpha}{dr} + 8\left(\frac{d\lambda}{dr}\right)^2 - \frac{14}{r}\frac{d\lambda}{dr} +$$
$$6\frac{e^{2\alpha+4\lambda}}{M_p^2}\sum_j \rho_j - 3p_j, \qquad (51)$$

where Equation (49) is the Hamiltonian constraint, which could be rewritten in a form containing no second derivatives using Equations (50) and (51):

$$\mathcal{H} = \frac{e^{2\alpha-4\lambda}}{2r^2}\left(-3r^2\left(\frac{d\alpha}{dr}\right)^2 + 4r\frac{d\alpha}{dr}\left(r\frac{d\lambda}{dr}-1\right) - \left(r\frac{d\lambda}{dr}-1\right)^2 + e^{6\lambda}\right) + \frac{3e^{4\alpha}}{M_p^2}\sum_j p_j = const. \qquad (52)$$

Each kind of substance has to satisfy

$$\frac{dp_j}{dr} + (p_j + \rho_j)\frac{d\alpha}{dr} = 0. \qquad (53)$$

A vacuum solution of Equations (49)–(51) corresponding to the point massive particle was considered in [9] where an absence of evidence for a horizon was demonstrated. Let us consider another solution, corresponding to the substance of a radiation-type filling all the space. This particular solution is written as

$$\alpha(r) = \ln r - \frac{1}{6}\ln 7, \quad \lambda(r) = \frac{1}{6}\ln 7, \qquad (54)$$

and, under (45), it follows from (53):

$$\frac{d}{dr}\left(\rho e^{4\alpha}\right) = 0, \quad \rho = \frac{1}{2}r^{-4}7^{-1/3}, \qquad (55)$$

if we use (54) and (49) with $const = 0$ in the right hand side of Equation (49). Here, ρ is measured in the terms of $r_g^{-2}M_p^{-2}$, and r is measured in the units of r_g, which is not a gravitational radius, but some arbitrary spatial scale. It should be noted that, for (45), Equations (50) and (51) look similar to those for an empty space, whereas Equation (49) could also be considered as that for an empty space, but with $const \neq 0$. Thus, in the CUM metric of the FVT where the Hamiltonian constraint is satisfied up to some constant, one could alternatively consider the F-vacuum polarization solution similar to that for an empty space, but with some non-zero value of $const$ in Equations (49) and (52).

Since the solution (55) is singular, it could be treated as unphysical. To obtain a realistic model, one has to consider at least two substances: a compact object in the center consisting of a substance with a constant energy density and a substance with the radiation equation of state (42). We must emphasize the importance of such a dense kernel for obtaining non-singular vacuum polarization of F-type.

5.2. Equations in the Schwarzschild-Type Metric

It is more convenient to begin a consideration from the Schwarzschild-type metric [58]

$$ds^2 = B(R)dt^2 - A(R)dR^2 - R^2 d\Omega, \qquad (56)$$

where Equations (54) and (55) correspond to the well-known solution [58]

$$\rho_2(R) = \frac{1}{14R^2}, \qquad (57)$$

obeying the TOV Equation [59,60] for a radiation fluid

$$\rho_2' = -\frac{3\rho_2(m + 4\pi R^3 \rho_2/3)}{\pi R(R - \frac{3m}{2\pi})}, \qquad (58)$$

in all the spatial region $R \in (0, \infty)$, where $m(R)$ is defined by

$$m'(R) = 4\pi R^2 \rho_2. \tag{59}$$

Again, ρ_2 is measured in the terms of $r_g^{-2} M_p^2$, and R is measured in the units of r_g. The solutions (57) and (55) are singular at $R = 0$ and, thereby, non-physical. The situation changes cardinally in the presence of a core consisting of incompressible matter. More exactly, in the presence of incompressible matter of low density ρ_1, the corresponding solution remains singular. However, if $\rho_1 > \frac{1}{2}(\frac{8}{9})$, a solid ball in the metric (48) looks similar to a shell over r_g in the metric (56) [9] that is shown in Figure 1a. Here, we again imply the gravitational radius r_g as a measure of the distances, but calculate it taking into account only an incompressible matter. Such a matter occupies a region between R_i and R_f, where

$$R_f = \sqrt[3]{R_i^3 + \frac{1}{2\rho_1}} \tag{60}$$

in the units of r_g. Here the energy density ρ_1 is constant and measured in the terms of $r_g^{-2} M_p^2$, where the gravitational radius is defined as $r_g = \frac{3m_1}{2\pi M_p^2}$ and $m_1 = \frac{4}{3}\pi \rho_1 (R_f^3 - R_i^3)$. A compact object of such a type arising in FVT is known as "eicheon" [9] and replaces a black hole of GR. The appearance of eicheon in the center makes the solution (58) to be non-singular because it allows for setting the finite boundary conditions for radiation.

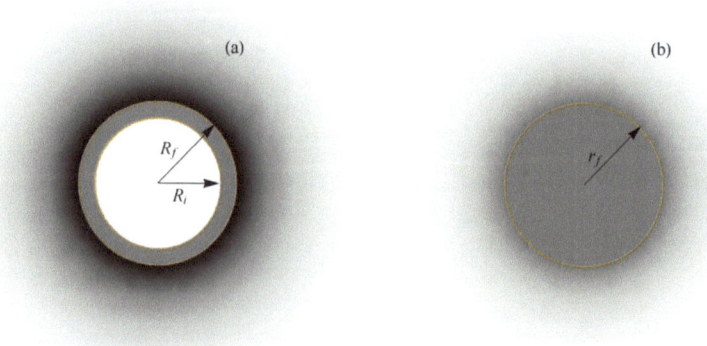

Figure 1. (a) Schematic picture of an eicheon in the metric (56), taking into account a vacuum polarization in the form of dark radiation; (b) an eicheon in the metric (48) looks similar to a solid sphere with a "dark radiation" of the finite energy density in the center.

To explain this, let us consider two fluids in the metric (56) obeying the TOV equations

$$p_1' = -\frac{3(p_1 + \rho_1)(m + 4\pi R^3(p_1 + \frac{p_2}{3}))}{4\pi R(R - \frac{3m}{2\pi})}, \tag{61}$$

$$p_2' = -\frac{3 p_2(m + 4\pi R^3(p_1 + \frac{p_2}{3}))}{\pi R(R - \frac{3m}{2\pi})}, \tag{62}$$

where the function $m(R)$ satisfies

$$m'(R) = 4\pi R^2 (\rho_1 + \rho_2). \tag{63}$$

For $\rho_1 > \frac{1}{2}(\frac{8}{9})$, the above equations hold for the internal range $R_i < R < R_f$, where $R_i > r_g$, and the border of a region, occupied by ρ_1, is defined through (60).

The pressure of incompressible fluid must turn to zero at the edge of the range filled by matter $R = R_f$, and it is a boundary condition for p_1. Then, one could set an amount of radiation at $R = R_f$ and solve the system of equations in a region of $\{R_i, R_f\}$ assuming $m(R_i) = 0$. A solution allows determining $m(R_f)$, and, using this value as an initial condition, one should solve the equation for the radiation fluid (58) in an outer region of $\{R_f, \infty\}$. The metric obtained by solving the equations is [58]

$$\frac{1}{B}\frac{dB}{dR} = -\frac{2}{p_1 + \rho_1}\frac{dp_1}{dR} = -\frac{2}{p_2 + \rho_2}\frac{dp_2}{dR}, \qquad (64)$$

$$\frac{d}{dR}\left(\frac{R}{A}\right) = 1 - 6R^2(\rho_1 + \rho_2). \qquad (65)$$

Comparing the metrics (46) and (56) leads to relation for the radial coordinates R and r [9]

$$\frac{dR}{dr} = \left(\frac{r}{R}\right)^2 \frac{B^{3/2}}{A^{1/2}}, \qquad (66)$$

where the dependencies $B(R(r))$ and $A(R(r))$ are implied. Equation (66) has to be integrated with the initial condition $R(0) = R_i$, which means that R_i in the metric (56) corresponds to $r = 0$ in the metric (46). Knowing $R(r)$ allows plotting $\rho_2(R)$ shown in Figure 2a as the r-dependent function $\rho_2(R(r))$ (Figure 2b).

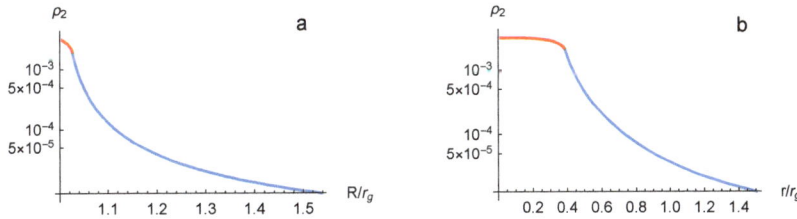

Figure 2. (a) ρ_2-nergy density of the vacuum polarization in a form of "dark radiation" in the coordinates $R > R_i$ calculated for the eicheon parameters $\rho_1 = 7M_p^2 r_g^{-2}$, $R_i = 1.001 r_g$, $R_f = 1.024 r_g$, $\rho_2(R_f) = 0.002 M_p^2 r_g^{-2}$. Red part of the curve corresponds to $R_i < R < R_f$, i.e., lies inside an eicheon. (b) ρ_2 calculated in the coordinates r of the metric (48). Red part of the curve corresponds to $0 < r < r_f$.

Let us consider the motion of a test particle on a circular orbit in the metric (56). The angular velocity on a circular orbit is calculated as [58]:

$$\frac{d\phi}{dt} = \sqrt{\frac{1}{2R}\frac{dB}{dR}}. \qquad (67)$$

A spatial interval followed by a particle along the circular orbit is given by $dl = Rd\phi = R\frac{d\phi}{dt}dt$. To obtain the rotation velocity observed by an observer situated at rest near the moving particle, one has to divide the spatial interval over the proper time $\sqrt{g_{00}}dt = \sqrt{B}dt$ of such an observer [61]:

$$v_{rot} = \frac{dl}{\sqrt{B}dt} = \sqrt{\frac{R}{2B}\frac{dB}{dR}} = \sqrt{-\frac{R}{p_2 + \rho_2}\frac{dp_2}{dR}} = \frac{1}{2}\sqrt{\frac{R}{\rho_2}\frac{d\rho_2}{dR}}. \qquad (68)$$

A qualitative example of the general form of the numerical solution for the rotation velocity is shown in Figure 3. Although the shape of the curve resembles observational data, the asymptotic of the rotation curve corresponds to $v_{rot} \sim 1/\sqrt{2} \approx 0.71$. This very large velocity (in units of speed of light) corresponds to asymptotic value $\rho_2 \sim R^{-2}$ in (57), whereas, in the reality, the rotation velocities of galaxies are $v_{rot} \sim 100 - 300$ km/s ~ 0.001. To obtain smaller velocities, one has to diminish the density of radiation in the center of eicheon, i.e., at $r = 0$ in the metric (48) or $R = R_i$ in the metric (56). For central radiation density of $\rho_2 = 4.6 \times 10^{-27}\ M_p^2 r_g^{-2} = 9.6 \times 10^{-24}$ g/cm^3, one has the rotation curve shown in Figure 4. That is a pure "dark radiation" contribution without the galaxy bulge or disk. It increases linearly with the distance and corresponds to the rising part of the general curve shown in Figure 3. In the logarithmic scale, one could see (Figure 5) together the contribution of the eicheon of the mass of $4.2 \times 10^6\ M_\odot$ in the center of the Milky Way (the left side of the curve) and the impact of the dark radiation (the right side of the curve), whereas the effects of the galactic bulge and disk responsible for the intermediate region are not taken into account. However, it is expected that bulge and disk attraction will influence the F-type vacuum polarization in such a way that the curve in Figure 4 will be not pure linear but slightly bent. We do not gain insight into such details because our goal is to show that the F-type vacuum polarization could arise only around a "sewed" black hole, i.e., around eicheon.

We emphasize that the presented consideration is heuristic because, although the linear system for the perturbation and the eikonal approximation for vacuum polarization seems reasonable, we use its results in the nonlinear TOV model. Another thing is that we set the density of radiation (the F-type vacuum polarization) in the center of eicheon, i.e., at $r = 0$, of $R = R_i$ empirically but not calculate it from the first principles, i.e., we use only the equation of state obtained from the calculations in the eikonal approximation.

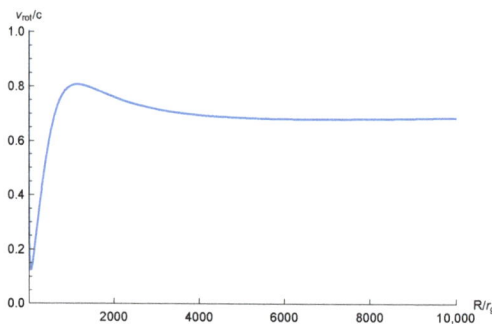

Figure 3. The general form of a model rotational curve for the eicheon parameters specified in the caption to Figure 2.

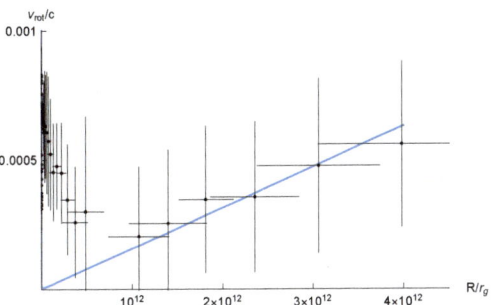

Figure 4. The rotational curve for the eicheon parameters $\rho_1 = 100 M_p^2 r_g^{-2}$, $R_i = 1.0001 r_g$ and $\rho_2(R_f) = 4 \times 10^{-27} M_p^2 r_g^{-2}$, where r_g is defined by an eicheon mass. In the physical units, $\rho_1 = 100 \frac{3c^6}{16\pi G^3 m^2} \approx 2.1 \times 10^5$ g/cm^3. The points and error bars correspond to the Milky Way rotational curve from [62].

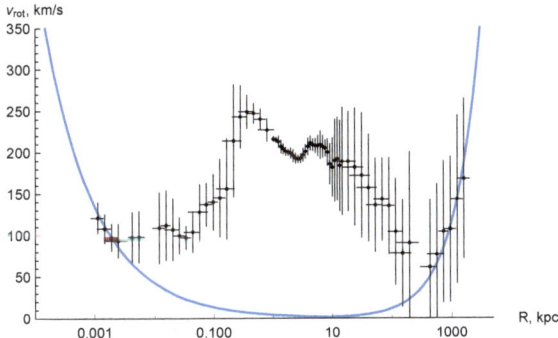

Figure 5. The rotational curve of eicheon with the mass of Sgr A* with taking into account the vacuum polarization of F-type. The logarithmic scale is used and the points correspond to the Milky Way rotational curve from [62]. The eicheon parameters are given in the caption to Figure 4.

6. Vacuum Polarization of Φ-Type

In Sections 3 and 4, the linear system of Equations (20)–(23) and (41) was deduced, which describes the evolution of perturbation by taking into account vacuum polarization (see Equation (41) and Appendix A for an example of an exact solution). Galaxy formation is a complex nonlinear process that develops over cosmological time scales. Generally, the linear system is insufficient to describe the galaxy evolution. However, one could create a heuristic picture, setting an approximate profile of matter near the galaxy center, and obtain a gravitational potential produced by vacuum polarization obeying the linear equations. Below we will discuss that the observed galaxy halo could originate from a very fast (compared to the cosmological times) change of the galactic nucleus mass. We will neglect a cosmological evolution assuming $\alpha(\eta) = 0$. This reduces the above system of the equations to

$$-12\Phi_q - 3F_q'' + q^2 F_q = 0, \tag{69}$$

$$-9\Phi_q'' - 9q^2\Phi_q + q^4 F_q + \frac{9}{M_p^2}\left(\sum_i \delta\wp_{ki} - 3\delta\Pi_{qi}\right) = 0. \tag{70}$$

$$\delta\wp_{qv} - 3\delta\Pi_{qv} = \frac{N_{sc}}{8\pi^2} k_{max}^2 (\Phi_q'' + q^2 \Phi_q), \tag{71}$$

where the last equation holds for the vacuum polarization of Φ-type and is denoted by $i = v$. Substituting Φ_q from Equation (69), and $\delta\wp_{qv} - 3\delta\Pi_{qv}$ from Equation (71) into Equation (70) gives the equation

$$3\left(k_{max}^2 - 8\pi^2 M_p^2\right)\left(3F_q'''' + 2q^2 F_q''\right) - q^4 F_q\left(3N_{sc}k_{max}^2 + 8\pi^2 M_p^2\right) = 288\pi^2 \delta\wp_{qeff}(\eta), \quad (72)$$

where an effective "density" of all the substances except vacuum is denoted as

$$\delta\wp_{qeff}(\eta) = \sum_{i\neq v}\delta\wp_{ki} - 3\delta\Pi_{qi}. \quad (73)$$

Equation (72) allows for developing a simple model: setting profile and time dependencies of the quantity $\wp_{qeff}(\eta)$ empirically determines the metric perturbation F_q and Φ_q using (69).

Let us, for simplicity, take $\wp_{qeff}(\eta)$ in the form of

$$\wp_{qeff}(\eta) = m\, Z(q) e^{\eta/T}, \quad (74)$$

where m is a "mass" of the object at $\eta = 0$, $Z(q)$ is a form-factor and T is some typical time of the "mass" growth. The model implies some rapid processes such as accretion occurring around the massive object, i.e., around the galaxy nucleus. Substitution of the expression (74) into Equation (72) allows finding $F_q(\eta) = \tilde{F}_q e^{\eta/T}$, where

$$\tilde{F}_q = -\frac{288\pi^2 T^4 m Z(q)}{3N_{sc}k_{max}^2(q^4 T^4 - 2q^2 T^2 - 3) + 8\pi^2 M_p^2(q^2 T^2 + 3)^2}, \quad (75)$$

and Equation (69) give $\Phi_q(\eta) = \check{\Phi}_q e^{\eta/T}$:

$$\check{\Phi}_q = -\frac{24\pi^2 T^2(q^2 T^2 - 3) m Z(q)}{3N_{sc}k_{max}^2(q^4 T^4 - 2q^2 T^2 - 3) + 8\pi^2 M_p^2(q^2 T^2 + 3)^2}. \quad (76)$$

At $T \to \infty$, the corresponding static limit is

$$\check{\Phi}_q = -\frac{24\pi^2 m Z(q)}{(3N_{sc}k_{max}^2 + 8\pi^2 M_p^2) q^2}, \quad (77)$$

which implies that the vacuum polarization leads to the renormalization (increasing) of the Planck mass, i.e., decreasing the gravitational constant. In particular, using the value (11) obtained from the cosmological observations [10] gives

$$M_{p\,ren}^2 = \left(1 + \frac{54 N_{sc}}{\pi^2(2 + N_{sc})}\right) M_p^2, \qquad G_{ren} = G / \left(1 + \frac{54 N_{sc}}{\pi^2(2 + N_{sc})}\right). \quad (78)$$

It seems that the vacuum polarization, in some sense, acts similar to antigravitation, and the gravitational constant G_{ren} appearing in Newton's law has to differ from the gravitational constant G in the Friedmann equations for a uniform universe. Although the gravitational constant's renormalization does not influence the cosmological balance of the different kinds of matter expressed in the units of the critical density $M_p^2 H^2$, it should be taken into account for comparison with the directly measured (for instance, utilizing luminosity) density. Numerically, $N_{sc} = 2$ gives $G_{ren} \approx 0.27\, G$.

Invariant Potentials and Rotational Curves

Astrophysicists express the results of observations in terms of gauge-invariant quantities, which are not dependent on a system of coordinates. The potentials $\Phi(\eta, x)$ and

$F(\eta, x)$ are not invariant relatively to the infinitesimal transformations of coordinates and time of the following type

$$t = \eta + \xi_1(\eta, x), \qquad r = x + \nabla \xi_2(\eta, x), \tag{79}$$

where $\xi_1(\eta, x)$ and $\xi_2(\eta, x)$ are some small functions. Usually, the potentials $\Phi_{inv}(\eta, x)$ and $\Psi_{inv}(\eta, x)$ are introduced [63–65] which are invariant relatively transformations (79). The potentials correspond to the metric

$$ds^2 = a^2(\eta)\Big((1 + 2\Phi_{inv}(\eta, x))d\eta^2 - (1 - 2\Psi_{inv}(\eta, x))\delta_{ij}dx^i dx^j\Big) \tag{80}$$

and are expressed through Φ and F as

$$\Phi_{q\,inv}(\eta) = \Phi_q(\eta) + \frac{a'(\eta)F_q'(\eta) + a(\eta)F_q''(\eta)}{2a(\eta)} = \Phi_q + \frac{F_q}{2T^2}, \tag{81}$$

$$\Psi_{q\,inv}(\eta) = -\frac{a'(\eta)F_q'(\eta)}{2a(\eta)} - \Phi_q(\eta) + \frac{1}{6}q^2 F_q(\eta) = -\Phi_q(\eta) + \frac{1}{6}q^2 F_q, \tag{82}$$

where the final equalities at the right-hand side of (81), (82) hold for our case of $a = const$, and $\Phi, F \sim \exp(\eta/T)$. Using (75), (76) gives

$$\tilde{\Phi}_{q\,inv} = -\frac{24\pi^2 T^2 (q^2 T^2 + 3) m\, Z(q)}{3 N_{sc}\, k_{max}{}^2 (q^4 T^4 - 2q^2 T^2 - 3) + 8\pi^2 M_p{}^2 (q^2 T^2 + 3)^2}, \tag{83}$$

and $\tilde{\Psi}_{q\,inv} = \tilde{\Phi}_{q\,inv}$. Thus, we obtained the Fourier transformation of the time-dependent gravitational potential $\Phi_{q\,inv} = \tilde{\Phi}_{q\,inv} e^{\eta/T}$, allowing us to define

$$\Phi_{inv}(x, \eta) = \frac{e^{\eta/T}}{(2\pi)^3} \int \tilde{\Phi}_{q\,inv}\, e^{iqx} d^3 q. \tag{84}$$

To obtain a concrete empirical formula, one has to set the form factor $Z(q)$, for instance, using the Gaussian profile $\delta\tilde{\rho}_{eff}(x) = \pi^{-3/2} m\, D^{-3} e^{-x^2/D^2}$. The spatial dependence of the potential (84) at the present time, i.e., $\eta = 0$, allows us to find the rotational velocity dependence on the spatial coordinate

$$v_{rot}(r) = \sqrt{-r\frac{d\Phi_{inv}(r)}{dr}}. \tag{85}$$

Here, the potential (84) is time-dependent, and actually, there are no pure rotational curves because the radial velocities are present. Here, for an estimation, we discuss only tangential velocity. The parameters m, D and $Z(q)$ are adopted to produce a typical rotational curve without an DM (blue curve in Figure 6), then vacuum polarization produces a halo corresponding to black curve in Figure 6.

The rotational curve has some similarities with the conventional picture at $N_{sc} = 2$, but in the conventional picture, the contribution of the galactic nucleus, bulge and disk are taken into account. We include all these components into the Gaussian form factor of galactic baryonic skeleton and call it "nucleus" in our oversimplified picture. Then, we permit it to increase (or decrease) with time and obtain vacuum polarization caused by this process.

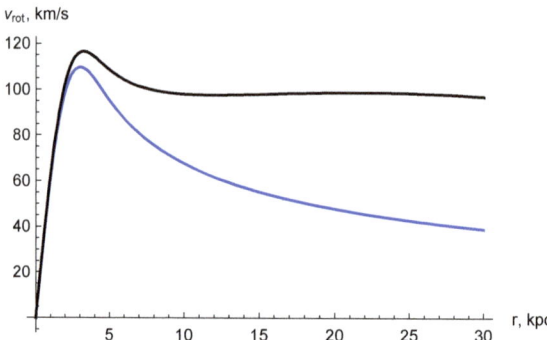

Figure 6. The lower blue curve corresponds to the contribution of a galactic nucleus of baryonic matter, including specifically nucleus, bulge and disk. The upper black curve takes the vacuum Φ-polarization into account. The form factor of a galaxy nuclei is taken as $Z(q) - \exp(-\lambda q^2)$, $\lambda = 1$, the accretion rate is of $T = 10$, i.e., 10 kPc, which corresponds to 32,000 years. The number of the minimally coupled scalar fields is of $N_{sc} = 2$, and $k_{max}^2 = 8 M_p^2 \pi^2 \frac{98}{100}$ is assumed.

7. Conclusions

We have considered two types of vacuum polarization corresponding to the F- and Φ-types of metric perturbations in the CUM frame.

The F-type of spatially non-uniform vacuum polarization has the radiation-type equation of state. In the first order on perturbations, it is impossible to determine a form of the static gravitational potential around an astrophysical object. In the frameworks of a nonlinear heuristic model using the TOV equations for matter and radiation, it was found that the solution, which is non-singular at $r = 0$, only arises if an eicheon is present in the galaxy's center. Eicheon is an analog of the black hole in GR and looks similar to an empty "nut" in the Schwarzschild-type metric. From this point of view, we assume that DM, as a vacuum polarization, arises only in the galaxies having an eicheon (i.e., a "black hole-like" object) in the center. Namely, the eicheon conjecture allows us to convert a singular solution for pure radiation into a non-singular physical one. Galaxies without an eicheon in the center (e.g., diffuse galaxies) do not have a DM halo[4].

Under the oversimplified assumption of an isolated galaxy, the dark halo, in terms of a test particle's rotation velocity, always increases with the distance from the galaxy's center. Decreasing the halo could occur only due to a violation of the galaxy's isolation, i.e., at the distance of $\sim$2 Mpc. It should be noted that the Andromeda galaxy is only 0.7 Mpc away. Generally, the galaxies tend to form clusters. These evident facts urge the development of a model of interacting galaxies with vacuum polarization.

For the Φ-type of vacuum polarization, the renormalization of the gravitational constant (or Planck mass) has been found. This means that the gravitational constant found in the Earth, the Solar System, and galaxy observations is not equal (approximately four times less) to the gravitational constant used in cosmology to describe a spatially uniform universe. This fact does not influence the balance of the different kinds of matter in cosmology if one measures them in $M_p^2 H^2$. Nevertheless, it increases the directly counted matter contribution fourfold, i.e., the luminous baryonic matter has to contribute 3.7-times stronger into the cosmological Friedmann equations.

The second effect of the Φ-type polarization is the creation of the dark halo in the non-stationary process. It is found that the time-dependent evolving mass of the galaxy nuclei produces the gravitational potential of the dark halo-type. This point urges a more careful observational investigation of the possible non-stationary origin of the dark halo. However, the required time for the galaxy nuclei mass growth seems very short: $\sim$32,000 years. In such a situation, clarifying the physical status of the possible accretion of vacuum energy and vacuum condensates discussed in [68–70] is very desirable. In particular, it was shown in [68,69] that accretion of substance with the equation of state of $p = -\rho$ (e.g.,

Higgs or QCD condensates) decreases a black hole mass, while accretion of the ordinary substance with radiation equation of state increases a black hole mass.

Investigations of these processes in the CUM metric with the applications to an eicheon are waiting. However, one may suggest some scenarios of a galaxy center evolution. Accretion by an eicheon could be more complicated than a traditional black hole. At some stage, eicheon could accrete more "dark radiation", increasing its mass, but at some stage, it could accrete more condensates, decreasing its mass. One could associate this with the fast processes with a typical time of ~32,000 years. Both growth of the galaxy's center mass and its lowering produce a halo. Thus, a galaxy center is reminiscent of "Alice from Wonderland" [71], which takes a bit of a mushroom from one side and rises, then takes a bit from another side and shrinks. These processes can interlace in a galaxy center.

To summarize, it is possible to obtain an equation of the state of vacuum polarization, which is some kind of "æther". It is challenging to find the "amount" of æther because it depends on the object's entire history due to the nonlocality of the vacuum state on the curved background. Here, we have adjusted this "amount" to astrophysical observations. Thus, the obtained final results have, in some sense, a heuristic nature.

Author Contributions: Concepts and methodology are developed by S.L.C. and V.L.K.; software, S.L.C.; validation, writing and editing, S.L.C. and V.L.K. All authors have read and agreed to the published version of the manuscript.

Funding: This research received no external funding.

Institutional Review Board Statement: Not applicable.

Informed Consent Statement: Not applicable.

Data Availability Statement: Not applicable.

Acknowledgments: We are gratefull to Yoshiaki Sofue for a permission to use the results of his observation of Milky Way rotational curve in Figures 4 and 5.

Conflicts of Interest: The authors declare no conflict of interest.

Appendix A

We emphasize that the presented consideration is heuristic because, although the linear system for the perturbation and the eikonal approximation for vacuum polarization seem trustable, we use its results in the nonlinear TOV model. Another point is that we empirically set the density of radiation (the vacuum polarization of F-type) in the center of an eicheon, i.e., at $r = 0$, of $R = R_i$. That is, we do not calculate it from the first principles, i.e., we use only the equation of state from the eikonal calculations.

Let us consider the system of Equations (18)–(23) for an empty space–time with the vacuum polarization of F-type in the form of radiation fluid. For $e^{4\alpha}\rho = const$, the constant in Equation (6) can be chosen in such a way that there is no evolution of the scale factor, i.e., $\alpha = 0$ (a static universe).

For the substance obeying (45), Equations (23) and (22) are reduced to

$$-\delta\wp'_{q\,vF} + q^2 v_{q\,vF} = 0, \tag{A1}$$

$$\delta\Pi_{q\,vF} + v'_{q\,vF} = 0, \tag{A2}$$

and have the solution

$$\delta\wp_{q\,vF} = c_1 \sin\frac{q\eta}{\sqrt{3}} + c_2 \cos\frac{q\eta}{\sqrt{3}}, \tag{A3}$$

$$v_{q\,vF} = \frac{c_2 \cos\left(\frac{\eta q}{\sqrt{3}}\right) - c_1 \sin\left(\frac{\eta q}{\sqrt{3}}\right)}{\sqrt{3}q}. \tag{A4}$$

Let us also place into this universe some amount of a dust matter $\delta\wp_{qm}$ obeying $\delta\Pi_{qm} = 0$ and without a uniform component, i.e., $\Pi_m = 0$, $\wp_m = 0$. The complete solution of the system (18)–(23) takes the form

$$\delta\wp_{qm}(\eta) = -\frac{1}{36}q^4 M_p^2(c_6\eta + c_5), \tag{A5}$$

$$v_{qm} - \frac{c_6}{36}q^2 M_p^2, \tag{A6}$$

$$F_q(\eta) = c_6\eta + c_5 - 3q^{-4}M_p^{-2}\left(\sin\left(\frac{\eta q}{\sqrt{3}}\right)\left(c_4 q^2 M_p^2 + 2\sqrt{3}c_1\eta q + 15c_2\right) + \cos\left(\frac{\eta q}{\sqrt{3}}\right)\left(q\left(c_3 q M_p^2 - 2\sqrt{3}c_2\eta\right) + 15c_1\right)\right), \tag{A7}$$

$$\Phi_q(\eta) = \frac{q^2}{12}(c_6\eta + c_5) - \frac{1}{2M_p^2 q^2}\left(6\sin\left(\frac{\eta q}{\sqrt{3}}\right)\left(c_4 M_p^2 q^2 + 2\sqrt{3}c_1\eta q + 9c_2\right) + \cos\left(\frac{\eta q}{\sqrt{3}}\right)\left(q\left(c_3 M_p^2 q - 2\sqrt{3}c_2\eta\right) + 9c_1\right)\right). \tag{A8}$$

Then, in accordance with (38), the energy density for a F-vacuum polarization is expressed approximately as

$$\delta\wp_{vF}(\eta, x) = \frac{1}{2}\sum_k -\frac{k\nabla\Theta_k}{k} + \Theta'_k, \tag{A9}$$

which gives

$$\delta\wp_{qvF}(\eta) = -2\pi k_{max}^4 \int_{-\infty}^{\eta}\left(\left(9 - 4q^2(\eta - \tau)^2\right)\sin(q(\eta - \tau)) + q(\eta - \tau)\left(q^2(\eta - \tau)^2 - 9\right)\cos(q(\eta - \tau))\right)\frac{F_q(\tau)}{3q(\eta - \tau)^4}d\tau. \tag{A10}$$

If we consider this equation as an additional equation to the system (18)–(23), we can find that the constants c_1, c_2, c_3, c_4, c_6 have to be zero and only c_5 term is permitted because

$$\int_{-\infty}^{\eta} \frac{\left((9 - 4q^2(\eta - \tau)^2)\sin(q(\eta - \tau)) + q(\eta - \tau)\left(q^2(\eta - \tau)^2 - 9\right)\cos(q(\eta - \tau))\right)}{3q(\eta - \tau)^4}d\tau = 0.$$

Thus, the static gravitational potential

$$\Phi_q = \frac{q^2 c_5(q)}{12} \tag{A11}$$

of arbitrary form (because c_5 could be function of q) is permitted in the framework of a linear system of the equations considered.

Notes

[1] The CUM metric implies a preferred time foliation of space–time. Using the CUM metric per se does not predict some visible effects in the Solar System and all satellite experiments if their results are expressed in a gauge invariant way. At the same time, the use of the UV-cutoff at k_{max} implies the Lorentz invariance violation. In the local particle physics experiments, it leads to effects of the order of $\sim \varepsilon/k_{max} \sim \varepsilon/M_p$, where ε is the typical energy of a particle, but certainly does not produce some restrictions for Earth and satellite experiments. However, as it will be shown below, the consideration of vacuum physics using CUM and k_{max} could produce observable effects in a galaxy scale.

[2] We consider only scalar perturbations because the vector and tensor perturbations do not perturb the matter.

[3] For instance, see a DM vacuum model with the equation of state "running" from radiation-type to dark energy-type [20].

[4] The diffuse galaxy NGC1052-DF2 [66] seems to contain no DM, whereas another diffuse galaxy Dragonfly 44 is supposed to contain a lot of DM [67]. However, for the last, we do not know for definite whether or not there is an eicheon in its center.

References

1. Weinberg, S. The cosmological constant problem. *Rev. Mod. Phys.* **1989**, *61*, 1. [CrossRef]
2. Peebles, P.J.E.; Ratra, B. The cosmological constant and dark energy. *Rev. Mod. Phys.* **2003**, *75*, 559. [CrossRef]
3. Mostepanenko, V.M.; Klimchitskaya, G.L. Whether an Enormously Large Energy Density of the Quantum Vacuum Is Catastrophic. *Symmetry* **2019**, *11*, 314 . [CrossRef]
4. Unruh, W.; Wald, R. Information loss. *Rep. Progr. Phys.* **2017**, *80*, 092002. [CrossRef] [PubMed]
5. Chakraborty, S.; Lochan, K. Black Holes: Eliminating Information or Illuminating New Physics? *Universe* **2017**, *3*, 55. [CrossRef]
6. Mizner, C.W.; Thorne, K.; Wheeler, J.A. *Gravitation*; Freeman: San Francisco, CA, USA, 1973; Volume 1.
7. Landau, L.D.; Lifshitz, E. *The Classical Theory of Fields*; Butterworth-Heinemann: Oxford, UK, 1975; Volume 2.
8. Cherkas, S.L.; Kalashnikov, V.L. An approach to the theory of gravity with an arbitrary reference level of energy density. *Proc. Natl. Acad. Sci. Belarus Ser. Phys. Math.* **2019**, *55*, 83. [CrossRef]
9. Cherkas, S.L.; Kalashnikov, V.L. Eicheons instead of Black holes. *Phys. Scr.* **2020**, *95*, 085009. [CrossRef]
10. Haridasu, B.S.; Cherkas, S.L.; Kalashnikov, V.L. A reference level of the Universe vacuum energy density and the astrophysical data. *Fortschr. Phys.* **2020**, *68*, 2000047. [CrossRef]
11. Cherkas, S.L.; Kalashnikov, V.L. Universe driven by the vacuum of scalar field: VFD model. In Proceedings of the International Conference "Problems of Practical Cosmology", Saint Petersburg, Russia, 23–27 June 2008; pp. 135–140. [CrossRef]
12. Iorio, L. Solar system planetary orbital motions and dark matter. *J. Cosmol. Astropart. Phys.* **2006**, *2006*, 002. [CrossRef]
13. Freese, K. Review of Observational Evidence for Dark Matter in the Universe and in upcoming searches for Dark Stars. *EAS Publ. Ser.* **2009**, *36*, 113–126. [CrossRef]
14. Oks, E. Brief review of recent advances in understanding dark matter and dark energy. *New Astron. Rev.* **2021**, *93*, 101632. [CrossRef]
15. Weinberg, D.H.; Bullock, J.S.; Governato, F.; Kuzio de Naray, R.; Peter, A.H. Cold dark matter: Controversies on small scales. *Proc. Natl. Acad. Sci. USA* **2015**, *112*, 12249–12255. [CrossRef]
16. de Martino, I.; Chakrabarty, S.S.; Cesare, V.; Gallo, A.; Ostorero, L.; Diaferio, A. Dark matters on the scale of galaxies. *Universe* **2020**, *6*, 107. [CrossRef]
17. Bertone, G.; Hooper, D.; Silk, J. Particle dark matter: Evidence, candidates and constraints. *Phys. Rep.* **2005**, *405*, 279–390. [CrossRef]
18. Buchmueller, O.; Doglioni, C.; Wang, L.T. Search for dark matter at colliders. *Nat. Phys.* **2017**, *13*, 217–223. [CrossRef]
19. Aprile, E.; Abe, K.; Agostini, F.; Maouloud, S.A.; Althueser, L.; Andrieu, B.; Angelino, E.; Angevaare, J.R.; Antochi, V.C.; Martin, D.A.; et al. Search for New Physics in Electronic Recoil Data from XENONnT. *arXiv* **2022**, arXiv:2207.11330.
20. Albareti, F.; Cembranos, J.; Maroto, A. Vacuum energy as dark matter. *Phys. Rev. D* **2014**, *90*, 123509. [CrossRef]
21. Hajdukovic, D.S. Quantum vacuum and dark matter. *Astrophys. Space Sci.* **2012**, *337*, 9–14. [CrossRef]
22. Penner, A.R. Gravitational anti-screening as an alternative to dark matter. *Astrophys. Space Sci.* **2016**, *361*, 124. [CrossRef]
23. Hajdukovic, D. On the gravitational field of a point-like body immersed in a quantum vacuum. *Mon. Not. R. Astron. Soc.* **2019**, *491*, 4816–4828. [CrossRef]
24. Fiscaletti, D. About dark matter as an emerging entity from elementary energy density fluctuations of a three-dimensional quantum vacuum. *J. Theor. Appl. Phys.* **2020**, *14*, 203–222. [CrossRef]
25. Penner, A.R. A relativistic mass dipole gravitational theory and its connections with AQUAL. *Class. Quant. Grav.* **2022**, *39*, 075001. [CrossRef]
26. Blanchet, L.; Le Tiec, A. Model of dark matter and dark energy based on gravitational polarization. *Phys. Rev. D* **2008**, *78*, 024031. [CrossRef]
27. Chardin, G.; Dubois, Y.; Manfredi, G.; Miller, B.; Stahl, C. MOND-like behavior in the Dirac–Milne universe. *Astron. Astrophys.* **2021**, *652*, A91. [CrossRef]
28. Huang, K. *A Superfluid Universe*; World Scientific: Singapore, 2016.
29. Sbitnev, V.I. Hydrodynamics of the physical vacuum: Dark matter is an illusion. *Mod. Phys. Lett. A* **2015**, *30*, 1550184. [CrossRef]
30. Zloshchastiev, K.G. An alternative to dark matter and dark energy: Scale-dependent gravity in superfluid vacuum theory. *Universe* **2020**, *6*, 180. [CrossRef]
31. Hamber, H.; Liu, S. On the quantum corrections to the Newtonian potential. *Phys. Lett. B* **1995**, *357*, 51–56. [CrossRef]
32. Bonanno, A.; Reuter, M. Renormalization group improved black hole spacetimes. *Phys. Rev. D* **2000**, *62*, 043008. [CrossRef]
33. Ward, B. Quantum corrections to Newton's law. *Mod. Phys. Lett. A* **2002**, *17*, 2371–2381. [CrossRef]
34. Kirilin, G.G.; Khriplovich, I.B. Quantum power correction to the Newton law. *J. Exp. Theor. Phys.* **2002**, *95*, 981–986. [CrossRef]
35. Satz, A.; Mazzitelli, F.D.; Alvarez, E. Vacuum polarization around stars: Nonlocal approximation. *Phys. Rev. D* **2005**, *71*, 064001. [CrossRef]
36. Morley, T.; Winstanley, E.; Taylor, P. Vacuum polarization on topological black holes with Robin boundary conditions. *Phys. Rev. D* **2021**, *103*, 045007. [CrossRef]
37. Birrell, N.D.; Davis, P.C.W. *Quantum Fields in Curved Space*; Cambridge University Press: Cambridge, UK, 1982.
38. Arnowitt, R.; Deser, S.; Misner, C.W. Republication of: The dynamics of general relativity. *Gen. Rel. Grav.* **2008**, *40*, 1997. [CrossRef]

39. Cherkas, S.L.; Kalashnikov, V.L. Quantization of the inhomogeneous Bianchi I model: Quasi-Heisenberg picture. *Nonlin. Phenom. Complex Syst.* **2015**, *18*, 1–14. [CrossRef]
40. Cherkas, S.L.; Kalashnikov, V.L. Structure of the compact astrophysical objects in the conformally-unimodular metric. *J. Belarusian State Univ. Phys.* **2020**, *3*, 97–111. [CrossRef]
41. Cherkas, S.L.; Kalashnikov, V.L. Determination of the UV cut-off from the observed value of the Universe acceleration. *J. Cosmol. Astropart. Phys.* **2007**, *01*, 028. [CrossRef]
42. Cherkas, S.L.; Kalashnikov, V.L. The equation of vacuum state and the structure formation in universe. *Vestn. Brest Univ. Ser. Fiz.-Mat.* **2021**, *1*, 41–59. (In Russian). [CrossRef]
43. Visser, M. Lorentz Invariance and the Zero-Point Stress-Energy Tensor. *Particles* **2018**, *1*, 138–154. [CrossRef]
44. Visser, M. The Pauli sum rules imply BSM physics. *Phys. Lett. B* **2019**, *791*, 43–47. [CrossRef]
45. Workman, R.L.; Burkert, V.D.; Crede, V.; Klempt, E.; Thoma, U.; Tiator, L.; Agashe, K.; Aielli, G.; Allanach, B.C.; Amsler, C.; Particle Data Group. Review of Particle Physics. *Prog. Theor. Exp. Phys.* **2022**, *2022*, 083C01. [CrossRef]
46. Cherkas, S.; Kalashnikov, V. Dark-Energy-Matter from Vacuum owing to the General Covariance Violation. *Nonlin. Phenom. Complex Syst.* **2020**, *23*, 332–337. [CrossRef]
47. Mattingly, D. Modern tests of Lorentz invariance. *Liv. Rev. Relat.* **2005**, *8*, 5. [CrossRef]
48. Amelino-Camelia, G. Quantum-spacetime phenomenology. *Liv. Rev. Relat.* **2013**, *16*, 5. [CrossRef]
49. Bluhm, R.; Yang, Y. Gravity with Explicit Diffeomorphism Breaking. *Symmetry* **2021**, *13*, 660. [CrossRef]
50. Anber, M.M.; Aydemir, U.; Donoghue, J.F. Breaking diffeomorphism invariance and tests for the emergence of gravity. *Phys. Rev. D* **2010**, *81*, 084059. [CrossRef]
51. Mavromatos, N.E. On CPT symmetry: Cosmological, quantum-gravitational and other possible violations and their phenomenology. In *Beyond the Desert 2003*; Springer: Berlin/Heidelberg, Germany, 2004; pp. 43–72.
52. Nilsson, N.A. Aspects of Lorentz and CPT Violation in Cosmology. Ph.D. Thesis, National Centre for Nuclear Research, Otwock-Świerk, Poland, 2020.
53. Cherkas, S.L.; Kalashnikov, V.L. Plasma perturbations and cosmic microwave background anisotropy in the linearly expanding Milne-like universe. In *Fractional Dynamics, Anomalous Transport and Plasma Science*; Skiadas, C.H., Ed.; Springer: Cham, Switzerland, 2018; Chapter 9. [CrossRef]
54. Dirac, P.A.M. Is there an Aether? *Nature* **1951**, *168*, 906–907. [CrossRef]
55. Cherkas, S.L.; Kalashnikov, V.L. Wave optics of quantum gravity for massive particles. *Phys. Scr.* **2021**, *96*, 115001. [CrossRef]
56. Czyz, W.; Maximon, L. High energy, small angle elastic scattering of strongly interacting composite particles. *Ann. Phys. (NY)* **1969**, *52*, 59–121. [CrossRef]
57. Kabat, D.; Ortiz, M. Eikonal quantum gravity and planckian scattering. *Nucl. Phys. B* **1992**, *388*, 570–592. [CrossRef]
58. Weinberg, S. *Gravitation and Cosmology: Principles and Applications of the General Theory of Relativity*; John Wiley & Sons: New York, NY, USA, 1972.
59. Tolman, R.C. Static Solutions of Einstein's Field Equations for Spheres of Fluid. *Phys. Rev.* **1939**, *55*, 364. [CrossRef]
60. Oppenheimer, J.R.; Volkoff, G.M. On Massive Neutron Cores. *Phys. Rev.* **1939**, *55*, 374. [CrossRef]
61. Rahaman, F.; Nandi, K.; Bhadra, A.; Kalam, M.; Chakraborty, K. Perfect fluid dark matter. *Phys. Lett. B* **2010**, *694*, 10–15. [CrossRef]
62. Sofue, Y. Rotation Curve and Mass Distribution in the Galactic Center—From Black Hole to Entire Galaxy. *Publ. Astron. Soc. Jpn.* **2013**, *65*, 118. [CrossRef]
63. Riotto, A. Inflation and the Theory of Cosmological Perturbations. In Proceedings of the Summer School on Astroparticles Physics and Cosmology, Trieste, Italy, 17 June–5 July 2002; pp. 317–417.
64. Hu, W. Covariant Linear Perturbation Formalism. In Proceedings of the Summer School on Astroparticles Physics and Cosmology, Trieste, Italy, 17 June–5 July 2002; pp. 147–185.
65. Mukhanov, V. *Physical Foundations of Cosmology*; Cambridge University Press: Cambridge, UK, 2005.
66. van Dokkum, P.; Danieli, S.; Cohen, Y.; Merritt, A.; Romanowsky, A.J.; Abraham, R.; Brodie, J.; Conroy, C.; Lokhorst, D.; Mowla, L.; et al. A galaxy lacking dark matter. *Nature* **2018**, *555*, 629–632. [CrossRef]
67. van Dokkum, P.; Abraham, R.; Brodie, J.; Conroy, C.; Danieli, S.; Merritt, A.; Mowla, L.; Romanowsky, A.; Zhang, J. A high stellar velocity dispersion and 100 globular clusters for the ultra-diffuse galaxy Dragonfly 44. *Astr. J. Lett.* **2016**, *828*, L6. [CrossRef]
68. Babichev, E.; Dokuchaev, V.; Eroshenko, Y. Black Hole Mass Decreasing due to Phantom Energy Accretion. *Phys. Rev. Lett.* **2004**, *93*, 021102. [CrossRef]
69. Babichev, E.O. The Accretion of Dark Energy onto a Black Hole. *J. Exp. Theor. Phys.* **2005**, *100*, 528. [CrossRef]
70. Sun, C.-Y. Dark Energy Accretion onto a Black Hole in an Expanding Universe. *Comm. Theor. Phys.* **2009**, *52*, 441–444. [CrossRef]
71. Carroll, L. *Alice's Adventures in Wonderland*; Princeton University Press: Princeton, NJ, USA, 2015.

Article

Equivalence Principle in Classical and Quantum Gravity

Nikola Paunković [1,†] and Marko Vojinović [2,*,†]

1. Instituto de Telecomunicações and Departamento de Matemática, Instituto Superior Técnico, Universidade de Lisboa, Avenida Rovisco Pais 1, 1049-001 Lisboa, Portugal
2. Institute of Physics, University of Belgrade, Pregrevica 118, 11080 Belgrade, Serbia
* Correspondence: vmarko@ipb.ac.rs
† These authors contributed equally to this work.

Abstract: We give a general overview of various flavours of the equivalence principle in classical and quantum physics, with special emphasis on the so-called weak equivalence principle, and contrast its validity in mechanics versus field theory. We also discuss its generalisation to a theory of quantum gravity. Our analysis suggests that only the strong equivalence principle can be considered fundamental enough to be generalised to a quantum gravity context since all other flavours of equivalence principle hold only approximately already at the classical level.

Keywords: equivalence principle; general relativity; quantum gravity

1. Introduction

Quantum mechanics (QM) and general relativity (GR) are the two cornerstones of modern physics. Yet, merging them together in a quantum theory of gravity (QG) is still elusive despite the nearly century-long efforts of vast numbers of physicists and mathematicians. While the majority of the attempts were focused on trying to formulate the full theory of quantised gravity, such as string theory, loop quantum gravity, non-commutative geometry, and causal set theory, to name a few, a number of recent studies embraced a rather more modest approach by exploring possible consequences of basic features and principles of QM and GR, and their status, in a tentative theory of QG. Acknowledging that the superposition principle, as a defining characteristic of any quantum theory, must be featured in QG as well, led to a number of papers studying gravity-matter entanglement [1–7], genuine indefinite causal orders [8–15], quantum reference frames [16–20] and deformations of Lorentz symmetry [21–25], to name a few major research directions. Exploring spatial superpositions of masses, and in general gravitational fields, led to the analysis of the status of various versions of the equivalence principle, and their exact formulations in the context of QG. In particular, in [26], it was shown that the weak equivalence principle (WEP) should generically be violated in the presence of a specific type of superpositions of gravitational fields, describing small quantum fluctuations around a dominant classical geometry. On the other hand, a number of recent studies propose generalisations of WEP to QG framework (see for example [16,20,27–31]), arguing that it remains satisfied in such scenarios, a result *seemingly* at odds with [26] (for details, see the discussion from Section 5).

The modern formulation of WEP is given in terms of a *test particle* and it's *trajectory*: it is a *theorem* within the mathematical formulation of GR stating that the trajectory of a test particle satisfies the so-called geodesic equation [32–46], while it is *violated* within the context of QG, as shown in [26]. In this paper, we present a brief overview of WEP in GR and a critical analysis of the notions of particle and trajectory in both classical and quantum mechanics, as well as in the corresponding field theories. Our analysis demonstrates that WEP, as well as all other flavours of the equivalence principle (EP) aside from the strong one (SEP), hold only approximately. From this we conclude that neither WEP nor any other flavour of EP (aside from SEP) can be considered a viable candidate for generalisation to the quantum gravity framework.

The paper is organised as follows. In Section 2, we give a brief historical overview of various flavours of the equivalence principle, with a focus on WEP. In Section 3, we analyse the notion of a trajectory in classical and quantum mechanics, while in Section 4 we discuss the notion of a particle in field theory and QG. Finally, in the Conclusion, we briefly review and discuss our results, and present possible future lines of research.

2. Equivalence Principle in General Relativity

The equivalence principle is one of the most fundamental principles in modern physics. It is one of the two cornerstone building blocks for GR, the other being the principle of general relativity. While its importance is well understood in the context of gravity, it is often underappreciated in the context of other fundamental interactions. In addition, there have been numerous studies and everlasting debates about whether EP holds also in quantum physics, if it should be generalised to include quantum phenomena or not, etc. Finally, EP has been historically formulated in a vast number of different ways, which are often not mutually equivalent, leading to a lot of confusion about the actual statement of the principle and its physical content [47–53]. Given the importance of EP, and the amount of confusion around it, it is important to try and help clarify these issues.

The equivalence principle is best introduced by stating its purpose—in its traditional sense, the purpose of EP is to *prescribe the interaction between gravity and all other fields in nature, collectively called matter* (by "matter" we assume not just fermionic and scalar fields, but also gauge vector bosons, i.e., nongravitational interaction fields). This is important to state explicitly since EP is often mistakenly portrayed as a property of gravity alone, without any reference to matter. In a more general, less traditional, and often not appreciated sense, the purpose of EP is to prescribe the interaction between *any gauge field* and all other fields in nature (namely fermionic and scalar matter, as well as other gauge fields, including gravity), which we will reflect on briefly in the case of electrodynamics below.

Given such a purpose, let us for the moment concentrate on the gravitational version of EP, and provide its modern formulation, as it is known and understood today. The statement of the equivalence principle is the following:

The equations of motion for matter coupled to gravity remain locally identical to the equations of motion for matter in the absence of gravity.

This kind of statement requires some unpacking and comments.

- When comparing the equations of motion in the presence and in the absence of gravity, the claim that they remain identical may naively suggest that gravity does not influence the motion of matter in any way whatsoever. However, on closer inspection, the statement is that the two sets of equations remain *locally* identical, emphasising that the notion of locality is a crucial feature of the EP. While equations of motion are already local in nature (since they are usually expressed as partial differential equations of finite order), the actual interaction between matter and gravity enters only when *integrating* those equations to find a solution (see Appendix A for a detailed example).
- In order to compare the equations of motion for matter in the presence of gravity to those in its absence, the equations themselves need to be put in a suitable form (typically expressed in general curvilinear coordinates, as tensor equations). The statement of EP relies on a theorem that this can always be achieved, first noted by Erich Kretschmann [54].
- Despite being dominantly a statement about the interaction between matter and gravity, EP also implicitly suggests that the best way to describe the gravitational field is as a property of the geometry of spacetime, such as its metric [55]. In that setup, EP can be reformulated as a statement of *minimal coupling* between gravity and matter, stating that equations of motion for matter may depend on the spacetime metric and its first derivatives, but not on its (antisymmetrised) second derivatives, i.e., the *spacetime curvature does not explicitly appear in the equations of motion for matter*.

- The generalisation of EP to other gauge fields is completely straightforward, by replacing the role of gravity with some other gauge field, and suitably redefining what matter is. For example, in electrodynamics, the EP can be formulated as follows:

    The equations of motion for matter coupled to the electromagnetic field remain locally identical to the equations of motion for matter in the absence of the electromagnetic field.

    This statement can also be suitably reformulated as the minimal coupling between the electromagnetic (EM) field and matter, i.e., coupling matter to the electromagnetic potential A_μ but not to the corresponding field strength $F_{\mu\nu} = \partial_\mu A_\nu - \partial_\nu A_\mu$. This is in fact the standard way the EM field is coupled to matter (see Appendix A for an illustrative example). Even more generally, the gauge field sector of the whole Standard Model of elementary particles (SM) is built using the minimal coupling prescription, meaning that the suitably generalised version of the EP actually prescribes the interaction between matter and all fundamental interactions in nature, namely strong, weak, electromagnetic and gravitational. In this sense, EP is a cornerstone principle for the whole fundamental physics, as we understand it today.

Of course, much more can be said about the statement of EP, its consequences, and various other details. However, in this work, our attention will focus on the so-called *weak equivalence principle* (WEP), which is a reformulation of EP applied to matter which consists of mechanical particles. To that end, it is important to understand various flavours and reformulations of EP that have appeared through history.

As with any deep concept in physics, EP has been expressed historically through a painstaking cycle of formulating it in a precise way, understanding the formulation, understanding the drawbacks of that formulation, coming up with a better formulation, and repeating. In this sense, EP, as quoted above, is a modern product of long and meticulous refinement over several generations of scientists. Needless to say, each step in that process made its way into contemporary physics textbooks, leading to a plethora of different formulations of EP that have accumulated in the literature over the years. This can bring about a lot of confusion about what EP actually states [47–50] since various formulations from old and new literature may often be not merely phrased differently, but in fact substantively inequivalent. To that end, let us comment on several of the most common historical statements of EP (for a more detailed historical overview and classification, see [56,57]), and their relationship with the modern version:

- *Equality of gravitational and inertial mass.* This is one of the oldest variants of EP, going back to Newton's law of universal gravitation. The statement claims that the "gravitational charge" of a body is the same as the body's resistance to acceleration, in the sense that the mass appearing on the left-hand side of Newton's second law of motion exactly cancels the mass appearing in Newton's gravitational force law on the right-hand side. This allows one to relate it to the modern version of EP, in the sense that a suitably accelerated observer could rewrite Newton's law of motion as the equation for a free particle, exploiting the cancellation of the "intertial force" and the gravitational force on the right-hand side of the equation. Unfortunately, this version of EP is intrinsically nonrelativistic, and applicable only in the context of Newtonian gravity since already in GR the source of gravity becomes the full stress-energy tensor of matter fields, rather than just the total mass. Finally, this principle obviously fails when applied to photons, as demonstrated by the gravitational bending of light.
- *Universality of free fall.* Going back all the way to Galileo, this statement claims that the interaction between matter and gravity does not depend on any intrinsic property of matter itself, such as its mass, angular momentum, chemical composition, temperature, or any other property, leading to the idea that gravity couples universally (i.e., in the same way) to all matter. Formulated from experimental observations by Galileo, its validity is related to the quality of experiments used to verify it. As we shall see below,

in a precise enough setting, one can experimentally observe direct coupling between the angular momentum of a body and spacetime curvature [32–46], invalidating the statement.

- *Local equality between gravity and inertia.* Often called Einstein's equivalence principle, the statement claims that a local and suitably isolated observer cannot distinguish between accelerating and being at rest in a uniform gravitational field. While this statement is much closer in spirit to the modern formulation of EP, it obscures the crucial aspect of the principle — coupling of matter to gravity. Namely, in this formulation, it is merely implicit that the only way an observer can *attempt to distinguish* gravity from inertia is by making local experiments using some form of *matter*, i.e., studying the equations of motion of matter in the two situations and trying to distinguish them by observing whether or not matter behaves differently. Moreover, the statement is often discussed in the context of mechanics, arguing that any given particle does not distinguish between gravity and inertia. This has two main pitfalls—first, the reliance on particles is very misleading (as we will discuss below in much more detail), and second, it implicitly suggests that gravity and inertia are the same phenomenon, which is completely false. Namely, inertia can be understood as a specific form of gravity, but a general gravitational field cannot be simulated by inertia, since inertia cannot account for tidal effects of inhomogeneous configurations of gravity.

- *Weak equivalence principle.* Stating that the equations of motion of particles do not depend on spacetime curvature, or equivalently, that the motion of a free particle is always a geodesic trajectory in spacetime, WEP is in fact an application of modern EP to mechanical point-like particles (i.e., test particles). One can argue that, as far as the notion of a point-like particle is a well-defined concept in physics, WEP is a good principle. Nevertheless, as we will discuss below in detail, the notion of a point-like particle is an idealisation that does not actually have any counterpart in reality, in either classical or quantum physics. Regarding a realistic particle (with nonzero size), WEP *never holds*, due to the explicit effect of gravitational tidal forces across the particle's size. In this sense, WEP can be considered at best an *approximate* principle, which can be assumed to hold only in situations where particle size can be approximated to zero.

- *Strong equivalence principle.* This version of the principle states that the equations of motion of all fundamental fields in nature do not depend on spacetime curvature (see [55], Section 16.2, page 387). To the best of our knowledge so far, fields are indeed the most fundamental building blocks in modern physics (such as SM), while the strength of the gravitational field is indeed described by spacetime curvature (as in GR). In this sense, the statement of SEP is actually an instance of EP applied to field theory, and as such equivalent to the modern statement of EP. So far, all our knowledge of natural phenomena is consistent with the validity of SEP.

As can be seen from the above review, various formulations of EP are both mutually inequivalent and have different domains of applicability. Specifically, only SEP holds universally, while all other flavours of EP hold only approximately. In the remainder of the paper, we focus on the study of WEP since recently it gained a lot of attention in the literature [20,27–29,31], primarily in the context of its generalisation to a "quantum WEP", and in the context of a related question of particle motion in a quantum superposition of different gravitational configurations, the latter being a scenario that naturally arises in QG. Since WEP is stated in terms of a test particle and its trajectory, in order to try and generalise it to the scope of QG one should first analyse these two notions in classical and quantum mechanics and field theory in more detail.

3. The Notion of Trajectory in Classical and Quantum Mechanics

A trajectory of a physical system in three-dimensional space is a set of points that form a line, usually parameterised by time. More formally, a trajectory is a set $\{(x(t), y(t), z(t)) \in$

$\mathbb{R}^3 | t \in [t_i, t_f] \subset \mathbb{R} \wedge t_i < t_f\}$, given by three smooth functions $x, y, z : \mathbb{R} \mapsto \mathbb{R}$. Depending on the nature of the system, the choice of points that form its trajectory may vary.

In classical mechanics, one often considers an ideal "point-like particle" localised in one spatial point $(x(t), y(t), z(t))$ at each moment of time t, in which case the choice of the points forming the system's trajectory is obvious. In the case of systems continuously spread over certain volumes ("rigid bodies", or "objects") or composite systems consisting of several point-like particles or bodies, it is natural to consider their centres of mass as points that form the trajectory. While this definition is natural, widely accepted, and formally applicable to any classical mechanical system, there are cases in which the very notion of a trajectory loses its intuitive, as well as useful, meaning.

Consider for example an electrical dipole, i.e., a system of two point-like particles with equal masses and opposite electrical charges, separated by the distance $\ell(t)$. As long as this distance stays "small" and does not vary significantly with time, the notion of a trajectory of a dipole, defined as the set of centres of mass of the two particles, does meet our intuition, and can be useful. Informally, if the trajectories of each of the two particles are "close" to each other, they can be approximated, and consequently represented, by the trajectory of the system's centre of mass. However, if the separate trajectories of the two particles diverge, one going to the "left", and another to the "right", one could hardy talk of a trajectory of such a composite system, although the set of locations of its centres of mass is still well defined. In fact, the dipole itself ceases to make physical sense when the distance between its constituents is large.

Moving to the realm of quantum mechanics, due to the superposition principle, even the ideal point-like particles do not have a well-defined position, which is further quantified by the famous Heisenberg uncertainty relations. Thus, the trajectory of point-like particles (and any system that in a given regime can be approximated to be point-like) is defined as a set of expectation values of the position operator. Like in the case of composite classical systems, here as well the definition of a trajectory of a point-like particle is mathematically always well defined, yet for a very similar reason is applicable only to certain cases. Namely, in order to give a useful meaning to the above definition of trajectory, the system considered must be *well localised*. Consider for example the double-slit experiment, in which the point-like particle is highly delocalised so that we say that *its trajectory is not well defined*, even though the set of the expectation values of the position operator is.

We see that, while in mechanics both the notions of a particle and its trajectory are rather straightforward and always well defined, the latter make sense only if our system is well localised in space (see for example [58], where the authors analyse the effects of wave-packet spreading to the notion of a trajectory).

4. The Notion of a Particle in Field Theory

While in classical mechanics a point-like particle is always well localised, we have seen that in the quantum case one must introduce an additional constraint in order for it to be considered localised—the particle should be represented by a wave-packet. The source for this requirement lies in the fact that quantum particles, although mechanical, are represented by a *wavefunction*. Thus, it is only to be expected that when moving to the realm of the field ontology, the notion of a particle becomes even more involved and technical.

In field theory, the fundamental concept is the *field*, rather than a particle. The notion of a particle is considered a derived concept, and in fact in QFT one can distinguish two vastly different phenomena that are called "particles".

The first notion of a particle is an elementary excitation of a free field. For example, the state
$$|\Psi\rangle = \hat{a}^\dagger(\vec{k})|0\rangle,$$
is called a *single particle state* of the field, or a *plane-wave-particle*. It has the following properties:

- It is an eigenstate of the *particle number operator* for the eigenvalue 1.

- It has a sharp value of the momentum $\vec{k}$, and corresponds to a completely delocalised plane wave configuration of the field.
- It has no centre of mass, and no concept of "position" in space since the "position operator" is not a well-defined concept for the field.
- States of this kind are said to describe *elementary particles*, understood as asymptotic free states of past and future infinity, in the context of the S-matrix for scattering processes. An example of a real scalar particle of this type would be the *Higgs particle*. For fields of other types (Dirac fields, vector fields, etc.) examples would be an *electron*, a *photon*, a *neutrino*, an asymptotically free *quark*, and so on. Essentially, all particles tabulated in the Standard Model of elementary particles are of this type.

Note that all the above notions are defined within the scope of free field theory, and do not carry over to interacting field theory. In other words, free field theory is a convenient idealisation, which does not really reflect realistic physics. One should therefore understand the concept of a plane-wave-particle in this sense, merely as a convenient mathematical approximation. Moreover, the particle number operator is not an invariant quantity, as demonstrated by the Unruh effect. We should also emphasise that in an interacting QFT, the proper way to understand the notion of a particle is as a localised wave-packet, interacting with its virtual particle cloud, which does have a position in space and whose momentum is defined through its group velocity. In this sense, the particle as a wave-packet could be better interpreted as a kink, discussed below.

The second notion of the particle in field theory is a bound state of fields, also called a *kink solution*. This requires an interacting theory since interactions are necessary to form bound states. This kind of configuration of fields has the following properties:

- It is not an eigenstate of the particle number operator, and the expectation value of this operator is typically different from 1.
- It is usually well localised in space, and does not have a sharp value of momentum.
- As long as the kink maintains a stable configuration (i.e., as long as it does not decay), one can in principle assign to it the concept of *size*, and as a consequence also the concepts of *centre of mass*, *position in space*, and *trajectory*. In this sense, a kink can play the role of a test particle.
- States of this kind are said to describe *composite particles*. Given an interacting theory such as the Standard Model, under certain circumstances quarks and gluons form bound states called a *proton* and a *neutron*. Moreover, protons and neutrons further form bound states called *atomic nuclei*, which together with electrons and photons form *atoms*, *molecules*, and so on.

For a kink, the notions of centre of mass, position in space and size are described only as classical concepts, i.e., as expectation values of certain field operators, such as the stress-energy tensor. Moreover, given the nonzero size of the kink, its centre of mass and position are not uniquely defined, even classically, since in relativity different observers would assign different points as the centre of mass.

Given the two notions of particles in QFT, one can describe two different corresponding notions of WEP. In principle, one first needs to apply SEP in order to couple the matter fields to gravity, at the fundamental level. Assuming this is completed, the motions of both the plane-wave-particles and kink particles can be derived from the combined set of Einstein's equations and matter field equations, without any appeal to any notion of WEP. In this sense, once the trajectory of the particle in the background gravitational field has been determined from the field equations, one can verify *as a theorem* whether the particle satisfies WEP or not.

Specifically, in the case of a matter field coupled to general relativity such that it locally resembles a plane wave, one can apply the WKB approximation to demonstrate that the wave 4-vector $k^\mu(x)$, orthogonal to the wavefront at its every point $x \in \mathbb{R}^4$, will satisfy a geodesic equation,

$$k^\mu(x)\nabla_\mu k^\lambda(x) = 0. \tag{1}$$

However, given that the plane-wave-particle is completely delocalised in space, the fact that the wave 4-vector satisfies the geodesic equation could hardly be interpreted as "the particle following a geodesic trajectory", and thus obeying WEP. Indeed, identifying the vector field orthogonal to the wavefront to the notion of "particle's trajectory" is at best an abuse of terminology.

Next, in the case of the kink particle coupled to general relativity, one assumes the configuration of the background gravitational field is such that the particle maintains its structure and that its size can be completely neglected. One can then apply the procedure given in [26,32–46] to demonstrate that the 4-vector $u^\mu(\tau)$, tangent to the kink's world line (i.e., its trajectory), will satisfy a geodesic equation ($\tau \in \mathbb{R}$ represents kink's proper time),

$$u^\mu(\tau)\nabla_\mu u^\lambda(\tau) = 0. \tag{2}$$

Thus, one concludes that the kink obeys WEP as a *theorem* in field theory, without the necessity to actually postulate it.

Note the crucial difference between Equations (1) and (2)—while the former features 4-dimensional variable x, the latter is given in terms of only 1-dimensional proper time τ. This reflects the fact that the plane-wave-particle is a highly delocalised object, with no well-defined position and trajectory, while the kink is a highly localised object, with a well-defined position and trajectory. As a consequence, WEP can be formulated only for the kink, and not for the plane-wave particle.

In the case of the kink, it is also important to emphasise that the zero-size approximation of the kink is crucial. Namely, without this assumption, the particle will feel the tidal forces of gravity across its size, effectively coupling its angular momentum $J^{\mu\nu}(\tau)$ to the curvature of the background gravitational field [32–46] (see also [59] for a more refined analysis of tidal effects). This will give rise to an equation of motion for the kink of the form

$$u^\mu(\tau)\nabla_\mu u^\lambda(\tau) = R^\lambda{}_{\mu\rho\sigma} u^\mu(\tau) J^{\rho\sigma}(\tau), \tag{3}$$

which features explicit coupling to curvature (absent from (2)) and thus fails to obey WEP. In this sense, for realistic kink solutions WEP is *always violated*, and can be considered to hold only as an approximation when the size of the particle can be completely neglected compared to the radius of curvature of the background gravitational field. If in addition the kink has negligible total energy, it can be used as a point-like test particle.

In the above discussion, while matter fields are described as quantum, using QFT, the background gravitational field is considered to be completely classical. It should therefore not be surprising that WEP may fail to hold if one allows the gravitational field to be quantum, such as matter fields, and one needs to revisit all steps of the above analysis from the perspective of QG. In fact, the case of the kink particle has been studied in precisely this scenario [26], and it has been shown that if the background gravitational field is in a specific type of quantum superposition of different configurations, the kink will fail to obey WEP even in the zero size approximation. Simply put, the equation of motion for the kink will contain extra terms due to the interference effects between superposed configurations of gravity, giving rise to an effective force that pushes the kink off the geodesic trajectory. Moreover, of course, similar to the case of classical gravity, the resulting conclusion is a *theorem*, which follows from the fundamental field equations of the theory. One of the assumptions of that theorem is that the field equations allow for kink solutions in the first place. Namely, it is entirely possible that in quantum gravity particles cannot be localised at all, as opposed to the classical case where such an approximation can be feasible. If that is the case, then one cannot even formulate (i.e., generalise from classical theory) the notion of WEP in a quantum gravity setup. However, one can instead assume that kink solutions do exist, as was performed in [26], where a particular superposition of gravitational fields was considered, describing small quantum fluctuations around a dominant classical geometry. It was then argued that such superpositions are compatible with the approximation of a well-defined localised particle (see the discussion around Equations (2.2) and (3.15), as well as Section 3.4 of that paper). As it turns out, even in such cases the trajectory of the kink fails to obey WEP. Therefore, the generalisations of WEP and other approximate versions of EP are not the best candidates for analysing the properties of quantum gravity.

Moreover, the assumption of a well-defined notion of a particle in the QG framework can also be supported from the point of view of nonrelativistic limit. Namely, in [4,5] an experiment was proposed in which the effects of QG fluctuations are expected to be observable, by measuring the motion of nonrelativistic particles. Furthermore, an extension of this experiment was also suggested [60], which aims to determine the potential difference between gravitational and inertial masses of those particles in such a setup. In fact, the relation between the two types of masses in the nonrelativistic limit has also been previously analysed in [26], predicting their difference due to quantum fluctuations of geometry. In this sense, the notion of a kink should make sense even in the QG setup, at least in the nonrelativistic limit.

For the case of the plane-wave-particle travelling through the superposed background of two gravitational field configurations, the analysis of the equation of motion for the wave-vector field $k^\mu(x)$, in the sense of [26,32–46], has not been performed so far (to the best of our knowledge). However, in principle, one can expect a similar interference term to appear in the WKB analysis, and give rise to a non-geodesic equation for the wave 4-vector as well. In this sense, it is to be expected that generically even the wavefronts of such plane-wave-particles would fail to obey WEP.

5. Conclusions and Discussion

In this paper, we give an overview of the equivalence principle and its various flavours formulated historically, with a special emphasis on the weak equivalence principle. We performed a critical analysis of the notions of particle and trajectory in various frameworks of physics, showing that the notion of a point-like particle and its trajectory are not always well defined. This in turn suggests that WEP might not be the best starting point for generalisation to QG, as we argue in more detail below.

As discussed in Section 4, in [26] it was shown that if superpositions of states of gravity and matter are allowed, WEP can be violated. It is important to note that the cases considered in [26] feature a specific type of superposition of three groups of states: the first consists of a single so-called dominant state—a classical state whose expectation values of the metric and the stress-energy tensors satisfy Einstein field equations; the second consists of states similar to the dominant one, with arbitrary coefficients; and the third consists of states quasi-orthogonal to the dominant one, but with negligible coefficients. Only then one may talk of a (well-localised) trajectory of the test particle in the overall superposed state and consequently about the straightforward generalisation of the classical WEP to the realm of QG. Considering that for the dominant state, being classical, the trajectory of the test particle follows the corresponding geodesic, we see that in the superposed state its trajectory would *deviate from the geodesic of the dominant state*, thus violating WEP. Note that, as discussed in Section 4, this deviation, in addition to classically weighted trajectories of the individual branches, also features purely quantum (i.e., off-diagonal) interference terms.

On the other hand, a number of recent studies propose generalisations of WEP to QG framework, arguing that it remains satisfied in such scenarios, a result *seemingly* at odds with [26]. For example, in [29–31], the authors consider superpositions of an arbitrary number of classical quasi-orthogonal states with arbitrary coefficients, arguing that since WEP is valid in each classical branch, it is valid in its superposition as well. If taken as a *definition* of what it means that a certain principle is satisfied in a superposition of different quantum states, then the above statement is manifestly true. As such, being a definition, it tells little about physics—it merely rephrases one statement ("principle A is separately satisfied in all branches of a superposition") with another, simpler ("principle A is satisfied in a superposition"). Namely, note that in [29,30], such a generalised version of EP plays no functional role in the analyses conducted in those papers. What does play a functional role is the statement of one version of classical EP (specifically, local equality between gravity and inertia) applied to each particular state in a superposition. All physically relevant (and otherwise interesting) conclusions of the two papers could be equally obtained without ever talking about the generalised EP. In addition, in [31] EP itself is not even the main focus of the paper, and its generalisation is just introduced in analogy to the analysis of the conservation laws, which is itself an interesting topic. On the other hand, in the case of weakly superposed gravitational fields, such as in proposed experiments [4,5], the violation of the equality of inertial and gravitational masses is to be expected [26,60]. Moreover, following the spirit of the above definition, one could be misled to conclude that the notions of the particle's position and trajectory are always well-defined, as long as they are defined in each (quasi-classical) branch of the superposition.

An alternative approach to the generalisation of EP to the quantum domain was proposed in [16,20,27,28]. In those works, the authors discuss the coupling of a spatially delocalised wave-particle to gravity, with the aim of generalising such a scenario to QG. To that end, they prove a theorem which essentially states that for such a delocalised wave-particle, even when it is entangled with the gravitational field, one can always find a quantum reference frame transformation, such that in the vicinity of a given spacetime point one has a locally inertial coordinate system. The theorem employs the novel techniques of quantum reference frames (QRF) to generalise to the quantum domain the well-known result from differential geometry, that in the infinitesimal neighbourhood of any spacetime point one can always choose a locally inertial coordinate system.

The authors then employ the theorem to generalise one flavour of EP to the quantum domain. Specifically, even if the wave-particle is entangled with the gravitational field, one can use the appropriate QRF transformation to switch to a locally inertial coordinate system, and then in that system "all the (nongravitational) laws of physics must take on their familiar non-relativistic form". Here, to the best of our understanding, the phrase "non-gravitational laws of physics" refers to the equations of motion for a quantum-mechanical wave-particle, while "non-relativistic form" means that these equations of motion take the same form as in special-relativistic context.

Our understanding is that the above wave-particle generalisation of EP lies somewhere "in between" mechanics and field theory, i.e., it is in a sense stronger than WEP, which discusses particles, but weaker than SEP, which discusses full-blown matter fields. Since it refers to wave-particles rather than kinks, our analysis of WEP and its reliance on the particle trajectory does not apply to this version of EP.

The methodology in [16,20,27,28] is that one should try to generalise even approximate flavours of EP, as a stopgap result in a bigger research programme, in the hope that they may still shed some light on QG. This is of course a legitimate methodology, and from that point of view these kinds of generalisations of EP to the quantum domain represent interesting results. Nevertheless, we also believe it would be preferable to formulate a generalisation of SEP, and in a way which does not appeal to reference frames at all, since that would be closer to the essence of the statement of EP, as discussed in Section 2.

To conclude, our analysis suggests that, instead of trying to generalise various approximate formulations of EP, one should rather talk of operationally verifiable statements regarding the (in)equality of gravitational and inertial masses, possible deviation from the geodesic motion of test particles, the universality of free fall, etc., and study other principles and their possible generalisations to QG, such as SEP (see Section 4.2 in [26]), background independence, quantum nonlocality, and so on.

Author Contributions: Investigation, N.P. and M.V.; writing—original draft preparation, N.P. and M.V.; writing—review and editing, N.P. and M.V. All authors have read and agreed to the published version of the manuscript.

Funding: NP's work was partially supported by SQIG—Security and Quantum Information Group of Instituto de Telecomunicações, by Programme (COMPETE 2020) of the Portugal 2020 framework [Project Q.DOT with Nr. 039728 (POCI-01-0247-FEDER-039728)] and the Fundação para a Ciência e a Tecnologia (FCT) through national funds, by FEDER, COMPETE 2020, and by Regional Operational Program of Lisbon, under UIDB/50008/2020 (actions QuRUNNER, QUESTS), Projects QuantumMining POCI-01-0145-FEDER-031826, PREDICT PTDC/CCI-CIF/29877/2017, CERN/FIS-PAR/0023/2019, QuantumPrime PTDC/EEI-TEL/8017/2020, as well as the FCT Estímulo ao Emprego Científico grant no. CEECIND/04594/2017/CP1393/CT000. MV was supported by the Ministry of Education, Science and Technological Development of the Republic of Serbia, and by the Science Fund of the Republic of Serbia, grant 7745968, "Quantum Gravity from Higher Gauge Theory 2021"—QGHG-2021. The contents of this publication are the sole responsibility of the authors and can in no way be taken to reflect the views of the Science Fund of the Republic of Serbia.

Institutional Review Board Statement: Not applicable.

Acknowledgments: The authors wish to thank Časlav Brukner, Flaminia Giacomini, Chiara Marletto and Vlatko Vedral for useful discussions. MV is also indebted to Milovan Vasilić and Igor Salom for clarifications regarding the notion of symmetry localisation.

Conflicts of Interest: The authors declare no conflict of interest.

Appendix A

Here, we provide a detailed example of the two applications of the EP. First, we discuss the gravitational EP and apply it to a real scalar field, giving all mathematical details and discussing various related aspects such as locality, symmetry localisation, and so on. Then, we turn to the application of the gauge field generalisation of EP, using electrodynamics as an example. We describe how one can couple matter to an EM field, mimicking the previous gravitational example, and emphasize the analogy between the gravitational and EM case at each step. Note also that the non-Abelian gauge fields can be studied in exactly the same way. Finally, we discuss the case of test particles, and the violation of the WEP in both gravitational and electromagnetic cases.

Throughout this section, we assume that the Minkowski metric $\eta_{\mu\nu}$ has signature $(-,+,+,+)$.

Appendix A.1. The Gravitational Case

Let us begin with an example of a real scalar field in Minkowski spacetime, and apply the equivalence principle by coupling it to gravity. The equation of motion in this case is the ordinary Klein–Gordon equation,

$$\left(\eta^{\mu\nu}\partial_\mu\partial_\nu - m^2\right)\phi(x) = 0\,. \tag{A1}$$

As it stands, it describes the free scalar field in Minkowski spacetime, in an inertial coordinate system. In order to couple it to gravity (in the framework of GR), we first rewrite this equation into an arbitrary curvilinear coordinate system, as

$$\left(\tilde{g}^{\mu\nu}\tilde{\nabla}_\mu\tilde{\nabla}_\nu - m^2\right)\phi(\tilde{x}) = 0\,. \tag{A2}$$

Here the covariant derivative $\tilde{\nabla}_\mu$ is defined in terms of the Levi-Civita connection,

$$\tilde{\Gamma}^\lambda{}_{\mu\nu} = \frac{1}{2}\tilde{g}^{\lambda\sigma}\left(\partial_\mu\tilde{g}_{\nu\sigma} + \partial_\nu\tilde{g}_{\mu\sigma} - \partial_\sigma\tilde{g}_{\mu\nu}\right), \tag{A3}$$

which is in turn defined in terms of the curvilinear Minkowski metric $\tilde{g}_{\mu\nu}$. Note that the tilde symbol denotes the fact that this metric has been obtained by a coordinate transformation $\tilde{x}^\mu = \tilde{x}^\mu(x)$ from the Minkowski metric in an inertial coordinate system, $\eta_{\mu\nu}$,

$$\tilde{g}_{\mu\nu} = \frac{\partial x^\rho}{\partial \tilde{x}^\mu}\frac{\partial x^\sigma}{\partial \tilde{x}^\nu}\eta_{\rho\sigma}\,, \tag{A4}$$

and, therefore, if one were to evaluate the Riemann curvature tensor using $\tilde{g}_{\mu\nu}$ and $\tilde{\Gamma}^\lambda{}_{\mu\nu}$, according to the equation

$$R^\lambda{}_{\rho\mu\nu} = \partial_\mu \tilde{\Gamma}^\lambda{}_{\rho\nu} - \partial_\nu \tilde{\Gamma}^\lambda{}_{\rho\mu} + \tilde{\Gamma}^\lambda{}_{\sigma\mu}\tilde{\Gamma}^\sigma{}_{\rho\nu} - \tilde{\Gamma}^\lambda{}_{\sigma\nu}\tilde{\Gamma}^\sigma{}_{\rho\mu}, \tag{A5}$$

one would obtain that $R^\lambda{}_{\mu\nu\rho} = 0$ at every point in spacetime since transforming into a different coordinate system cannot induce the curvature of spacetime.

Now one can apply EP (in this example specifically SEP) in order to couple the scalar field to gravity. The statement of SEP is that, in the presence of a gravitational field (i.e., in curved spacetime), the equation of motion for the scalar field should locally retain the same form as in the absence of the gravitational field (i.e., in flat spacetime). Since Equation (A2) depends only on the field at a given spacetime point and its first and second derivatives at the same point, the equation is in fact local—it is defined within an infinitesimal neighbourhood of a single point. Given this, EP states that the corresponding equation of motion in the presence of gravity should have precisely the same form:

$$\left(g^{\mu\nu}\nabla_\mu\nabla_\nu - m^2\right)\phi(x) = 0. \tag{A6}$$

The absence of the tilde now denotes the fact that the covariant derivative ∇_μ is defined in terms of a generic Levi-Civita connection $\Gamma^\lambda{}_{\mu\nu}$ which is in turn defined in terms of a generic metric $g_{\mu\nu}$, which does not necessarily satisfy (A4). In other words, EP postulates that the Equation (A6) now holds even in curved spacetime since for a generic metric and connection, the Riemann curvature tensor need not be equal to zero everywhere. The interaction between the scalar field and gravity, as postulated by EP and implemented in Equation (A6), is also known in the literature as the *minimal coupling* prescription [61].

In order to convince oneself that the preparation step of transforming (A1) to (A2) is trivial in the sense that it does not introduce any substantial modification of (A1), one can additionally demonstrate that (A6) is in fact locally equivalent to (A1) as well. Namely, according to a theorem in differential geometry (see for example the end of Chapter 85 in [62]), at any specific spacetime point x_0 one can choose the locally inertial coordinate system, in which the generic metric $g_{\mu\nu}$, the corresponding connection $\Gamma^\lambda{}_{\mu\nu}$ and consequently also the covariant derivative ∇_μ take their usual Minkowski values,

$$g_{\mu\nu}(x_0) = \eta_{\mu\nu}, \qquad \Gamma^\lambda{}_{\mu\nu}(x_0) = 0, \qquad \nabla_\mu\big|_{x=x_0} = \partial_\mu, \tag{A7}$$

so that in the infinitesimal neighbourhood of the point x_0 Equation (A6) obtains the form precisely equal to (A1).

However, note that when *integrating* (A6), one must take into account that spacetime is curved since integration is a nonlocal operation, and the locally inertial coordinate system cannot eliminate spacetime curvature. Therefore, the *solutions* of (A6) will in general be *different* from solutions of (A1), indicating the physical interaction of the scalar field with gravity, despite the fact that the form of the equation of motion is identical in both cases.

Another thing that should be emphasised is that EP itself is not a mathematical theorem, but rather a principle with physical content, since it can be either satisfied or violated. Specifically, we could have prescribed a different coupling of the scalar field to gravity, such that in curved spacetime its equation of motion takes for example the form

$$\left(g^{\mu\nu}\nabla_\mu\nabla_\nu - m^2 + R^2 + K^2\right)\phi(x) = 0, \tag{A8}$$

where $R \equiv R^{\mu\nu}{}_{\mu\nu}$ and $K \equiv R^{\mu\nu\rho\sigma}R_{\mu\nu\rho\sigma}$ are the curvature scalar and Kretschmann invariant, respectively. This equation is not equivalent to (A2) and there is no coordinate system in which it can take the form (A1) since R and K are invariants. In this sense, (A8) is an example of a scalar field coupled to gravity such that EP is violated. This type of interaction between matter and gravity is also known in the literature as *non-minimal coupling* [61].

Finally, we should note that the transformation from (A1) to (A2) amounts to what is also known in the literature as *symmetry localisation* [61]. In particular, one can verify that (A1) remains invariant with respect to the group $\mathbb{R}^4$ of global translations,

$$x^\mu \to \tilde{x}^\mu = x^\mu + \zeta^\mu, \qquad (\zeta \in \mathbb{R}^4), \tag{A9}$$

while (A2) remains invariant with respect to the group $Diff(\mathbb{R}^4)$ of spacetime diffeomorphisms, obtained by localisation of the translational symmetry group,

$$x^\mu \to \tilde{x}^\mu = x^\mu + \zeta^\mu(x) \equiv \tilde{x}^\mu(x), \tag{A10}$$

which represent general curvilinear coordinate transformations, used in (A4). One can explicitly verify that all three Equations (A2), (A6) and (A8) remain invariant with respect to local translations (A10) while describing no coupling to gravity, coupling to gravity that satisfies EP, and coupling to gravity that violates EP, respectively. In this sense, contrary to a common misconception (often stated in the literature) that symmetry localisation gives rise to interactions, one can say that the process of symmetry localisation *does not* introduce nor prescribe interactions in any way whatsoever. In particular, a direct counterexample is the Equation (A4), which manifestly *does* obey local translational symmetry, while it *does not* give rise to any interaction whatsoever (see below for the analogous counterexample in electrodynamics).

Appendix A.2. The Electromagnetic Case

Let us begin with an example of a Dirac field in Minkowski spacetime, and apply the generalised equivalence principle by coupling it to the EM field. The equation of motion in this case is the ordinary Dirac equation,

$$(i\gamma^\mu \partial_\mu - m)\psi(x) = 0, \tag{A11}$$

where γ^μ are standard Dirac gamma matrices, satisfying the anticommutator identity of the Clifford algebra $\{\gamma^\mu, \gamma^\nu\} = -2\eta^{\mu\nu}$. As it stands, Equation (A11) describes the free Dirac field, not coupled to an EM field in any way. Note that it is invariant with respect to global $U(1)$ transformations, defined as

$$\psi \to \psi' = e^{-i\lambda}\psi, \qquad e^{-i\lambda} \in U(1), \qquad \lambda \in \mathbb{R}. \tag{A12}$$

In order to couple it to standard Maxwell electrodynamics, we first rewrite this equation into a form which is invariant with respect to local $U(1)$ transformations,

$$\psi \to \psi' = e^{-i\lambda(x)}\psi, \qquad \partial_\mu \to \tilde{D}_\mu = \partial_\mu + i\partial_\mu \lambda(x), \tag{A13}$$

so that the equation takes the form

$$(i\gamma^\mu \tilde{D}_\mu - m)\psi(x) = 0, \tag{A14}$$

Note that here, $\tilde{D}$ denotes the covariant derivative with respect to the "pure gauge" connection

$$\tilde{A}_\mu \equiv \partial_\mu \lambda(x), \tag{A15}$$

where $\lambda(x)$ denotes the arbitrary gauge function. Moreover, note that (A11) is analogous to (A1), (A14) is analogous to (A2), while the global and local $U(1)$ gauge transformations (A12) and (A13) are EM analogues of the global and local spacetime translations (A9) and (A10) from the gravitational case. Finally, note that if one were to evaluate the electromagnetic Faraday field strength tensor using $\tilde{A}_\mu$ from (A15), according to the equation

$$F_{\mu\nu} = \partial_\mu \tilde{A}_\nu - \partial_\nu \tilde{A}_\mu, \tag{A16}$$

one would obtain that $F_{\mu\nu} = 0$ at every point in spacetime since the potential that is a pure gauge cannot induce an EM field. Here (A16) is analogous to (A5).

Once the Dirac equation in the form (A14) is in hand, one can apply the electromagnetic generalisation of EP in order to couple the Dirac field to an EM field. The statement of EP, in this case, is that in the presence of an EM field, the equation of motion for the Dirac field should locally retain the same form as in the absence of the EM field. Since Equation (A14) depends only on the field at a given spacetime point and its first derivatives at the same point, it is therefore defined within an infinitesimal neighbourhood of a single point—in other words, it is local. Given this, electromagnetic EP states that the corresponding equation of motion in the presence of EM field should have precisely the same form (the analogue of (A6)):

$$(i\gamma^\mu \mathcal{D}_\mu - m)\psi(x) = 0. \tag{A17}$$

The absence of the tilde now denotes the fact that the covariant derivative $\mathcal{D}_\mu \equiv \partial_\mu + iA_\mu$ is defined in terms of a generic $U(1)$ connection A_μ which does not necessarily satisfy (A15), but does obey the usual gauge transformation rule,

$$A_\mu \to A'_\mu = A_\mu + \partial_\mu \lambda(x). \tag{A18}$$

In other words, electromagnetic EP postulates that the Equation (A17) holds even in the presence of an EM field since for a generic connection A_μ the Faraday tensor may not be equal to zero everywhere. The interaction between the Dirac field and the EM field as postulated by the electromagnetic EP and implemented in Equation (A17) is again known in the literature as the *minimal coupling* prescription [61,63].

If one wishes to convince oneself that the preparation step of transforming (A11) to (A14) is trivial in the sense that it does not introduce any substantial modification of (A11), one can additionally demonstrate that (A17) is in fact locally equivalent to (A11). To do this, at any specific spacetime point x_0 one can choose the following $U(1)$ gauge,

$$\lambda(x) = -A_\mu(x_0)x^\mu, \tag{A19}$$

so that, according to (A18)

$$A'_\mu(x) = A_\mu(x) - \partial_\mu(A_\nu(x_0)x^\nu) \quad \Rightarrow \quad A'_\mu(x_0) = 0, \quad \mathcal{D}_\mu\Big|_{x=x_0} = \partial_\mu. \tag{A20}$$

This choice of gauge is the EM analogue of the choice of a locally inertial coordinate system (A7). Substituting this into the primed version of (A17) and evaluating the whole equation at $x = x_0$, it reduces precisely to the form (A11) in the infinitesimal neighbourhood at that point, despite the presence of nonzero EM field.

Again note that when *integrating* (A17), one must take into account that the EM field is nonzero since integration is a nonlocal operation, and the choice of gauge (A19) eliminates the EM potential from (A17) only at the point x_0, while the Faraday tensor is gauge invariant. Therefore, the *solutions* of (A17) will in general be *different* from solutions of (A11), indicating the physical interaction of the Dirac field with EM field, despite the fact that the form of the equation of motion for the Dirac field is identical in both cases.

As in the case of gravity, we should emphasise that the electromagnetic EP is not a mathematical theorem, but rather a principle with physical content, since it can be either satisfied of violated. Specifically, we could have prescribed a different coupling of the Dirac field to electrodynamics, such that in the presence of an EM field its equation of motion takes for example the form (analogue of (A8))

$$(i\gamma^\mu \mathcal{D}_\mu - m + I_1 + I_2)\psi(x) = 0, \tag{A21}$$

where $I_1 \equiv F^{\mu\nu}F_{\mu\nu}$ and $I_2 \equiv \varepsilon^{\mu\nu\rho\sigma}F_{\mu\nu}F_{\rho\sigma}$ are the two fundamental invariants of the Faraday tensor. This equation is not equivalent to (A14), and there exists no local $U(1)$ gauge in which it could take the form (A11), since I_1 and I_2 are invariants. In this sense, (A21) is an example of a Dirac field coupled to the EM field such that the electromagnetic EP is violated. This is also known in the literature as *non-minimal coupling* [61,63].

Finally, we should also note that the transformation from (A11) to (A14) amounts to what is also known in the literature as *symmetry localisation* [61,63]. Specifically, one can explicitly verify that all three Equations (A14), (A17) and (A21) remain invariant with respect to local $U(1)$ gauge transformations, while describing no coupling to an EM field, coupling to an EM field that satisfies the electromagnetic EP, and coupling to an EM field that violates electromagnetic EP, respectively. In this sense, one can again say that the process of symmetry localisation *does not* introduce nor prescribe interactions in any way whatsoever. In the case of electrodynamics and other gauge theories, this is quite often misrepresented in literature—the step of symmetry localisation is silently joined together with the step of applying the electromagnetic version of EP; thus, in the end, giving rise to an interacting theory, and the resulting presence of the interaction is then mistakenly attributed to the localisation of symmetry, rather than to the application of EP. Similar to the gravitational case above, the equation of motion (A14) is an explicit counterexample to such an attribution, since it *does* have local $U(1)$ symmetry, but *does not* have any interaction with an EM field.

Appendix A.3. The Test Particle Case

The last topic we should address is the context in which the statement of electromagnetic EP is compatible with the existence of the Lorentz force law, acting on charged test particles. Namely, one often distinguishes the motion of a test particle in a gravitational field from a motion of a test particle in an EM field, by comparing the geodesic Equation (2)

$$u^\mu(\tau)\nabla_\mu u^\lambda(\tau) = 0, \qquad (A22)$$

where u^μ is the 4-velocity of the test particle, with the Lorentz force equation

$$u^\mu(\tau)\nabla_\mu u^\lambda(\tau) = \frac{q}{m} F^{\lambda\rho} u_\rho(\tau), \qquad (A23)$$

where q/m is the charge-to-mass ratio of a test particle moving in an external EM field, described by the Faraday tensor $F_{\mu\nu}$. A typical conclusion one draws from this comparison is that the interaction with the EM field gives rise to a "real force", while the interaction with the gravitational field does not.

However, it is highly misleading to compare (A22) to (A23) in the first place. Namely, as we have discussed in detail in Section 4, in field theory the notion of a particle can be defined only approximately, and this applies equally for electrodynamics as well as for gravity. Specifically, given the example discussed above, of a Dirac field coupled to an EM field via Equation (A17), we have seen that in the infinitesimal neighbourhood of a given point x_0 one can completely gauge away any presence of the coupling to EM field from (A17). In this sense, the notion of a test particle that satisfies (A23) cannot be identified with an idealised point-particle, that has exactly zero size. Instead, the realistic test particle is a wave-packet configuration of a Dirac field (a kink), and as such has a small but nonzero size. As it evolves, the different parts of the wave-packet are subject to interaction with the EM potential A_μ at *different* points of spacetime, giving rise to an effective non-minimal coupling with the Faraday tensor $F_{\mu\nu}$. This is completely analogous to the case of a test particle with small but nonzero size interacting with spacetime curvature due to tidal forces—both effects are equally nonlocal since both kinks have nonzero size. On the other hand, a test particle that satisfies (A22) represents an idealised point-particle (a leading order approximation in the multipole expansion of the matter field), i.e., a kink which thus has precisely zero size.

In this sense, the Lorentz force Equation (A23) rather ought to be compared with the Papapetrou Equation (3),

$$u^\mu(\tau)\nabla_\mu u^\lambda(\tau) = R^\lambda{}_{\mu\rho\sigma} u^\mu(\tau) J^{\rho\sigma}(\tau). \qquad (A24)$$

Indeed, one can see quite a reasonable analogy between (A23) and (A24). There are of course small technical differences due to the precise nature of the coupling to various moments of the kink, but nevertheless, the two equations are strikingly similar. Given this, while one can still draw the conclusion that the interaction of a kink with the EM field gives rise to a "real force", one can draw precisely the same conclusion for the interaction of a kink with the gravitational field. There is no distinction between gravity and the other gauge interactions at this level—all four interactions in nature (strong, weak, electromagnetic and gravitational) are equally "real". In addition, all four interactions satisfy EP at the fundamental field theory level (i.e., in the sense of strong generalised EP), while at the level of mechanics, a corresponding weak generalised EP is manifestly violated in all four cases.

References

1. Kay, B.S. Decoherence of macroscopic closed systems within Newtonian quantum gravity. *Class. Quantum Gravity* **1998**, *15*, L89. [CrossRef]
2. Oniga, T.; Wang, C.H.T. Quantum gravitational decoherence of light and matter. *Phys. Rev. D* **2016**, *93*, 044027. [CrossRef]
3. Bruschi, D.E. On the weight of entanglement. *Phys. Lett. B* **2016**, *754*, 182. [CrossRef]
4. Bose, S.; Mazumdar, A.; Morley, G.W.; Ulbricht, H.; Toroš, M.; Paternostro, M.; Geraci, A.A.; Barker, P.F.; Kim, M.S.; Milburn, G. Spin entanglement witness for quantum gravity. *Phys. Rev. Lett.* **2017**, *119*, 240401. [CrossRef]
5. Marletto, C.; Vedral, V. Gravitationally Induced Entanglement between Two Massive Particles is Sufficient Evidence of Quantum Effects in Gravity. *Phys. Rev. Lett.* **2017**, *119*, 240402. [CrossRef]
6. Marletto, C.; Vedral, V. When can gravity path-entangle two spatially superposed masses? *Phys. Rev. D* **2018**, *98*, 046001. [CrossRef]
7. Paunković, N.; Vojinović, M. Gauge protected entanglement between gravity and matter. *Class. Quantum Gravity* **2018**, *35*, 185015. [CrossRef]
8. Oreshkov, O.; Costa, F.; Brukner, Č. Quantum correlations with no causal order. *Nat. Commun.* **2012**, *3*, 1092. [CrossRef]
9. Araújo, M.; Branciard, C.; Costa, F.; Feix, A.; Giarmatzi, C.; Brukner, Č. Witnessing causal nonseparability. *New J. Phys.* **2015**, *17*, 102001. [CrossRef]
10. Vilasini, V. An Introduction to Causality in Quantum Theory (and Beyond). Master's Thesis, ETH, Zürich, Switzerland, 2017.
11. Oreshkov, O. Time-delocalized quantum subsystems and operations: On the existence of processes with indefinite causal structure in quantum mechanics. *Quantum* **2019**, *3*, 206. [CrossRef]
12. Paunković, N.; Vojinović, M. Causal orders, quantum circuits and spacetime: Distinguishing between definite and superposed causal orders. *Quantum* **2020**, *4*, 275. [CrossRef]
13. Vilasini, V.; Colbeck, R. General framework for cyclic and fine-tuned causal models and their compatibility with space-time. *Phys. Rev. A* **2022**, *106*, 032204. [CrossRef]
14. Vilasini, V.; Renner, R. Embedding cyclic causal structures in acyclic spacetimes: No-go results for process matrices. *arXiv* **2022**, arXiv:2203.11245.
15. Ormrod, N.; Vanrietvelde, A.; Barrett, J. Causal structure in the presence of sectorial constraints, with application to the quantum switch. *arXiv* **2022**, arXiv:2204.10273.
16. Giacomini, F.; Castro-Ruiz, E.; Brukner, Č. Quantum mechanics and the covariance of physical laws in quantum reference frames. *Nat. Commun.* **2019**, *10*, 494. [CrossRef] [PubMed]
17. Vanrietvelde, A.; Höhn, P.A.; Giacomini, F.; Castro-Ruiz, E. A change of perspective: Switching quantum reference frames via a perspective-neutral framework. *Quantum* **2020**, *4*, 225. [CrossRef]
18. Krumm, M.; Höhn, P.A.; Müller, M.P. Quantum reference frame transformations as symmetries and the paradox of the third particle. *Quantum* **2021**, *5*, 530. [CrossRef]
19. Ahmad, S.A.; Galley, T.D.; Höhn, P.A.; Lock, M.P.E.; Smith, A.R.H. Quantum Relativity of Subsystems. *Phys. Rev. Lett.* **2022**, *128*, 170401. [CrossRef]
20. de la Hamette, A.C.; Kabel, V.; Castro-Ruiz, E.; Brukner, Č. Falling through masses in superposition: Quantum reference frames for indefinite metrics. *arXiv* **2021**, arXiv:2112.11473.
21. Colladay, D.; Kostelecký, V.A. Lorentz-violating extension of the standard model. *Phys. Rev. D* **1998**, *58*, 116002. [CrossRef]
22. Kostelecký, V.A.; Russell, N. Data tables for Lorentz and CPT violation. *Rev. Mod. Phys.* **2011**, *83*, 11–31.
23. Amelino-Camelia, G. Particle-dependent deformations of Lorentz symmetry. *Symmetry* **2012**, *4*, 344–378. [CrossRef]
24. Amelino-Camelia, G.; Palmisano, M.; Ronco, M.; D'Amico, G. Mixing coproducts for theories with particle-dependent relativistic properties. *Int. J. Mod. Phys. D* **2020**, *29*, 2050017. [CrossRef]
25. Torri, M.D.C.; Antonelli, V.; Miramonti, L. Homogeneously Modified Special relativity (HMSR). *Eur. Phys. J. C* **2019**, *79*, 808.
26. Pipa, F.; Paunković, N.; Vojinović, M. Entanglement-induced deviation from the geodesic motion in quantum gravity. *J. Cosmol. Astropart. Phys.* **2019**, *2019*, 57. [CrossRef]
27. Giacomini, F.; Brukner, Č. Einstein's Equivalence principle for superpositions of gravitational fields. *arXiv* **2020**, arXiv:2012.13754.

28. Giacomini, F.; Brukner, Č. Quantum superposition of spacetimes obeys Einstein's equivalence principle. *AVS Quantum Sci.* **2022**, *4*, 015601.
29. Marletto, C.; Vedral, V. On the testability of the equivalence principle as a gauge principle detecting the gravitational t^3 phase. *Front. Phys.* **2020**, *8*, 176.
30. Marletto, C.; Vedral, V. Sagnac interferometer and the quantum nature of gravity. *J. Phys. Commun.* **2021**, *5*, 051001.
31. Marletto, C.; Vedral, V. The quantum totalitarian property and exact symmetries. *AVS Quantum Sci.* **2022**, *4*, 015603. [CrossRef]
32. Einstein, A.; Infeld, L.; Hoffmann, B. The Gravitational Equations and the Problem of Motion. *Ann. Math.* **1938**, *39*, 65. [CrossRef]
33. Mathisson, M. Neue mechanik materieller systemes. *Acta Phys. Pol.* **1937**, *6*, 163.
34. Papapetrou, A. Spinning test-particles in general relativity, I. *Proc. R. Soc. A* **1951**, *209*, 248.
35. Tulczyjev, W. Equations of motion of rotating bodies in general relativity theory. *Acta Phys. Pol.* **1959**, *18*, 393.
36. Taub, A.H. Motion of Test Bodies in General Relativity. *J. Math. Phys.* **1964**, *5*, 112. [CrossRef]
37. Dixon, G. A covariant multipole formalism for extended test bodies in general relativity. *Nuovo Cim.* **1964**, *34*, 317. [CrossRef]
38. Dixon, G. Classical theory of charged particles with spin and the classical limit of the Dirac equation. *Nuovo Cim.* **1965**, *38*, 1616. [CrossRef]
39. Dixon, G. Dynamics of extended bodies in general relativity. I. Momentum and angular momentum. *Proc. R. Soc. A* **1970**, *314*, 499.
40. Dixon, G. Dynamics of extended bodies in general relativity - II. Moments of the charge-current vector. *Proc. R. Soc. A* **1970**, *319*, 509.
41. Dixon, G. The definition of multipole moments for extended bodies. *Gen. Relativ. Gravit.* **1973**, *4*, 199. [CrossRef]
42. Yasskin, P.B.; Stoeger, W.R. Propagation equations for test bodies with spin and rotation in theories of gravity with torsion. *Phys. Rev. D* **1980**, *21*, 2081. [CrossRef]
43. Nomura, K.; Shirafuji, T.; Hayashi, K. Spinning Test Particles in Spacetime with Torsion. *Prog. Theor. Phys.* **1991**, *86*, 1239. [CrossRef]
44. Nomura, K.; Shirafuji, T.; Hayashi, K. Semiclassical particles with arbitrary spin in the Riemann-Cartan space-time. *Prog. Theor. Phys.* **1992**, *87*, 1275. [CrossRef]
45. Vasilić, M.; Vojinović, M. Classical spinning branes in curved backgrounds. *JHEP* **2007**, *7*, 28. [CrossRef]
46. Vasilić, M.; Vojinović, M. Spinning branes in Riemann-Cartan spacetime. *Phys. Rev. D* **2008**, *78*, 104002. [CrossRef]
47. Accioly, A.; Paszko, R. Conflict between the Classical Equivalence Principle and Quantum Mechanics. *Adv. Stud. Theor. Phys.* **2009**, *3*, 65.
48. Longhi, S. Equivalence principle and quantum mechanics: Quantum simulation with entangled photons. *Opt. Lett.* **2018**, *43*, 226. [CrossRef]
49. Chowdhury, P.; Home, D.; Majumdar, A.S.; Mousavi, S.V.; Mozaffari, M.R.; Sinha, S. Strong quantum violation of the gravitational weak equivalence principle by a non-Gaussian wave packet. *Class. Quantum Gravity* **2012**, *29*, 025010. [CrossRef]
50. Rosi, G.; D'Amico, G.; Cacciapuoti, L.; Sorrentino, F.; Prevedelli, M.; Zych, M.; Brukner, Č.; Tino, G.M. Quantum test of the equivalence principle for atoms in coherent superposition of internal energy states. *Nat. Commun.* **2017**, *8*, 15529. [CrossRef]
51. Zych, M.; Brukner, Č. Quantum formulation of the Einstein Equivalence Principle. *Nat. Phys.* **2018**, *14*, 1027. [CrossRef]
52. Anastopoulos, C.; Hu, B.L. Equivalence principle for quantum systems: Dephasing and phase shift of free-falling particles. *Class. Quantum Gravity* **2018**, *35*, 035011. [CrossRef]
53. Hardy, L. Implementation of the Quantum Equivalence Principle. In *Progress and Visions in Quantum Theory in View of Gravity*; Finster, F., Giulini, D., Kleiner, J., Tolksdorf, J., Eds.; Springer International Publishing: Cham, Switzerland, 2020; pp. 189–220.
54. Kretschmann, E. Über den physikalischen Sinn der Relativitätspostulate, A. Einsteins neue und seine ursprüngliche Relativitätstheorie. *Ann. Phys.* **1918**, *358*, 575–614.
55. Misner, C.W.; Thorne, K.S.; Wheeler, J.A. *Gravitation*; W. H. Freeman and Co.: San Francisco, CA, USA, 1973.
56. Okon, E.; Callender, C. Does Quantum Mechanics Clash with the Equivalence Principle—And Does it Matter? *Eur. J. Phil. Sci.* **2011**, *1*, 133. [CrossRef]
57. Casola, E.D.; Liberati, S.; Sonego, S. Nonequivalence of equivalence principles. *Am. J. Phys.* **2015**, *83*, 39. [CrossRef]
58. Viola, L.; Onofrio, R. Testing the equivalence principle through freely falling quantum objects. *Phys. Rev. D* **1997**, *55*, 455–462. [CrossRef]
59. Plyatsko, R. Gravitational ultrarelativistic spin-orbit interaction and the weak equivalence principle. *Phys. Rev. D* **1998**, *58*, 084031.
60. Bose, S.; Mazumdar, A.; Schut, M.; Toroš, M. Entanglement witness for the weak equivalence principle. *arXiv* **2022**, arXiv:2203.11628.
61. Blagojević, M. *Gravitation and Gauge Symmetries*; Institute of Physics Publishing: Bristol, UK, 2002.
62. Landau, L.D.; Lifshitz, E.M. *The Classical Theory of Fields*, 4th ed.; Butterworth-Heinemann: Oxford, UK, 1980.
63. Peskin, M.E.; Schroeder, D.V. *An Introduction to Quantum Field Theory*; Addison-Wesley Publishing Co.: Boston, USA, 1995.

Article

A New Perspective on Doubly Special Relativity

J. M. Carmona [1,2], J. L. Cortés [1,2], J. J. Relancio [2,3,*] and M. A. Reyes [1,2]

[1] Departamento de Física Teórica, Universidad de Zaragoza, 50009 Zaragoza, Spain
[2] Centro de Astropartículas y Física de Altas Energías (CAPA), Universidad de Zaragoza, 50009 Zaragoza, Spain
[3] Departamento de Matemáticas y Computación, Universidad de Burgos, 09001 Burgos, Spain
* Correspondence: jjrelancio@ubu.es

Abstract: Doubly special relativity considers a deformation of the special relativistic kinematics parametrized by a high-energy scale, in such a way that it preserves a relativity principle. When this deformation is assumed to be applied to any interaction between particles, one faces some inconsistencies. In order to avoid them, we propose a new perspective where the deformation affects only the interactions between elementary particles. A consequence of this proposal is that the deformation cannot modify the special relativistic energy–momentum relation of a particle.

Keywords: quantum gravity; doubly special relativity; relative locality; relativistic deformed kinematics

1. Introduction

The aim of this work is to present a new proposal to introduce a departure from the notions of spacetime and locality in special relativity (SR). The first reference to the possible limitations of these notions goes back to Einstein himself (1905) [1]: *"We shall not here discuss the inexactitude which lurks in the concept of simultaneity of two events at approximately the same place, which can only be removed by an abstraction"*.

This departure might be very relevant in the development of theories going beyond SR trying to put in the same scheme general relativity (GR) and quantum field theory (QFT). Indeed, a possible source of inconsistencies that impedes the unification of these two theories is the role that spacetime plays in them. While in QFT spacetime is given as a rigid static framework in which propagation and interactions can be described, in GR spacetime is understood as a dynamic deformation of the Minkowski spacetime modeled by energy–momentum sources. Both theories are expected to be limiting cases of a quantum gravity theory (QGT), where spacetime would reveal its quantum nature, leading to a completely new structure. This is the case of some proposals for a QGT, such as string theory [2–4], loop quantum gravity [5,6], causal dynamical triangulations [7], or causal set theory [8–10]. However, the correct fundamental spacetime structure is still unknown to date.

An alternative, bottom-up, approach to QGT may come from trying to incorporate some generic expected modifications of the classical spacetime to our present, low-energy theories, as a way to try to capture residual quantum gravity effects. Such modifications include, most notably, departures from the symmetries and from the standard notion of locality of SR and can indeed have a well-defined and testable phenomenology [11]. Two main possibilities have emerged in this quantum spacetime phenomenology [12], affecting the standard kinematics of SR in a different way: either Lorentz invariance is broken, which implies the existence of a privileged system of reference [13,14], or the symmetries of SR are deformed, as in the so-called "Doubly Special Relativity" (DSR) scenario, so that a relativity principle is still present [15,16].

DSR is an attractive theoretical possibility, since it represents a step from SR which is analogous to the one that SR made from Galilean relativity, and it dismisses the need of a privileged system of reference. The kinematics of DSR models is characterized by two different ingredients that may appear in a generic deformation: a modification of the dispersion relation of SR (MDR) and a modification in the composition law of momenta (MCL), which is no longer additive, therefore affecting the conservation of energy and momentum in processes. The MDR adds new terms to the quadratic relation between the energy and momentum of a particle in SR, which are proportional to powers of the energy of the particle divided by a high-energy scale Λ, usually considered as the Planck energy. The MCL for two particles differs from the simple addition in terms involving products of momenta of both particles and again, the inverse of the high-energy scale Λ, since we want to recover SR when this scale goes to infinity. While an MDR may be absent in specific DSR models (so that the dispersion relation is that of SR), an MCL is always present, as well as deformed Lorentz transformations, in order to keep the relativity principle [17]. The MCL can be related [18] to the coproduct operation in the formalism of Hopf algebras [19], being the κ-Poincaré example [20,21] the most studied model. In the particular case of κ-Poincaré, the MCL is noncommutative but associative, so that the composition law of a multiparticle system is completely defined by gathering momenta in pairs.

However, the implementation of the deformed kinematics of DSR leads to several problems of consistency of the theory. In particular, a deformation of the usual relativistic expression of the dispersion relation leads to the so-called "soccer-ball" problem [22,23]: the terms proportional to the inverse of the high-energy scale (even when considering a Planckian energy) produce a huge contribution when considering a macroscopic object. This problem is not exclusive to the MDR; it also affects the MCL, whose nonlinear terms would give enormous contributions to the total energy of a macroscopic object, when expressed in terms of the energy of its constituents. Although some proposals have been pointed out to solve this problem [23–25], they are not free from issues, as we discuss in Section 2.

Another consistency problem of DSR has to do with the apparent nonlocal influences that a nonadditive composition law seems to suggest [26]. This "spectator problem" is more evident when one tries to analyze the translational symmetry of a process that includes several interactions [27–29]. The generator of translations, the total momentum defined with the MCL, contains momenta of particles that do not directly participate in one of the interactions, but that appear in other interactions that form part of the process. The translational invariance can then only be consistently implemented if one knows the whole sequence of causally connected vertices for the particles involved in every interaction. This means that a complete knowledge of the content of the Universe should be given for the description of every process, which impedes us having a physical description of reality.

In this work, we present a solution to these problems of consistency by introducing a new perspective on how the deformed kinematics of DSR plays a role in processes of particles. Essentially, we propose that DSR should be seen as a way to go beyond the standard locality of interactions present in relativistic quantum field theories that is still compatible with relativistic invariance. Our proposal is described in Section 2 and, in particular, it leads us to conclude that modified composition laws should appear only whenever elementary particles are involved in an interaction, while, at the same time, their dispersion relation should be the same as in special relativity. As we see in Section 3, this new perspective on DSR has specific phenomenological implications that can be distinguished from those considered in the standard interpretation of DSR. We end with a brief summary in Section 4 and give some technical details on a modified composition law compatible with the present proposal in Appendix A.

2. DSR as a Way to Go Beyond Local Interactions

We present in this work a new interpretation of DSR as a way to go beyond local interactions in a quantum relativistic theory. In relativistic quantum field theory (RQFT), interactions are local in the sense that they are defined by products of more than two fields in a spacetime point, which is the way one introduces interactions in a Lagrangian density of the theory of quantum fields. When one uses the plane-wave expansion for the fields, the locality of the interactions automatically leads to linear relations between the momenta defining the plane-wave expansion when one considers the integral over spacetime of the Lagrangian density (action).

One can identify DSR with a nonlinear modification of the sum of momenta (composition of momenta) depending on an energy scale Λ. This can be implemented in the field theory framework as a deformation of the product of fields (see [30] and references therein). However, this mathematical formalism has not yet provided a systematic treatment of the effects of the deformation in the calculation of observables in particle processes. In any case, what is clear is that the nonlinear relations between momenta that are obtained through that procedure imply a loss of the locality of interactions. Interactions in DSR do not define a point in spacetime but a region of size $\ell \sim (1/\Lambda)$. This is the ingredient that we take as the key point for the proposed new interpretation of DSR.

We have explored nature at the microscopic level up to scales of the order of $(\text{TeV})^{-1}$ with collisions of particles in the highest energy accelerators. We have not seen any sign of deviations from the locality of interactions; then, we conclude that if DSR is realized in nature, the energy scale Λ of DSR should be larger than 1 TeV.

The interactions involving a system whose size is much larger than the scale ℓ are not be affected by the transition from SR to DSR. This argument reduces the search for observable effects of DSR to the interactions of what we identify at present as elementary particles (leptons, quarks and mediators of the interactions in the standard model of particle physics).

The above perspective of DSR can be contrasted with the standard interpretation in which DSR is identified as the necessary modification in SR to make relativistic invariance compatible with a (minimum) length scale. One finds that the linear Lorentz transformations between different frames in SR have to be replaced by nonlinear transformations. This leads to a deformation of the energy–momentum relation of a particle in SR and also to a deformation of the relations imposed by the energy–momentum conservation associated with the translational invariance. When one considers a classical model for the interaction of particles based on the deformed energy–momentum conservation, one finds that the notion of locality of SR (interactions associated to the intersection of the worldlines of the particles) becomes an observer-dependent property; absolute locality is replaced by relative locality [31,32]. The deviations from locality depend on the observer when one considers a family of observers related by translations.

The search for observable effects in the standard interpretation of DSR follows different rules from those of the proposal in this work. One usually interprets the deformation of SR as a signal of a theory of quantum gravity, identifying the energy scale of the deformation Λ with the Planck scale. The limitation on the energies one can reach in accelerators, or even in high-energy astrophysics, makes any effect of DSR in the interaction of particles unobservable. One has to look for the amplification of the effects of DSR in the propagation of particles over astrophysical distances due to the modification of the energy–momentum relation and, correspondingly, of the velocity of propagation. If one changes the choice of energy–momentum variables, one will have a different energy dependence of the velocity of propagation; in fact, one can always choose energy–momentum variables to make the velocity of propagation energy-independent. In this case, however, translations generated by these variables act nontrivially in spacetime and, when one compares two observers separated by astrophysical distances, one finds once more the amplification of the effects of DSR [33].

In the new perspective of DSR as a small departure from the locality of interactions, there are no effects of DSR in the propagation of a free particle, and then no amplification when one considers astrophysical distances. The only way to find observable effects is to assume that one can have interactions of particles with energies approaching the scale Λ of the deformation.

In order to illustrate the differences between the standard interpretation of DSR and the new perspective proposed in this work, we consider a process with two interactions in regions of spacetime separated by a large distance.

2.1. Production, Propagation and Detection of a Very High Energy Particle

We consider a process with an interaction producing a particle which propagates and is detected by a second interaction, as shown in Figure 1.

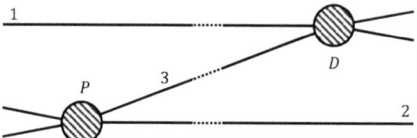

Figure 1. A particle, labelled '3', is produced in an interaction (*P*), propagates over a large distance, and is detected in a second interaction (*D*). Particles labelled by '1' and '2' only participate in the interactions *D* and *P*, respectively.

We assume that each interaction is an elastic scattering of two particles. From a kinematic point of view, the interaction produces a change of the momenta of the particles. This change of momenta is determined in SR by a conservation law. The effect of DSR is to modify this conservation law through a deformed composition of momenta and then, to change the possible momenta of the particles in the final state of the process with respect to their values in SR. The relativistic invariance of the deformed kinematics generally requires a modification of the energy–momentum relation of the particles compatible with the modification of the composition of momenta, since both modifications are related by the so-called "golden rules" [28,34]. However, there are examples of modifications of the composition of momenta compatible with the energy–momentum relation of SR, such as the classical basis [35] of κ-Poincaré and the well-known Snyder kinematics [36].

In previous works [27–29] on DSR, the kinematics of the process in Figure 1 was studied based on a modified implementation of the translational symmetry. One assumed the conservation of the total momentum of the three-particle system defined in terms of the modified composition of their momenta, which could be taken before the first interaction, between the two interactions, or after the second interaction. One could consider that the exchange of momenta in each interaction involves the momenta of three particles. In fact, in the algebraic interpretation of DSR, based on a κ-deformed Poincaré Hopf algebra, the composition of momenta is determined by the coproduct of momentum operators, which makes no reference to spacetime and then, the momentum of particle 1 could be modified by the interaction *P* even if this particle is far away from the spacetime region where the interaction takes place [26]. Even if one assumes that this is not the case, nothing tells us that the conservation of the total momentum of the three-particle system before and after the interaction *P* will lead to a relation between momenta independent of the momentum of particle 1. The total momentum of the three-particle system before the first interaction will be the result of the composition of the momentum of particle 1 and the momentum variables of the other two particles participating in the interaction *P* whose momenta change due to the interaction. However, nothing fixes the ordering of the momentum variables in the expression of the total momentum before and after the interaction. In particular, one could have a case where the conservation of the total momentum can be expressed as

$$p^{(1)} \oplus p^{(2)} \oplus p^{(3)} = p^{(2)'} \oplus p^{(1)} \oplus p^{(3)'}, \tag{1}$$

where the symbol $\oplus$ denotes the modified composition between momenta. An example of such MCL is given in Appendix A. The relation between the variables $(p^{(2)}, p^{(3)})$ and $(p^{(2)'}, p^{(3)'})$ depends on $p^{(1)}$ in this case.

The same applies to the transition due to the detection of the particle, which can depend on the momentum of particle 2. In this case, one cannot treat both interactions independently, even if the produced and detected particle is propagating over a very large distance. Therefore, DSR leads to a violation of the cluster property of interactions, which is at the basis of RQFT [37].

The previous argument also raises doubts on the consistency of considering the process in Figure 1 as an isolated system. The interactions producing particle 1 and the two particles in the initial state of the interaction P could affect the kinematics of the process and should be included. One could even go further and consider a fourth particle which does not participate in any of the two interactions. The conservation of the total momentum of the four-particle system, including this new particle, would lead to an exchange of momenta between the particles depending on the momentum of the fourth particle. This is what is known as the spectator problem [27–29].

The main idea in the new perspective of DSR is to associate the deformation to a deviation from the locality of the interactions characterized by a length scale $\ell \sim (1/\Lambda)$, where Λ is the energy scale of DSR. When one considers an interaction of particles which are seen as elementary when explored at distances larger than the scale ℓ, one sees an effect of the deviation from the locality of interactions; the change of momenta due to the interaction is determined by the kinematics of DSR. On the other hand, when one considers an interaction with composite particles whose size is much larger than the scale ℓ, one can neglect the deviations from locality, and the change of momenta produced by the interaction is determined by the kinematics of SR.

Let us assume that in Figure 1, particles 2 and 3 are elementary but particle 1 is a composite particle, so that the interaction producing particle 3 is between elementary particles, and the interaction in its detection involves a composite particle. A possible relation for the change of momenta in the interaction D is

$$p^{(1)} - p^{(1)'} = p^{(3)'} \oplus \hat{p}^{(3)}, \qquad (2)$$

where $\hat{p}^{(3)}$ is the antipode (see Appendix A) of the momentum of particle 3 before its detection. We selected in Equation (2) one of the two possible orderings of the composition of momentum variables on the right-hand side. The left-hand side is a difference of momentum variables of the composite particle 1, whose kinematics is not affected by the deformation of SR. Then, they transform linearly under Lorentz transformations, and this implies that the composition of momentum variables on the right-hand side also has to transform linearly.

The assumption that one can have interactions with different kinematics eliminates the arbitrariness in DSR associated with the different choices of energy–momentum variables. The composition of two momentum variables as well as a single momentum variable have to transform linearly under Lorentz transformations. Therefore, there is no modification of the dispersion relation with respect to SR and there is no signal of the deformation in the propagation of a particle. This agrees with the interpretation of DSR as a modification of the interaction between elementary particles instead of an effect in the propagation of particles in a quantum spacetime.

2.2. Deviation from Locality in DSR

Another consequence of the standard interpretation of DSR as a modified implementation of the translational symmetry generated by the total momentum of the three-particle system is that there is no observer that sees the two interactions in Figure 1 as local. Different observers related by a translation have a different origin of spacetime. An observer whose origin is close to the region where the interaction P is produced sees this interaction as approximately local but sees deviations from locality in the interaction D proportional to the distance between the regions of the two interactions. The same happens for an observer whose origin is close to the interaction D, who sees a deviation from locality in the interaction P proportional to the distance from it. One refers to this property as relative locality [31,32], which defines DSR in the sense that it reflects the effect in spacetime of the modification of the composition of momenta. In the standard interpretation of DSR, where one cannot treat the two interactions independently, the idea of relative locality plays an important role to understand that there is no conflict between the deviations from locality in DSR and the very precise experimental tests of locality [38].

In the new perspective of DSR proposed in this work, where one can consider the two interactions independently, one can directly use in all experimental tests of locality an observer whose origin is within the region where the interaction that detects the particle takes place. One can ignore the interaction P producing the particle, and one does not need to refer to the idea of relative locality in order to see that there is no conflict between DSR and the very precise tests of locality.

The standard interpretation of DSR is formulated in the framework of a classical model of interaction of particles based on worldlines, while in the new perspective proposed in this work, the deformation is introduced in the relation between the momenta associated with a product of quantum fields. Due to this difference of frameworks used in the formulation of the transition from SR to DSR, one has a different perspective on the loss of the notion of absolute locality present in SR.

2.3. Solution of the Spectator and Soccer-Ball Problems in the New Perspective of DSR

If one associates the effects of DSR to a deviation from the locality of the interactions between elementary particles, it is natural to determine the kinematics of each interaction by the momenta of the particles participating in it. This means that the change of momenta due to the interaction P is independent of the momentum of particle 1, which does not participate in the interaction. The same argument can be applied to the change of momenta due to the interaction D, leading us to conclude that it is independent of the momentum of particle 2. Then, both interactions can be treated independently, according to the cluster property, and each of them can be treated as taking place in an isolated system. As a consequence, the new perspective of DSR solves the spectator problem and treats the process as three independent steps: an interaction P producing particle 3, the propagation of the free particle 3 and an interaction D where particle 3 is detected, just as in the case of SR.

Another potential inconsistency of the original proposal of DSR comes when one considers the kinematics of macroscopic systems. For a microscopic system in DSR, one can understand the small deviations from SR as a consequence of the small ratio of the energies of the particles and the energy scale which parametrizes DSR. Then, however, if the modified composition of momenta is the same for any particle, including a macroscopic system, one would have very large corrections to the kinematics of SR, in obvious conflict with our observations at the macroscopic level. The solution that has been proposed for this paradox (soccer-ball problem [22,23,25]) is that the scale of energy which parametrizes the modification of the kinematics of a macroscopic system is much larger than the energy scale of DSR at the microscopic level. More specifically, one argues that the scale of deformation is proportional to the number of constituents of the macroscopic object by considering an approximation where those constituents all move with the same velocity [23]. Identifying those constituents with the atoms of a macroscopic object, one finds an effect on the kinematics of the macroscopic system of the same order as the correction to the kinematics of the atoms. The smallness of the energy of each atom compared with the energy scale of DSR solves the soccer-ball problem.

The previous argument can be criticized from different perspectives. Why should we identify the atoms as the constituents of the macroscopic system instead of the elementary particles (electrons, quarks and gluons)? Were we to identify these elementary particles as "constituents", how many of them would there be in, for example, a proton? How good is the approximation to consider all the constituents as a rigid system?

In the new perspective of DSR proposed in this work, a macroscopic system is a composite system with a size much larger than the scale $\ell \sim (1/\Lambda)$ of the deviation from the locality of the interactions of elementary particles. Then, the interactions of macroscopic systems are not affected by DSR and, according to the relativity principle, the energy–momentum relation of SR applies to a macroscopic system. The soccer-ball problem is solved as a direct consequence of the assumption that DSR only applies to the interactions of elementary particles.

We note that there is some similarity between this solution to the soccer-ball problem and the one offered in ref. [24], where this "problem" is seen as a "case of mistaken identity", distinguishing between the composition law, which is relevant in the modification of interactions in a noncommutative spacetime (then modifying the standard notion of locality) and the (additive) total momentum of a multiparticle system. However, ref. [24] does not discuss how one should consider the interaction between macroscopic particles; in the present proposal, we deduce that they should be described by a standard composition law between momenta.

3. Observable Effects of DSR

In contrast to the case of the Lorentz invariance violation (LIV), where one can have observable effects of the deviation from SR in the kinematics of processes at energies much smaller than the energy scale of the LIV [11,13,14], the relativistic invariance of the deformation of the kinematics of SR leads to a suppression of any effect in the kinematics of a process by powers of the ratio of the energy of the particles and the energy scale of DSR. If this scale is of the order of the Planck scale, as usually assumed when one considers the quantum structure of spacetime as the origin of the deformation of the kinematics, the possibilities to observe the effects of DSR are reduced to the problematic access to the details of the initial state of the Universe or the final stages of the evaporation of a black hole.

A completely different phenomenology of DSR is based on time delays of massless particles. In this case, even if the modification in the velocity of propagation of a particle with respect to SR is proportional to the ratio of the energy of the particle and the energy scale of DSR, there is an amplification effect, proportional to the long distance that astroparticles travel from the source to our detectors on Earth, which could be measured by current experiments. Therefore, a lot of papers were devoted to this possible effect in the context of DSR [33,39–45].

In the perspective of DSR discussed in this work, since the dispersion relation is the one of SR, time delays are absent [33,42]. This means that the strong constraints on the high-energy scale based on the possibility of time delays [46–52] do not apply in our scheme.

This opens up an attractive alternative from a phenomenological perspective, by keeping an open mind about the value of the energy scale of DSR. From this point of view, one can see which bounds are on such scale from the lack of signals of a departure from the prediction of SR in the kinematics of processes of particles, and which places are best to look for a first signal of DSR. The best candidates are the interactions involved in very high energy astroparticle physics [53,54]. Since the effects of the deformation are restricted to the interactions of elementary particles in the new perspective of DSR proposed in this work, one is led to identify the observations of the elementary cosmic messengers (gamma rays and neutrinos) at the highest energies as the best window to look for observable effects of DSR. A difficulty one faces in this program are the uncertainties in the predictions of SR due to the limited knowledge of the astrophysical processes in which those messengers are involved, which have to be disentangled from the possible effects of DSR on the interactions of those messengers affecting their observation at the highest energies.

Another alternative is to look for the interactions of particles at the highest energies in laboratory experiments, where one is free from the astrophysics uncertainties, but the energies one can reach are lower, and then one has to identify a smaller effect. This leads to concentrate on the most precise observations (Z-line shape is a good example) or on the experiments involving the highest possible energies (Large Hadron Collider at CERN and a future higher pp collider) [55].

4. Summary

We proposed a new way to introduce a relativistic deformation of the kinematics of SR. In this proposal, the deformation only applies to the interaction of elementary particles. What is elementary depends on the energy (and associated length) scale of the interaction considered. In this way, we can restrict the effects of the deformation to those particles which we have not been able to identify as composite particles. Thus, we obtain a new perspective of DSR where the different potential inconsistencies (soccer-ball and spectator problems, consistency with tests of locality) are automatically solved.

In order to explore the consequences of this proposal, we have to incorporate the effects of DSR in a model for the interaction of particles that gives a generalization of the result of the perturbative treatment in RQFT for cross sections and decay widths compatible with the deformed relativistic invariance of DSR. This will be the subject of future work.

Author Contributions: Writing—original draft preparation, review and editing, J.M.C., J.L.C., J.J.R. and M.A.R. All authors contributed equally to the present work. All authors have read and agreed to the published version of the manuscript.

Funding: This work is supported by Spanish grants PID2021-126078NB-C21 (FEDER/AEI) and DGIID-DGA no. 2020-E21-17R. The work of M.A.R. was supported by MICIU/AEI/FSE (FPI grant PRE2019-089024). This work has been partially supported by Agencia Estatal de Investigación (Spain) under grant PID2019-106802GB-I00/AEI/10.13039/501100011033, by the Regional Government of Castilla y León (Junta de Castilla y León, Spain) and by the Spanish Ministry of Science and Innovation MICIN and the European Union NextGenerationEU/PRTR. The authors would like to acknowledge the contribution of the COST Action CA18108 "Quantum gravity phenomenology in the multi-messenger approach".

Data Availability Statement: No new data were created or analyzed in this study. Data sharing is not applicable to this article.

Conflicts of Interest: The authors declare no conflict of interest.

Appendix A. Modified Composition of Momenta

A relativistic deformation of SR in which the dispersion relation and Lorentz transformations in the one-particle system are not deformed is provided within the mathematical framework of Hopf algebras by the classical basis of κ-Poincaré [35]. The coproduct of the momentum operators defines a composition law of momenta

$$(p \oplus q)_0 = p_0 \Pi(q) + \Pi^{-1}(p)\left(q_0 + \frac{\vec{p} \cdot \vec{q}}{\Lambda}\right), \qquad (p \oplus q)_i = p_i \Pi(q) + q_i, \qquad \text{(A1)}$$

with

$$\Pi(k) = \frac{k_0}{\Lambda} + \sqrt{1 + \frac{k_0^2 - \vec{k}^2}{\Lambda^2}}, \qquad \Pi^{-1}(k) = \left(\sqrt{1 + \frac{k_0^2 - \vec{k}^2}{\Lambda^2}} - \frac{k_0}{\Lambda}\right)\left(1 - \frac{\vec{k}^2}{\Lambda^2}\right)^{-1}, \qquad \text{(A2)}$$

where Λ plays the role of the high-energy scale deforming the kinematics. From this composition law of momenta, one defines the antipode (inverse element of the composition) $\hat{p}$ of a momentum p satisfying $(\hat{p} \oplus p) = (p \oplus \hat{p}) = 0$. For the classical basis, it reads

$$\hat{p}_0 = -p_0 + \frac{\vec{p}^2}{\Lambda}\Pi^{-1}(p), \qquad \hat{p}_i = -p_i \Pi^{-1}(p). \qquad \text{(A3)}$$

The nontrivial coproduct of Lorentz generators provides a Lorentz transformation of the two-particle system compatible with the linear Lorentz transformation of the total momentum defined by the noncommutative composition law (A1).

This relativistic deformation of SR can also be obtained from a geometric perspective [56]. In fact, the step from SR to DSR can be understood as going from a flat Minkowski momentum space to a curved momentum space [57,58]. If one considers the metric [59]

$$g_{\mu\nu}(k) = \eta_{\mu\nu} + \frac{k_\mu k_\nu}{\Lambda^2} \quad (A4)$$

in a de Sitter maximally symmetric curved momentum space as a deformation of the flat Minkowski metric $\eta_{\mu\nu}$, one can see that the isometries of this metric reproduce the kinematics of the classical basis of κ-Poincaré. Note that, in particular, the absence of a deformation of the dispersion relation is compatible with the usual interpretation of the dispersion relation as given by (a function of the square of) the geometric distance from the origin ($k = 0$) to a point $k = p$ with the previous metric. On the other hand, the composition law of momenta (A1) is the result of a translation in momentum space with the components of one of the two momentum variables identified as the set of parameters of the translation.

References

1. Einstein, A. Zur Elektrodynamik bewegter Körper. *Annalen der Physik* **1905**, *322*, 891–921. [CrossRef]
2. Mukhi, S. String theory: A perspective over the last 25 years. *Class. Quant. Grav.* **2011**, *28*, 153001. [CrossRef]
3. Aharony, O. A Brief review of 'little string theories'. *Class. Quant. Grav.* **2000**, *17*, 929–938. [CrossRef]
4. Dienes, K.R. String theory and the path to unification: A Review of recent developments. *Phys. Rept.* **1997**, *287*, 447–525. [CrossRef]
5. Sahlmann, H. Loop Quantum Gravity—A Short Review. In Proceedings of the Proceedings, Foundations of Space and Time: Reflections on Quantum Gravity, Cape Town, South Africa, 5 August 2010; pp. 185–210.
6. Dupuis, M.; Ryan, J.P.; Speziale, S. Discrete gravity models and Loop Quantum Gravity: A short review. *SIGMA* **2012**, *8*, 052. [CrossRef]
7. Loll, R. Quantum gravity from causal dynamical triangulations: A review. *Class. Quantum Gravity* **2019**, *37*, 013002. [CrossRef]
8. Wallden, P. Causal Sets Dynamics: Review & Outlook. *J. Phys. Conf. Ser.* **2013**, *453*, 012023. [CrossRef]
9. Wallden, P. Causal Sets: Quantum Gravity from a Fundamentally Discrete Spacetime. *J. Phys. Conf. Ser.* **2010**, *222*, 012053. [CrossRef]
10. Henson, J. The Causal set approach to quantum gravity. In *Approaches to Quantum Gravity: Toward a New Understanding of Space, Time and Matter*; Oriti, D., Ed.; Cambridge University Press: Cambridge, UK, 2009; pp. 393–413.
11. Addazi, A.; Alvarez-Muniz, J.; Alves Batista, R.; Amelino-Camelia, G.; Antonelli, V.; Arzano, M.; Asorey, M.; Atteia, J.-L.; Bahamonde, S.; Bajardi, F.; et al. Quantum gravity phenomenology at the dawn of the multi-messenger era—A review. *Prog. Part. Nucl. Phys.* **2022**, *125*, 103948. [CrossRef]
12. Amelino-Camelia, G. Quantum-Spacetime Phenomenology. *Living Rev. Rel.* **2013**, *16*, 5. [CrossRef]
13. Mattingly, D. Modern tests of Lorentz invariance. *Living Rev. Rel.* **2005**, *8*, 5. [CrossRef] [PubMed]
14. Liberati, S. Tests of Lorentz invariance: A 2013 update. *Class. Quant. Grav.* **2013**, *30*, 133001. [CrossRef]
15. Amelino-Camelia, G. Testable scenario for relativity with minimum length. *Phys. Lett. B* **2001**, *510*, 255–263. [CrossRef]
16. Amelino-Camelia, G. Relativity in space-times with short distance structure governed by an observer independent (Planckian) length scale. *Int. J. Mod. Phys. D* **2002**, *11*, 35–60. [CrossRef]
17. Kowalski-Glikman, J.; Nowak, S. Noncommutative space-time of doubly special relativity theories. *Int. J. Mod. Phys. D* **2003**, *12*, 299–316. [CrossRef]
18. Carmona, J.M.; Cortes, J.L.; Relancio, J.J. Beyond Special Relativity at second order. *Phys. Rev. D* **2016**, *94*, 084008. [CrossRef]
19. Majid, S.; Ruegg, H. Bicrossproduct structure of kappa Poincare group and noncommutative geometry. *Phys. Lett. B* **1994**, *334*, 348–354. [CrossRef]
20. Majid, S. *Foundations of Quantum Group Theory*; Cambridge University Press: Cambridge, UK, 1995.
21. Majid, S. Meaning of noncommutative geometry and the Planck scale quantum group. *Lect. Notes Phys.* **2000**, *541*, 227–276.
22. Hossenfelder, S. Multi-Particle States in Deformed Special Relativity. *Phys. Rev. D* **2007**, *75*, 105005. [CrossRef]
23. Amelino-Camelia, G.; Freidel, L.; Kowalski-Glikman, J.; Smolin, L. Relative locality and the soccer ball problem. *Phys. Rev. D* **2011**, *84*, 087702. [CrossRef]
24. Amelino-Camelia, G. Planck-scale soccer-ball problem: A case of mistaken identity. *Entropy* **2017**, *19*, 400. [CrossRef]
25. Kowalski-Glikman, J.; Bevilacqua, A. Doubly special relativity and relative locality. *PoS* **2022**, *CORFU2021*, 322. [CrossRef]
26. Kowalski-Glikman, J. Introduction to doubly special relativity. *Lect. Notes Phys.* **2005**, *669*, 131–159. [CrossRef]
27. Carmona, J.M.; Cortes, J.L.; Mazon, D.; Mercati, F. About Locality and the Relativity Principle Beyond Special Relativity. *Phys. Rev. D* **2011**, *84*, 085010. [CrossRef]
28. Amelino-Camelia, G. On the fate of Lorentz symmetry in relative-locality momentum spaces. *Phys. Rev. D* **2012**, *85*, 084034. [CrossRef]
29. Gubitosi, G.; Heefer, S. Relativistic compatibility of the interacting κ-Poincaré model and implications for the relative locality framework. *Phys. Rev. D* **2019**, *99*, 086019. [CrossRef]

30. Arzano, M.; Kowalski-Glikman, J. Deformations of Spacetime Symmetries: Gravity, Group-Valued Momenta, and Non-Commutative Fields. *Lect. Notes Phys.* **2021**, *986*. [CrossRef]
31. Amelino-Camelia, G.; Freidel, L.; Kowalski-Glikman, J.; Smolin, L. The principle of relative locality. *Phys. Rev. D* **2011**, *84*, 084010. [CrossRef]
32. Amelino-Camelia, G.; Freidel, L.; Kowalski-Glikman, J.; Smolin, L. Relative locality: A deepening of the relativity principle. *Gen. Rel. Grav.* **2011**, *43*, 2547–2553. [CrossRef]
33. Carmona, J.M.; Cortés, J.L.; Relancio, J.J.; Reyes, M.A. Time delays, choice of energy-momentum variables, and relative locality in doubly special relativity. *Phys. Rev. D* **2022**, *106*, 064045. [CrossRef]
34. Carmona, J.M.; Cortes, J.L.; Mercati, F. Relativistic kinematics beyond Special Relativity. *Phys. Rev. D* **2012**, *86*, 084032. [CrossRef]
35. Borowiec, A.; Pachol, A. Classical basis for kappa-Poincare algebra and doubly special relativity theories. *J. Phys. A* **2010**, *43*, 045203. [CrossRef]
36. Battisti, M.V.; Meljanac, S. Scalar Field Theory on Non-commutative Snyder Space-Time. *Phys. Rev. D* **2010**, *82*, 024028. [CrossRef]
37. Weinberg, S. *The Quantum Theory of Fields. Volume 1: Foundations*; Cambridge University Press: Cambridge, UK, 2005.
38. Amelino-Camelia, G.; Matassa, M.; Mercati, F.; Rosati, G. Taming Nonlocality in Theories with Planck-Scale Deformed Lorentz Symmetry. *Phys. Rev. Lett.* **2011**, *106*, 071301. [CrossRef]
39. Freidel, L.; Smolin, L. Gamma ray burst delay times probe the geometry of momentum space. *arXiv* **2011**, arXiv:1103.5626.
40. Amelino-Camelia, G.; Loret, N.; Rosati, G. Speed of particles and a relativity of locality in κ-Minkowski quantum spacetime. *Phys. Lett. B* **2011**, *700*, 150–156. [CrossRef]
41. Amelino-Camelia, G.; Arzano, M.; Kowalski-Glikman, J.; Rosati, G.; Trevisan, G. Relative-locality distant observers and the phenomenology of momentum-space geometry. *Class. Quant. Grav.* **2012**, *29*, 075007. [CrossRef]
42. Carmona, J.M.; Cortes, J.L.; Relancio, J.J. Does a deformation of special relativity imply energy dependent photon time delays? *Class. Quant. Grav.* **2018**, *35*, 025014. [CrossRef]
43. Mignemi, S.; Rosati, G. Relative-locality phenomenology on Snyder spacetime. *Class. Quant. Grav.* **2018**, *35*, 145006. [CrossRef]
44. Carmona, J.M.; Cortés, J.L.; Relancio, J.J. Observers and their notion of spacetime beyond special relativity. *Symmetry* **2018**, *10*, 231. [CrossRef]
45. Carmona, J.M.; Cortes, J.L.; Relancio, J.J. Spacetime and deformations of special relativistic kinematics. *Symmetry* **2019**, *11*, 1401. [CrossRef]
46. Martinez, M.; Errando, M. A new approach to study energy-dependent arrival delays on photons from astrophysical sources. *Astropart. Phys.* **2009**, *31*, 226–232. [CrossRef]
47. Abramowski, A. et al. [HESS Collaboration]. Search for Lorentz Invariance breaking with a likelihood fit of the PKS 2155-304 Flare Data Taken on MJD 53944. *Astropart. Phys.* **2011**, *34*, 738–747. [CrossRef]
48. Vasileiou, V.; Jacholkowska, A.; Piron, F.; Bolmont, J.; Couturier, C.; Granot, J.; Stecker, F.W.; Cohen-Tanugi, J.; Longo, F. Constraints on Lorentz Invariance Violation from Fermi-Large Area Telescope Observations of Gamma-Ray Bursts. *Phys. Rev. D* **2013**, *87*, 122001. [CrossRef]
49. Ahnen, M.L. et al. [MAGIC Collaboration]. Constraining Lorentz invariance violation using the Crab Pulsar emission observed up to TeV energies by MAGIC. *Astrophys. J. Suppl.* **2017**, *232*, 9. [CrossRef]
50. Abdalla, H. et al. [MAGIC Collaboration]. The 2014 TeV γ-Ray Flare of Mrk 501 Seen with H.E.S.S.: Temporal and Spectral Constraints on Lorentz Invariance Violation. *Astrophys. J.* **2019**, *870*, 93. [CrossRef]
51. Acciari, V.A. et al. [MAGIC Collaboration]. Bounds on Lorentz invariance violation from MAGIC observation of GRB 190114C. *Phys. Rev. Lett.* **2020**, *125*, 021301. [CrossRef]
52. Du, S.S.; Lan, L.; Wei, J.J.; Zhou, Z.M.; Gao, H.; Jiang, L.Y.; Zhang, B.B.; Liu, Z.K.; Wu, X.F.; Liu, Z.K.; et al. Lorentz Invariance Violation Limits from the Spectral-lag Transition of GRB 190114C. *Astrophys. J.* **2021**, *906*, 8. [CrossRef]
53. Carmona, J.M.; Cortés, J.L.; Pereira, L.; Relancio, J.J. Bounds on Relativistic Deformed Kinematics from the Physics of the Universe Transparency. *Symmetry* **2020**, *12*, 1298. [CrossRef]
54. Carmona, J.M.; Cortés, J.L.; Relancio, J.J.; Reyes, M.A.; Vincueria, A. Modification of the mean free path of very high-energy photons due to a relativistic deformed kinematics. *Eur. Phys. J. Plus* **2022**, *137*, 768. [CrossRef]
55. Albalate, G.; Carmona, J.M.; Cortés, J.L.; Relancio, J.J. Twin Peaks: A possible signal in the production of resonances beyond special relativity. *Symmetry* **2018**, *10*, 432. [CrossRef]
56. Carmona, J.M.; Cortés, J.L.; Relancio, J.J. Relativistic deformed kinematics from momentum space geometry. *Phys. Rev. D* **2019**, *100*, 104031. [CrossRef]
57. Kowalski-Glikman, J. De sitter space as an arena for doubly special relativity. *Phys. Lett. B* **2002**, *547*, 291–296. [CrossRef]
58. Magueijo, J.; Smolin, L. Gravity's rainbow. *Class. Quant. Grav.* **2004**, *21*, 1725–1736. [CrossRef]
59. Franchino-Viñas, S.A.; Relancio, J.J. Geometrizing the Klein–Gordon and Dirac equations in doubly special relativity. *Class. Quant. Grav.* **2023**, *40*, 054001. [CrossRef]

Disclaimer/Publisher's Note: The statements, opinions and data contained in all publications are solely those of the individual author(s) and contributor(s) and not of MDPI and/or the editor(s). MDPI and/or the editor(s) disclaim responsibility for any injury to people or property resulting from any ideas, methods, instructions or products referred to in the content.

Article

Æther as an Inevitable Consequence of Quantum Gravity

Sergey Cherkas [1,*,†] and Vladimir Kalashnikov [2,†]

1 Institute for Nuclear Problems, Bobruiskaya 11, 220006 Minsk, Belarus
2 Department of Physics, Norwegian University of Science and Technology, 7491 Trondheim, Norway
* Correspondence: cherkas@inp.bsu.by
† These authors contributed equally to this work.

Abstract: The fact that quantum gravity does not admit an invariant vacuum state has far-reaching consequences for all physics. It points out that space could not be empty, and we return to the notion of an æther. Such a concept requires a preferred reference frame for describing universe expansion and black holes. Here, we intend to find a reference system or class of metrics that could be attributed to "æther". We discuss a vacuum and quantum gravity from three essential viewpoints: universe expansion, black hole existence, and quantum decoherence.

Keywords: vacuum energy; preferred reference frame; vacuum state; quantum gravity

1. Introduction

From the earliest times, people comprehended an emptiness as "Nothing", which consists of absolutely nothing, no matter, no light, nothing. Others are convinced that "nothing" is unthinkable and a space-time should always contain "something", i.e., to be "æther" [1]. From straightforward point of view, the æther represents some stationary "medium" mimicking some matter and needs a preferred reference frame in which it is at rest "in tote". After the development of the quantum field theory (QFT), it was found that a vacuum actually contains a number of virtual particle–antiparticle pairs appearing and disappearing during the time of $\Delta t \propto \frac{1}{m}$, where m is a particle mass. That leads to the experimentally observable effects such as anomalous electron magnetic moment, the Lamb shift of atomic levels [2], the Casimir effect [3], etc. However, although a vacuum is not empty, a "soup" of the virtual particle–antiparticle pairs is not æther because it does not prevent the test particles from moving freely due to the Lorentz invariance (LI) of a QFT vacuum, as it is illustrated in Figure 1. That implies rigid limits on a local LI violation, and the existence of a preferred reference frame in the framework of QFT [4]. However, considering gravity seems to insist on the æther existence and the preferred reference frame due to an absence of a vacuum state invariant relative general transformation of coordinates. That demands reconsidering an idea of æther [5]. A possibility of the LI violation was also considered within string theory and loop quantum gravity (see [6] and references herein), the Einstein-Æther [7], and Horava–Lifshitz [8] theories, and others (see [9,10] for phenomenological implications). It could also mention the CPT invariance violation [11], which manifests itself both under Minkowski's space-time [12,13] and in the presence of gravity [10,14].

Another argument for a preferred reference frame is the vacuum energy problem. If the zero-point energy is real, we need to explain why this energy does not influence a universe's expansion. One of the solutions is to modify the gravity theory. That may violate the invariance relative to the general transformation of coordinates. For example, the Five Vectors Theory (FVT) of gravity demonstrates such a violation, including a LI violation [15].

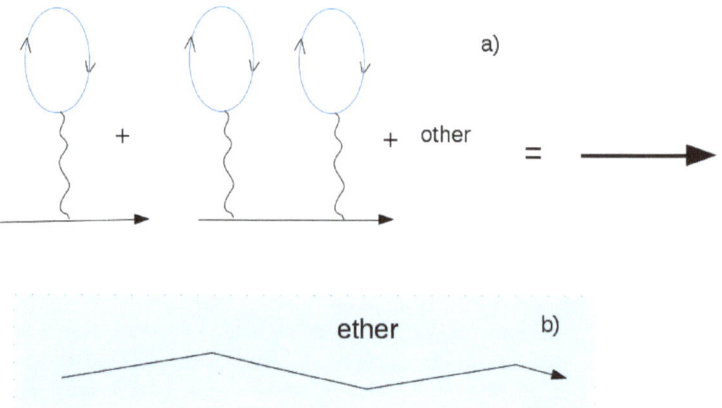

Figure 1. Illustration of vacuum influence on the particle propagation in (**a**) QFT, where the vacuum loops renormalize the mass and charge of the particle but do not prevent their free motion, and in (**b**) QG, where the æther fills space due to the absence of an invariant vacuum state.

2. Vacuum State and QG

The notion of a vacuum state originates from the ground state of a quantum oscillator. In QFT, the free fields are decomposed into a set of independent field oscillators by Fourier decomposition. Exited states of the oscillators are treated as particles, i.e., matter (both massive and massless). Introducing the interaction term leads to the renormalization of a particle mass and charge, but a one-particle state remains a one-particle state. Consequently, a one-particle wave packet moves freely with the constant envelope velocity, i.e., with no in vacuo dispersion [16]. That implies that even in the presence of perturbative interaction, one could still introduce a LI vacuum state in QFT.

Quantization of GR is too complicated to discuss a vacuum state. Nevertheless, let us consider a toy QG model regarding this issue. In this model only a spatially nonuniform scale factor represents gravity $a(\eta, \mathbf{r})$. It is certainly not a self-consistent approach within the GR frameworks [17]. Nevertheless, there exists a (1+1)– dimensional toy model [17] including a scalar fields $\phi(\tau, \sigma) = \{\phi_1(\tau, \sigma), \phi_2(\tau, \sigma)\dots\}$ and a scale factor $a(\tau, \sigma)$ described by the action

$$S = \int L \, d\tau = \frac{1}{2} \int \left(-a'^2 + (\partial_\sigma a)^2 + a^2 \left(\phi'^2 - (\partial_\sigma \phi)^2 \right) \right) d\sigma d\tau, \quad (1)$$

where τ is a time variable, σ is a spatial variable, and prime denotes differentiation with respect to τ. Here, like GR, the scalar fields evolve on the curved background $a(\tau, \sigma)$, which is, in turn, determined by the fields. The equations of motion is written as

$$\phi'' - \partial_\sigma^2 \phi + 2a' \phi' - 2\partial_\sigma \alpha \partial_\sigma \phi = 0, \quad (2)$$

$$a'' - \partial_\sigma^2 a + a'^2 - (\partial_\sigma a)^2 + \phi'^2 - (\partial_\sigma \phi)^2 = 0. \quad (3)$$

The relevant Hamiltonian and momentum constraints, written in terms of momentums $\pi(\tau, \sigma) \equiv \frac{\delta L}{\delta \phi'(\tau,\sigma)} = a^2 \phi'$, $p_a(\tau, \sigma) \equiv -\frac{\delta L}{\delta a'(\tau,\sigma)} = a'$ is

$$\mathcal{H} = \frac{1}{2} \left(-p_a^2 + \frac{\pi^2}{a^2} + a^2 (\partial_\sigma \phi)^2 - (\partial_\sigma a)^2 \right) = 0, \quad (4)$$

$$\mathcal{P} = -p_a \partial_\sigma a + \pi \partial_\sigma \phi = 0, \quad (5)$$

and obey the constraint evolution similar to GR [17]:

$$\partial_\tau \mathcal{H} = \partial_\sigma \mathcal{P}, \quad (6)$$

$$\partial_\tau \mathcal{P} = \partial_\sigma \mathcal{H}. \tag{7}$$

2.1. Quasi-Heisenberg Quantization and a Region of Small Scale Factor: Absence of Vacuum State

It is believed that our universe originates from a singularity in which a scale factor equals zero. Let us consider a region of small scale-factors first. In this region, it is convenient to use the quasi-Heisenberg picture [18], in which the setting of the initial conditions for operators at the initial moment allows quantization of the equations of motion. In the vicinity of small-scale factors, kinetic energy terms dominate over potential ones [17,18] so that the equations of motion (2) and (3) reduce to

$$\hat{\boldsymbol{\phi}}'' + 2\hat{\alpha}'\hat{\boldsymbol{\phi}}' \approx 0, \tag{8}$$

$$\hat{\alpha}'' + \hat{\alpha}'^2 + \hat{\boldsymbol{\phi}}'^2 \approx 0. \tag{9}$$

The solutions of Equations (8) and (9) for two scalar fields $\phi_1(\tau,\sigma)$, $\phi_2(\tau,\sigma)$ under initial conditions, discussed in Appendix A, are written as

$$\hat{\phi}_1(\tau,\sigma) = -\frac{i}{\pi_1}\int_{-\infty}^{\infty}\theta(\sigma-\sigma')S\left(k(\sigma')\partial_{\sigma'}\frac{\delta}{\delta k(\sigma')}\right)d\sigma' +$$

$$\frac{\pi_1}{2\sqrt{\pi_1^2+k^2(\sigma)}}\ln\left(1+2e^{-2\alpha_0}\sqrt{\pi_1^2+k^2(\sigma)}\,\tau\right),$$

$$\hat{\phi}_2(\tau,\sigma) = i\frac{\delta}{\delta k(\sigma)} + \frac{k(\sigma)}{2\sqrt{\pi_1^2+k^2(\sigma)}}\ln\left(1+2e^{-2\alpha_0}\sqrt{\pi_1^2+k^2(\sigma)}\,\tau\right),$$

$$\hat{\alpha}(\tau,\sigma) = \alpha_0 + \frac{1}{2}\ln\left(1+2e^{-2\alpha_0}\sqrt{\pi_1^2+k^2(\sigma)}\,\tau\right),$$

where the notations are given in the Appendix A.

As one can see, the scalar fields and the logarithm of the scale factor have monotonic behavior with time. It means that there are no oscillators in the vicinity of small-scale factors and no possibility of defining a vacuum state. In this situation, a quantum state is described by the momentum wave packet $C[k(\sigma)]$ as it is discussed in the Appendix A.

The difference in behavior in the vicinity of small-scale factors and at the epoch of the quantum oscillators occurrence was known long ago from analysis of the Wheeler–DeWitt equation solutions for the Gowdy model [19]. Therefore, we simply illustrate this fact in terms of asymptotic solutions of operator equations.

2.2. String-like Quantization within the Intermediate Region

From the previous subsection one can see that the operator equations of motion are not the oscillator equations in the vicinity of $a\sim 0$, which does not allow defining a vacuum state. A question arises: Could we define a vacuum state when the fields begin to oscillate, and quantum oscillators arise? In this region, the fields obey nonlinear wave Equations (2) and (3), which could not be solved analytically. That complicates using the quasi-Heisenberg picture, and to obtain some analytical results, we will use bosonic string quantization [20,21]. The action (1) could be rewritten in the reparametrization invariant form of a string on the curved background [17]

$$S = \frac{1}{2}\int d^2\xi\sqrt{-g}\,g^{\alpha\beta}(\xi)\partial_\alpha X^A\partial_\beta X^B G_{AB}(X(\xi)), \tag{10}$$

where $\xi = \{\tau,\sigma\}$, $X^A = \{a,\phi_1,\phi_2,\dots\}$, and the metric tensors $g_{\mu\nu}$, $G_{AB}(X)$ are in the form of

$$g = \begin{pmatrix} -N^2+N_1^2 & N_1 \\ N_1 & 1 \end{pmatrix}, \qquad G = \begin{pmatrix} 1 & 0 & 0 & \dots \\ 0 & -a^2 & 0 & \dots \\ 0 & 0 & -a^2 & \dots \\ \dots & \dots & \dots & \dots \end{pmatrix}.$$

The particular gauge for the lapse $N = 1$ and shift $N_1 = 0$ functions results in (1). The metric tensor $g_{\alpha\beta}(\xi)$ describes an intrinsic geometry of a (1+1)-dimensional manifold, i.e., a (1+1)-dimensional space-time, and it is an analog of the four-dimensional metric of general relativity. $G_{AB}(X(\xi))$ represents a geometry of the external space unifying scale factor and scalar fields and has no direct physical meaning here. The system (10) manifests an invariance relative to the reparametrization of the variables τ, σ, which is analog of the general coordinate transformation in GR. The transformations of coordinates $\tilde{\tau} = \tilde{\tau}(\tau, \sigma)$, $\tilde{\sigma} = \tilde{\sigma}(\tau, \sigma)$ imply transition to another reference frame for an observer who "lives on a string".

For obtaining a vacuum state, the key point is fixing the gauge by taking $g_{\mu\nu}$ in the form of Minkowski's metric by setting $N = 1$, $N_1 = 0$, which simplifies the action (10) to the form

$$S = \frac{1}{2} \int d\sigma d\tau G^{AB}(-\partial_\tau X_B \partial_\tau X_A + \partial_\sigma X_B \partial_\sigma X_A). \tag{11}$$

The momentum

$$P^A = \frac{\delta S}{\delta(\partial_\tau X_A)} = -\partial_\tau X^A = -G^{AB}\partial_\tau X_A \tag{12}$$

and the variable X_A obey the canonical commutation relations

$$[\hat{P}_A(\tau, \sigma), \hat{X}_B(\tau, \sigma')] = iG_{AB}\delta(\sigma - \sigma'). \tag{13}$$

As a zero-order approximation, one may take G to be equal to $\mathcal{G}$, where

$$\mathcal{G} = \begin{pmatrix} 1 & 0 & \ldots \\ 0 & -1 & \ldots \\ \ldots & \ldots & \ldots \end{pmatrix}.$$

Then, it could be possible to develop the perturbation theory on $G - \mathcal{G}$. In zero-order, X_A satisfies the wave equation

$$\hat{X}''_A - \partial_\sigma^2 X_A = 0, \tag{14}$$

and the commutation relations (13) can be realized using creation and annihilation operators

$$\hat{X}_A = \sum_{k=-\infty}^{\infty} \frac{1}{\sqrt{2k}} \left(a_{kA} e^{ik\sigma - i|k|\tau} + a_{kA}^+ e^{-ik\sigma + i|k|\tau} \right), \tag{15}$$

$$\hat{P}_A = \sum_{k=-\infty}^{\infty} i\sqrt{\frac{k}{2}} \left(-a_{kA} e^{ik\sigma - i|k|\tau} + a_{kA}^+ e^{-ik\sigma + i|k|\tau} \right), \tag{16}$$

where a_{kA}, a_{kB}^+ obey

$$[a_{kA}, a_{qB}^+] = -\mathcal{G}_{AB}\delta_{k,q}. \tag{17}$$

Thus, only when the gauge is fixed by $N = 1$, $N_1 = 0$, it is possible to define a vacuum state by $a_{kA}|0> = 0$. This vacuum state is not gauge-invariant because the dynamic variable X_A satisfies the wave Equation (14) in only this gauge (and in zero-order on $G - \mathcal{G}$). Moreover, one could see a problem with the definition of the Fock space of quantum states. Actually, Equation (17) leads to $[a_{k0}, a_{k0}^+] = -1$. That means that the state $a_{k0}^+|0>$ has a negative norm $<0|a_{k0}a_{k0}^+|0> = -1$. To avoid the negative norms, the string theory uses additional conditions on the physical Fock states $|>$:

$$\hat{\mathcal{L}}_f | > = 0, \tag{18}$$

where $\hat{\mathcal{L}}_f = \int (\hat{P}_A(\tau, \sigma) + \partial_\sigma \hat{X}_A(\tau, \sigma))^2 f(\sigma) d\sigma$, and $f(\sigma)$ is an arbitrary function. Operators $\hat{\mathcal{L}}_f$ obey the Virasoro algebra. It should be noted that the definition of the Virasoro operators includes the normal ordering [20–22], but it is beyond the concept of our work. If one accepts the feasibility of using the normal ordering, then the vacuum energy problem does not exist at all. However, we intend to refrain from discussing the status of excluding anomalies in the string theory here.

2.3. Towards a Classical Background

In Section 2.1, it is shown that there is no vacuum state in the vicinity of a small scale-factor because of an absence of field oscillators. In principle, the quasi-Heisenberg picture could be used for the description of the subsequent evolution, but it could be done only numerically because solving the operator equations with the initial conditions is complicated. Instead, we have used a string-like quantization described in Section 2.2. That allows an analytical consideration of the vacuum state, but it is only half of the problem because a further investigation of the perturbation series on $G - \mathcal{G}$ is needed. Moreover, the trouble with the negative norm of the states can be solved based on the Virasoro algebra by the transition to the $D = 26$ dimension in the string theory [20–22]. The general conclusion for us is that the vacuum state is not gauge-invariant and is defined in a single gauge $N = 1$, $N_1 = 0$. We could not make some other physical predictions for this region. However, one could put forward a hypothesis that in the presence of multiple scalar fields, a scale factor acquires monotonic behavior in time and could be considered classically finally. Such a situation is studied in the next section and allows for obtaining a number of physical predictions.

3. Vacuum Energy Problem as a Criterion for Finding the Preferred Reference Frame

The more straightforward problem is to define the vacuum state on a classical background space-time. Even in this case, the exact vacuum state exists only for some particular space-time. In other cases, the vacuum state has only an approximate meaning [23]. The observer moving with acceleration straightforwardly [24] or circularly [25] in Minkowski's space-time will detect quanta of the fields. That means that, although an observer could be in a resting coordinate system, the quantum fields are not in a vacuum state.

Nevertheless, a vacuum state could be defined, for example, in the slowly expanding universe, where a solution to the vacuum energy problem could serve as a criterion for choosing a preferred reference frame. The solution implies avoidance of the enormous zero-point energy density of the quantum fields affecting the universe's expansion. To do this, a class of conformally unimodular (CUM) metrics has been introduced [15]:

$$ds^2 \equiv g_{\mu\nu}dx^\mu dx^\nu = a^2(1 - \partial_m P^m)^2 d\eta^2 - \gamma_{ij}(dx^i + N^i d\eta)(dx^j + N^j d\eta), \quad (19)$$

where $x^\mu = \{\eta, x\}$, η is a conformal time, γ_{ij} is a spatial metric, $a = \gamma^{1/6}$ is a locally defined scale factor, and $\gamma = \det \gamma_{ij}$. The interval (19) is similar formally to the ADM one [26], but the lapse function is taken in the form of $a(1 - \partial_m P^m)$, where P^m is a three-dimensional vector, and ∂_m is a conventional partial derivative.

Using the restricted class of the metrics (19), the theory [15] has been suggested in which the Hamiltonian constraint is not necessarily zero but equals some constant. Such a theory is known as the Five Vectors theory (FVT) of gravity [15], because the interval (19) contains two 3-vectors P, N and, moreover, spatial metric can be decomposed into a set of three triads $\gamma_{ij} = e_{ia}e_{ja}$, where index a enumerates vectors of the triads e_a.

This theory satisfies the strong equivalence principle (EP) because no additional tensor fields appear.[1] Nevertheless, in contrast to GR, where the lapse and shift are arbitrary functions fixing the gauge, the restrictions $\partial_n(\partial_m N^m) = 0$ and $\partial_n(\partial_m P^m) = 0$ arise in FVT. The Hamiltonian $\mathcal{H}$ and momentum $\mathcal{P}_i$ constraints in the particular gauge $P^i = 0$, $N^i = 0$ obey the constraint evolution equations [15]:

$$\partial_\eta \mathcal{H} = \partial_i \left(\tilde{\gamma}^{ij} \mathcal{P}_j \right), \quad (20)$$

$$\partial_\eta \mathcal{P}_i = \frac{1}{3}\partial_i \mathcal{H}, \quad (21)$$

where $\tilde{\gamma}_{ij} = \gamma_{ij}/a^2$ is a matrix with a unit determinant. Equations (20) and (21) admit adding some constant to $\mathcal{H}$ and, in the FVT frame, it is not necessary that $\mathcal{H} = 0$, but $\mathcal{H} = const$ is also allowed. That solves the problem of the main part of the zero-point energy density.

Let us consider a spatially uniform, isotropic, and a flat universe with the metric

$$ds^2 = a(\eta)^2(d\eta^2 - dx^2), \qquad (22)$$

which belongs to a class of (19). Using the Pauli hard cutoff of the 3-momentums k_{max} [30,31] reduces the zero-point energy density calculated in the metric (22) to

$$\rho_v = \frac{(N_{boson} - N_{ferm})}{4\pi^2 a^4} \int_0^{k_{max}} k^2 \sqrt{k^2 + a^2 m^2} dk \approx \frac{(N_{boson} - N_{ferm})}{16\pi^2} \left(\frac{k_{max}^4}{a^4} + \frac{m^2 k_{max}^2}{a^2} + \frac{m^4}{8} \left[1 + 2\ln\left(\frac{m^2 a^2}{4k_{max}^2}\right) \right] + \ldots \right), \qquad (23)$$

where, for simplicity, bosons and fermions of equal masses are considered.

The main part of this energy density $\sim \frac{k_{max}^4}{a^4}$ scales as radiation, and it has to cause an extremely fast universe expansion in the frame of GR. This result contradicts the observations [32]. In our approach, a constant in the Hamiltonian constraint [15] compensates this main part of zero-point energy and makes it unobservable.[2]

The remaining parts in (23) are also huge but assuming the sum rules for masses of bosons and fermions (the condensates should be taken into account, as well) would provide a mutual compensation for these terms [31,34]. Of course, all spectrum of the particles in nature, including unknown now, should be taken into account. The empirical cutoff of momentums k_{max} is used in (23), with the hope that some fundamental basis will be found for that in the future (e.g., like a noncommutative geometry [35–37]), and will provide the UV completions of QG without a renormalization.

Equation (19) determines the preferred reference frame ensuring an æther existence and an absence of dipole anisotropy of the cosmic microwave background (CMB) [38]. Otherwise, the question arises: What is the physical foundation of the frame where CMB is in a rest "in tote", i.e., does not have a dipole component [39]?

4. Cosmological Consequences of Residual Vacuum Energy

Other contributors to the vacuum energy density are the terms depending on the derivatives of the universe expansion rate [34,40,41]. Sum rules cannot remove these terms, but they have the correct order of $\rho_v \sim M_p^2 H^2$, where H is the Hubble constant, and allow explaining the accelerated expansion of the universe. These energy density and pressure are [34,40,41]:

$$\rho_v = \frac{a'^2}{2a^6} M_p^2 S_0, \qquad p_v = \frac{M_p^2 S_0}{a^6} \left(\frac{1}{2} a'^2 - \frac{1}{3} a'' a \right), \qquad (24)$$

where, $S_0 = \frac{k_{max}^2}{8\pi^2 M_p^2}$ is determined by the UV cut-off of the comoving momenta and the reduced Planck mass $M_p = \sqrt{\frac{3}{4\pi G}}$ is implied. The energy density and pressure of vacuum (24) satisfy a continuity equation

$$\rho_v' + 3\frac{a'}{a}(\rho_v + p_v) = 0, \qquad (25)$$

and, in the expanding universe, are related to the equation of state $p_v = w\rho_v$, as Figure 2 (upper panel) illustrates. Using this equation of vacuum state leads to the cosmological Vacuum Fluctuations Domination (VFD) model [40–42]. According to VFD the universe behavior at early times, when the scale factor was small, is as freely rolling, i.e., without any deceleration or acceleration, but it is accelerated at a late time. The deceleration parameter $q(a) = -\frac{a''a}{a'^2} + 1$ is shown in Figure 2 (lower panel) [42]. The discovery of an accelerated universe expansion was a big surprise [43]. However, if the above view of a vacuum is true, a stage preceding the acceleration should be Milne-like, i.e., linear in a cosmic time. The Milne-like universes have been much discussed again recently [44–50].

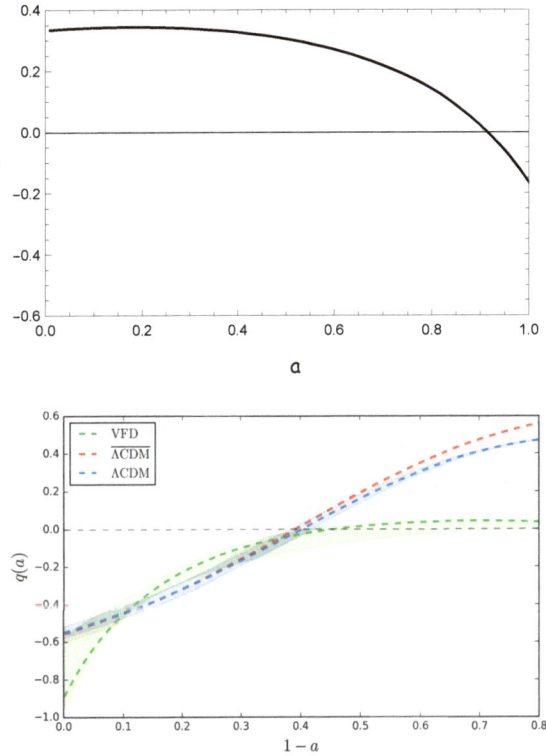

Figure 2. (**Upper panel**). Equation of the vacuum state in dependence on the universe scale factor a. (**Lower panel**) Deceleration parameter $q(a)$ and the corresponding dispersion channels for the VFD model (24) and two versions of ΛCDM model.

4.1. Nucleosynthesis in the Milne-like universe

Nucleosynthesis in a slowly expanding universe was considered earlier [51–53]. Here, we present our calculation for the VFD model, which has a Milne-like stage, as shown in Figure 2, corresponding to the region where the deceleration parameter q is close to zero. The calculations have been performed with the PRIMAT code (version 0.1.1) [54,55] including 423 nuclear reactions. The results of the calculations are presented in Table 1 and Figure 3. For comparison, the results for the standard cosmological model are also shown.

As expected, there is a very low rate of neutrons during a period of helium formation in the VFD model (see Figure 3b). That is because an equilibrium between protons and neutrons is shifted towards a neutron decay during the slow universe expansion. Nevertheless, a small amount of neutrons during a long time can create a necessary amount of helium if baryonic density $\Omega_b \approx 0.76$. From analysis of Supernovae Type-Ia, Cosmic Chronometers, and Gamma-ray bursts, it was also found that $\Omega_m \approx 0.87$ for the VFD model [42].[3] It means that there is no need for any ad hoc dark matter in the VFD cosmological model because $\Omega_m \approx \Omega_b$. Moreover, as it was conjectured in [56], spatially nonuniform vacuum polarization should be taken into account in the dynamics of the structure formation.

On the other hand, there is a lot of time for the growth of inhomogeneities in the VFD model [34,57], and the nonlinear regime begins soon after the last scattering surface. That allows the suggestion that almost all the baryonic matter collapses into eicheons [58], which replace the black holes in FVT. There are no strong constraints on the abundance of black holes in a region of mass $M \sim 10^{13}$–10^{19} $M_\odot$ [59], and it is possible that the matter concentrates namely in this region.

In the VFD model, there is no cosmological deuterium production. The amount of lithium is less than that in the ΛCDM, that alleviates the lithium overproduction problem of the standard cosmological model. The amount of CNO is 10^7 times greater compared to ΛCDM, but it does not contradict the observations [60,61].

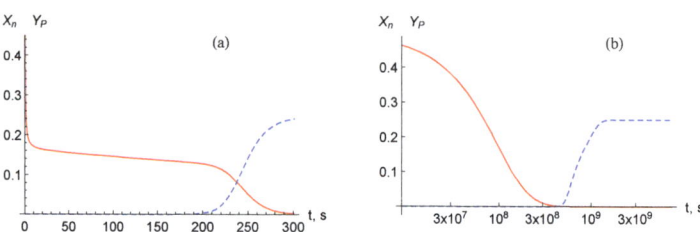

Figure 3. Dependencies of relative abundances of neutron and ^{4}He on cosmic time $dt = ad\eta$ are given by red and blue curves, respectively, (**a**) for standard cosmological model, (**b**) for the VFD model.

Table 1. Final abundances of light elements in the ΛCDM model at $\Omega_b = 0.049$ and the VFD model at $\Omega_b = 0.87$.

	ΛCDM	VFD
H	0.75	0.75
$Y_p = 4Y_{He}$	0.25	0.25
D/H $\times\ 10^5$	2.6	$<10^{-25}$
^{3}He/H $\times\ 10^5$	1.1	$<10^{-8}$
T/H $\times\ 10^8$	7.9	$<10^{-32}$
(^{7}Li + ^{7}Be)/H $\times\ 10^{10}$	5.7	2.1
^{6}Li/H $\times\ 10^{14}$	1.2	$<10^{-25}$
^{9}Be/H $\times\ 10^{19}$	9.2	$<10^{-34}$
^{10}B/H $\times\ 10^{21}$	2.9	$<10^{-8}$
^{11}B/H $\times\ 10^{16}$	3.3	$<10^{-10}$
CNO/H $\times\ 10^{16}$	8.0	5.6×10^7

It is widely believed that deuterium is produced only cosmologically in ΛCDM, but for the VFD model, the most plausible and direct way is to create necessary deuterium by beams of antineutrino arising during a collapse [62] before the formation of eicheons in the range of $M \sim 10^{13}$–10^{19} $M_\odot$. Their formation is unavoidable in the slowly expanding cosmologies because there is much time for the collapse of inhomogeneities in contrast to the standard cosmological model. Indeed, the matter stored in the supermassive eicheons is not related to "dark matter" observed in rotational curves of galaxies because the latter could be explained by the vacuum polarization [56]. Other mechanisms of non-cosmological deuterium production are also discussed [63].

4.2. Notes about Cosmic Microwave Background in the Slowly Expanding Cosmological Models

By this time, there are no trustable studies of the CMB background for slowly expanding cosmological models, and only some heuristic calculations exist [64]. The main question is the origin of a scale corresponding to the first peak in the CMB spectrum and the origin of the baryon acoustic oscillations (BAO) ruler [65]. In the standard cosmological model, this is the sound horizon's size at the recombination moment. For the Milne-like cosmology, these quantities must be different [65,66]. Apart from this, the sound horizon for the Milne-like flat universe is vast and cannot be a scale, which determines the position of the first CMB peak. Let us hypothesize that the width of the last scattering surface [64] could be such a scale for VFD.[4] In this light, the mechanisms of perturbation growth during a recombination period are of interest [67]. As for the BAO ruler, it has to be determined by the complex nonlinear process in the slowly evolving cosmologies and is not related directly to the scale corresponding to the first peak of CMB.

5. Size of Eicheon

The concept of æther considered in this work is based on postulating the preferred coordinate frame, namely, CUM. One more consequence of this hypothesis is a replacement of the black hole solutions of GR by the so-called "eicheons". In Ref. [58], the spherically symmetric solution of the Einstein equations in the CUM metrics (19) was analyzed, and it was found that the finite pressure solution exists for an arbitrarily large mass. As a result, there are no compact objects with an event horizon,[5] because an "eicheon" appears instead of a black hole [58].[6]

In Ref. [58], we have turned from the CUM metrics (27) to Schwarzschild-like in order to demonstrate that a compact object looks like a hollow sphere with a radius greater than that of Schwarzschild (see Figure 4).

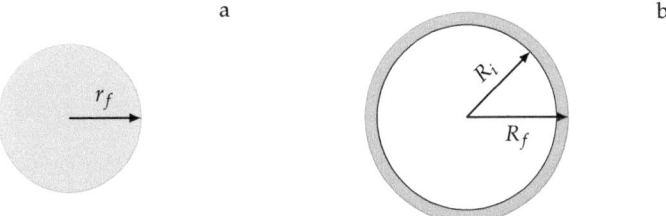

Figure 4. (**a**) A compact object of uncompressible fluid with the radius of r_f in the CUM metrics (27) looks as a shell (**b**) with the boundaries $r_g < R_i < R_f$ in Schwarzschild's type metric, where r_g is a Schwarzschild's radius.

Here, we intend to calculate the radius of a compact object of constant density in the CUM metrics depending on maximum pressure and density. For a spherically symmetric space-time, the CUM metrics (19) is reduced to

$$ds^2 = a^2(d\eta^2 - \tilde{\gamma}_{ij}dx^i dx^j) = e^{2\alpha}\left(d\eta^2 - e^{-2\lambda}(dx)^2 - (e^{4\lambda} - e^{-2\lambda})(xdx)^2/r^2\right), \quad (26)$$

where $r = |x|$ and α, λ are the functions of r. In the spherical coordinates, Equation (26) looks as

$$ds^2 = e^{2\alpha}\left(d\eta^2 - dr^2 e^{4\lambda} - e^{-2\lambda}r^2\left(d\theta^2 + \sin^2\theta d\phi^2\right)\right). \quad (27)$$

Let us compare (27) with Schwarzschild's type metrics

$$ds^2 = B(R)dt^2 - A(R)dR^2 - R^2\left(d\theta^2 + \sin^2\theta d\phi^2\right). \quad (28)$$

The difference between the metrics (26) and (28) is that the metric (28) suggests that the circumference equals $2\pi R$. However, there is no evidence for this fact in an arbitrary spherically symmetric space-time. For the metric (26), the circumference is not equal to $2\pi r$ in the close vicinity of a point-like mass. Coordinate transformation $t = \eta$, $R = R(r)$ relates the metrics (27) and (28), while their comparison gives:

$$B(R) = e^{2\alpha}, \quad (29)$$

$$R^2 = r^2 e^{-2\lambda + 2\alpha}, \quad (30)$$

$$A(R)\left(\frac{dR}{dr}\right)^2 = e^{4\lambda + 2\alpha}. \quad (31)$$

Using (29), (30) in (31) to exclude λ and α yields

$$\frac{dr}{dR} = \frac{R^2 A^{1/2}}{r^2 B^{3/2}}. \tag{32}$$

For an empty Schwarzschild space-time $A(R) = (1 - r_g/R)^{-1}$ and $B(R) = 1 - r_g/R$, whereas in the region filled by matter, $A(R)$ and $B(R)$ obey [72]

$$\frac{d}{dR}\left(\frac{R}{A}\right) = 1 - \frac{6}{M_p^2}\rho R^2, \tag{33}$$

$$\frac{1}{B}\frac{dB}{dR} = -\frac{2}{p+\rho}\frac{dp}{dR}, \tag{34}$$

where $r_g = \frac{3m}{2\pi M_p^2}$. Further, as in [58], we will consider a model of the constant density $\rho(R) = \rho_0$. In this case, Equations (33) and (34) can be integrated explicitly that gives

$$A = \frac{R}{R - r_g - 2\rho_0\left(R^3 - R_f^3\right)M_p^{-2}}, \tag{35}$$

$$B = \left(1 - \frac{r_g}{R_f}\right)\frac{\rho_0^2}{(p(R)+\rho_0)^2} \tag{36}$$

and one needs only to find a pressure $p(R)$, which obeys the Tolman–Volkov–Oppenheimer (TVO) equation

$$p'(R) = -\frac{3}{4\pi M_p^2 R^2}\mathcal{M}(R)\rho(R)\left(1 + \frac{4\pi R^3 p(R)}{\mathcal{M}(R)}\right)\left(1 + \frac{p(R)}{\rho(R)}\right)\left(1 - \frac{3\mathcal{M}(R)}{2\pi M_p^2 R}\right)^{-1}. \tag{37}$$

It is convenient to measure density and pressure in the units of $M_p^2 r_g^{-2}$, so that the mean density of Schwarzschild black hole $\rho_0 = m/(\frac{4}{3}\pi r_g^3)$ equals $1/2$, while the TOV limit $R_f < \frac{9}{8}r_g$ gives the value of $\rho_0 = \frac{1}{2}(\frac{8}{9})^3 \approx 0.35$. As for the eicheon radius in Schwarzschild's type metric, it equals $R_f = \sqrt[3]{R_i^3 + \frac{1}{2\rho_0}}$ in the units of r_g, where R_i is an inner radius, which determines maximum pressure. Using

$$\mathcal{M}(R) = \frac{4\pi}{3}\rho_0\left(R^3 - R_i^3\right), \tag{38}$$

for solving the TOV equation for pressure, it is possible to find B, and then solve (32) with the initial condition $r(R_i) = 0$ and find the eicheon radius $r_f = r(R_f)$ in the CUM metrics.

Let us plot (see Figure 5) the calculated radius of the eicheon in the CUM metrics in dependence on density ρ_0 and maximum pressure, that is, the pressure in the center of a solid ball in the metric (27). An approaching R_i to unity increases the maximal pressure. Actual density and pressure in the center of eicheon are defined by the extremal equation of state, which is the subject of future investigations. However, Figure 5 allows concluding that the pressure is considerably smaller than the energy density in a region of interest. That results in a straightforward analytic estimation of the eicheon radius. For the estimation, one could take pressure equal to some constant (e.g., $p(R) = \rho_0/10$) in (36), or even simply $p(R) = 0$. Then one could take $R_i = 1$, i.e., the Schwarzschild radius and integrate (32) to obtain

$$r_f = \sqrt[3]{3\int_1^{R_f}\frac{A^{1/2}}{B^{3/2}}R^2 dR} \approx \frac{\sqrt{3}\sqrt[3]{11}\rho_0^{1/6}}{2^{5/6}}, \tag{39}$$

where a small "thickness" of the eicheon surface $R_f - 1$ is used, R_f is expressed as $R_f = \sqrt[3]{1 + \frac{1}{2\rho_0}}$, and only asymptotic term of large ρ_0 is retained. In the ordinary units, the result reads

$$r_f = \frac{3\, 3^{5/6} \sqrt[3]{11}\, m^{4/3}\, \sqrt[6]{\rho_0}}{4\sqrt[6]{2}\, \pi^{4/3} M_p^3} \tag{40}$$

and it is slightly unexpected because the eicheon radius rises with the density that turns out to be a specific manifestation of the CUM geometry.[7] In particular, the eicheon of the Planck density $\rho_0 = M_p^4$, which is sometimes considered as a maximal density in nature [74] has a radius of $r_f \approx 0.8 \frac{1}{M_p} \left(\frac{m}{M_p}\right)^{4/3}$ in the CUM metrics. Looking at the last equation, one may assume that the large eicheons cannot be very dense. However, r_f given by (39) is not a physical distance but only points out a border of eicheon in the CUM metrics, whereas the physical distance is given by $l_{eiche} = \int_0^{r_f} e^{\alpha + 2\lambda} dr = \int_1^{R_f} A^{1/2} dR \approx \frac{5}{24\rho_0}$.

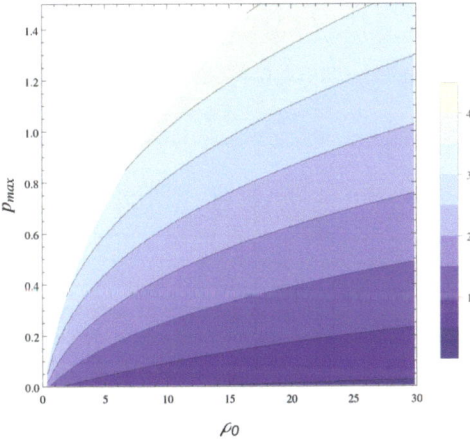

Figure 5. Dependence of the eicheon radius $r_f = r(R_f)$ in the CUM metrics, expressed in the units of gravitational radius, on the density and maximal pressure (i.e., pressure in the eicheon center). Pressure and density are in the units of $M_p^2 r_g^{-2}$.

Recently, many investigations explored the footprints of black holes manifesting themselves through star trajectories and a shadow in the accretion disks around the galaxy centers, gravitational lensing, and gravitational waves from the colliding compact objects (see footnote 3 on p. 15). These phenomena can be explained from the properties of both stationary and non-stationary metrics of Schwarzschild and Kerr types, where the radius of "source" objects is of the order of r_g. It seems reasonable to interpret these observations in the CUM framework and obtain an actual eicheon radius r_f using its equation of state.[8] The most informative study originates from collisions of ultracompact massive objects producing the gravitational waves observed by the existing and developing detectors. At this moment, direct astrophysical observations and, much less, analogous modeling cannot provide decisive evidence, which would rule out some alternative concepts of ultracompact massive objects without a horizon (nevertheless, see [76]). One could suggest that off-horizon properties of an eicheon are the same as for a black hole. However, near-horizon phenomena like gravitation wave emission in black hole collisions could be most informative [71,77,78] and the study of these phenomena in the framework of FVT is a matter for the future.

6. Decoherence of the Particles Due to Gravitational Potential Fluctuations

Here, we return to the consideration of a locally defined scale factor as an operator. One more implication of the CUM metrics and æther arises for a gravitational decoherence [79], which is the subject of the table-top quantum gravity experiments. In GR, it would not be possible to say that the vacuum fluctuations under Minkowski's space-time are small. Actually, for the small vacuum fluctuations, one could turn to the reference system where they are significant. That means the appearance of the so-called gauge waves, which are the consequence of the reference frame choice. By restricting the possible reference systems, it would be possible to reveal the actual vacuum fluctuation influencing the motion of the massive particles. The main fundamental question is: Does a massive particle lose its coherence due to interaction with æther? Under Minkowski's background, one could write for a locally defined scale factor:

$$a(\eta, r) = (1 + \Phi(\eta, r)), \tag{41}$$

According to [79], the correlator of the Fourier amplitudes for the gravitational potential in vacuum $\hat{\Phi}(\eta, r) = \sum_k \hat{\Phi}_k(\eta) e^{ikr}$ takes the form

$$S(\tau - \eta, k) = <0|\hat{\Phi}_k^+(\eta)\hat{\Phi}_k(\tau)|0> = \int_{-\infty}^{\infty} \tilde{S}(\omega, k) e^{i\omega(\tau - \eta)} d\omega, \tag{42}$$

where a spectral function $\tilde{S}(\omega, q)$ is approximately written as [79]

$$\tilde{S}(\omega, k) \approx \begin{cases} \frac{N_{all}}{32\pi^2 M_p^4}, & q < \omega < 2k_{max} \\ 0, & \text{otherwise.} \end{cases} \tag{43}$$

where $N_{all} = N_{boson} + N_{ferm}$ is a number of all degrees of freedom.

For a nonrelativistic massive point particle propagating among the fluctuations of the gravitational potential, the Fokker–Planck equation is:

$$\partial_\eta f_k(p) + i(E_{p+k/2} - E_{p-k/2}) f_k(p) = -iK_1 k \frac{\partial f_k}{\partial p} + 2iK_2 \, kp \, \Delta_p f_k(p) + 2iK_3 \, p_i k_j \frac{\partial^2 f_k}{\partial p_j \partial p_i}, \tag{44}$$

where $f_k(p)$ is a Fourier transform of the Wigner function $\tilde{f}(r, p) = \sum_k f_k(p) e^{ikr}$. In the first order on the constants K_1, K_2, K_3 it is possible to write:

$$\int f_k(p, \eta) f_{-k}(p, \eta) d^3 p d^3 k \approx 1 - (3K_1 + 3K_2 + 6K_3) \frac{\Gamma^2 \eta^2}{m}. \tag{45}$$

This means that the interaction with a vacuum produces decoherence manifesting itself in the decreasing of "purity" [79] of a particle state according to (45). From Equation (45), the decoherence time is estimated as

$$t_{dec} \approx \frac{1}{\Gamma} \sqrt{\frac{m}{3K_1 + 3K_2 + 6K_3}}, \tag{46}$$

and using the approximate expressions for the constants K_1, K_2, K_3, the decoherence length can be found [79]

$$L_{dec} \approx \frac{4M_p}{3\sqrt{3 N_{all} \pi m}} \frac{v}{\Gamma}, \tag{47}$$

where v is a particle velocity, m is a particle mass, $1/\Gamma$ is a localization length of the particle wave packet. That is, a point-like particle of mass $m \sim \frac{4M_p v}{3\sqrt{3N_{all}}\pi}$ loses coherence at a distance equal to the length of the wave packet $1/\Gamma$. It should be noted that interaction with the æther does not produce a particle scattering because the momentum distribution $f_0(p)$ does not change. Nevertheless, decoherence arises. That is a fundamental result implying Lorentz and Galilean invariance violation because one particle state becomes non-pure quantum state.

The difference in particle propagation in the QG and QFT is illustrated in Figure 1. The æther in QG originates from an absence of an invariant vacuum state. The last is not invariant relatively to the general transformation of coordinates and, in particular relative to the Lorentz transformation when it is considered as a subgroup of the general transformation of coordinates.

As regards the decoherence observation (47), such massive point particles are unknown. A real particle of large mass has a finite size, which restricts momentums transferred by the particle form factor: $q < 1/d$, where d is particle size. In this case, the following estimation arises [79]

$$L_{dec} \approx \frac{8\pi(M_p d)^2}{\sqrt{N_{all}}} \frac{v}{\Gamma}. \tag{48}$$

This quantity seems very large and unobservable. At the same time, the real particles are not rigid but have internal degrees of freedom and consist of a number of point particles, so more careful investigation is needed. Moreover, other possible fundamental mechanisms of decoherence also need investigation [80].

Recently, a gravitationally induced entanglement has attracted great attention (see e.g., [81–85]). There is no doubt that the nonrelativistic quantum mechanics holds for any interaction, including gravitational interaction in the form of the second Newton's law and any weak external gravitational field [86]. It is also no doubt that the gravitational waves of linearized gravity are fully analogous to electromagnetic waves and have to be quantized. Undoubtedly, the second Newton's law could be interpreted as an exchange by gravitons, like the Coulomb law can be interpreted as an exchange by photons. In contrast, the result (47) seems much less trivial because this fundamental decoherence implies the existence of æther with stochastic properties. Such an æther is absent in quantum electrodynamics due to the existence of the LI vacuum state. Implications of the æther in a photon sector of LI violation [87,88] have to be investigated.

7. Conclusions

To summarize, the CUM metrics gives a sustained basis for quantum gravity physics, cosmology, and physics of compact astrophysical objects. Although fascinating physics like closed time-like curves [89–91], time machines [92,93], wormholes [94], and Hawking radiation [95–97] are excluded in the CUM metrics, these metrics give a fresh impetus to investigate the real physical phenomena, including the structure formation [34], CMB [64], the structure of ultracompact astrophysical objects, and search for the decoherence QG effects and other QG consequences from the vacuum fluctuations of the gravitational potential. All these phenomena imply to single out the conformally unimodular metrics corresponding to a reference system, where the æther is at rest "in tote". Certainly, it suggests the æther existence per se. In QG, an æther is not simply some background but a thing that weaves all the physical phenomena into a whole quantum universe.

On the other hand, because black holes are absent in this theory, there is no actual "eraser" of the information in the CUM metrics. In other words, a wave function of some particular quantum system is only mixed to a more general wave function, including vacuum and, finally, all universe, and the universe's wave function seems not an idealization, but a reality conserving all information without any loss [98].

Author Contributions: Concepts and methodology are developed by S.C. and V.K.; software, S.C.; validation, writing and editing, S.C. and V.K. All authors have read and agreed to the published version of the manuscript.

Funding: This research received no external funding.

Data Availability Statement: Not applicable.

Conflicts of Interest: The authors declare no conflict of interest.

Appendix A. Quasi-Heisenberg Quantization

For simplicity, it is convenient to consider two scalar fields, ϕ_1 and ϕ_2, that correspond to a system with three degrees of freedom, including the logarithm of scale factor $\alpha = \ln a$. As a result, there is only one degree of freedom because the Hamiltonian and momentum constraints allow excluding two of them. Let us discuss a quantum picture of the system (1), (4) and (5). The quasi-Heisenberg picture suggests that one needs to define the commutation relations and initial values for operators at the initial moment and then permit the operator evolution according to the equation of motions. For quantization with the help of the Dirac brackets (see also [99]), one should set two additional gauge fixing conditions corresponding to the Hamiltonian and momentum constraints.

Let us take these conditions as

$$\hat{\alpha}(0,\sigma) = \alpha_0 = const, \tag{A1}$$

$$\partial_\sigma \hat{\pi}_1(0,\sigma) = 0, \tag{A2}$$

i.e., the logarithm of the scale factor and momentum $\hat{\pi}_1(0,\sigma) = \pi_1$ are c-number constants at the initial moment. Generally, that is some time-dependent gauge, which is known only at an initial moment. Then it is permissible for the commutation relations to evolve.

Dirac brackets could allow calculating the operator commutation relations at the initial moment, but the equivalent receipt is to set

$$\hat{\phi}_2(0,\sigma) \equiv \varphi(\sigma), \tag{A3}$$

$$\hat{\pi}_2(0,\sigma) \equiv -i\frac{\delta}{\delta\varphi(\sigma)}, \quad \hat{\phi}'_2(0,\sigma) = -i e^{-2\alpha_0}\frac{\delta}{\delta\varphi(\sigma)} \tag{A4}$$

and express other variables from constraints and gauge conditions to obtain

$$\hat{p}_\alpha(0,\sigma) = \sqrt{-\frac{\delta^2}{\delta\varphi^2(\sigma)} + \pi_1^2}, \quad \hat{\alpha}'(0,x) = e^{-2\alpha_0}\sqrt{-\frac{\delta^2}{\delta\varphi^2(\sigma)} + \pi_1^2}, \tag{A5}$$

$$\hat{\phi}_1(0,\sigma) = \frac{i}{\pi_1}\int_{-\infty}^{\infty} \theta(\sigma - \sigma') S\left(\frac{\delta}{\delta\varphi(\sigma')}\partial_{\sigma'}\varphi(\sigma')\right)d\sigma', \tag{A6}$$

$$\hat{\phi}'_1(0,\sigma) = e^{-2\alpha_0}\pi_1, \tag{A7}$$

where the symbol S denotes symmetrization of the noncommutative operators, i.e., $S(\hat{A}\hat{B}) = \frac{1}{2}(\hat{A}\hat{B} + \hat{B}\hat{A})$ or $S(\hat{A}\hat{B}\hat{C}) = \frac{1}{6}(\hat{A}\hat{B}\hat{C} + \hat{B}\hat{A}\hat{C} + \hat{A}\hat{C}\hat{B} + \ldots)$ and $\theta(\sigma)$ is a unit step function. Its appearance in (A6) is the only nontrivial moment that follows from calculation of the Dirac brackets [100], and we have introduced it here for expressing ϕ_1 from the momentum constraint (5).

The equations of motion (2), (3) should be considered as the operator equations with the initial conditions (A1), (A3)–(A7).

The second stage of quantization consists of building the Hilbert space where the quasi-Heisenberg operators act. This stage again begins from the classical Hamiltonian (4) and momentum (5) constraints. The momentum constraint and corresponding gauge condition (A2) are resolved relatively the variable ϕ_1 and its momentum π_1. Then, these quantities are substituted to the Hamiltonian constraint, which is then quantized and considered as the Wheeler–DeWitt equation in the vicinity of the small scale factor $a \sim 0$, i.e., $\ln a = \alpha \to -\infty$. In such a way, we come to

$$\left(\frac{\delta^2}{\delta\alpha(\sigma)} - \frac{\delta^2}{\delta^2\varphi(\sigma)} + \pi_1^2\right)\Psi[\alpha, V] = 0, \tag{A8}$$

where it is taken into account that π_1 is some constant. Space of the negative frequency solutions of the Equation (A8) constitutes the Hilbert space for the quasi-Heisenberg operators.

In the general case, the solution of Equation (A8) is of the form of the wave packet

$$\Psi[\alpha,\varphi] = \int C[k] \, e^{\int \left(-i\alpha(\sigma)\sqrt{\pi_1^2+k^2(\sigma)}+ik(\sigma)\varphi(\sigma)\right)d\sigma} \, \mathcal{D}k(\sigma), \tag{A9}$$

where only negative frequency solutions are taken and $\mathcal{D}k(\sigma)$ denotes a functional integration over $k(\sigma)$. The scalar product has a form [17,18,101]

$$<\Psi|\Psi> = iZ \prod_\sigma \int \left(\Psi^*[\alpha,\varphi] \hat{D}^{-1/2}(\sigma) \frac{\delta}{\delta\alpha(\sigma)} \Psi[\alpha,\varphi] \right.$$
$$\left. - \left(\hat{D}^{-1/2}(\sigma) \frac{\delta}{\delta\alpha(\sigma)} \Psi^*[\alpha,\varphi] \right) \Psi[\alpha,\varphi] \right) d\varphi(\sigma), \tag{A10}$$

where $\hat{D}(\sigma) = -\frac{\delta^2}{\delta\varphi^2(\sigma)} + \pi_1^2$ and Z is a normalization constant. The infinite product is taken over σ-points, and to be understood in a formal sense as representing the result of a limiting process based on a lattice in σ-space. The scalar product (A10) is independent of the choice of the hyperplane $\alpha(\sigma)$.

The mean value of an arbitrary operator can be evaluated as

$$<\Psi|\hat{A}[\alpha, -i\frac{\delta}{\delta\varphi(\sigma)}, \varphi(\sigma)]|\Psi> = iZ \prod_\sigma \int \left(\Psi^*[\alpha,\varphi] \hat{A}\, \hat{D}^{-1/2}(\sigma) \frac{\delta}{\delta\alpha(\sigma)} \Psi[\alpha,\varphi] \right.$$
$$\left. - \left(\hat{D}^{-1/2}(\sigma) \frac{\delta}{\delta\alpha(\sigma)} \Psi^*[\alpha,\varphi] \right) \hat{A}\, \Psi[\alpha,\varphi] \right) d\varphi(\sigma) \Bigg|_{\alpha(\sigma)=\alpha_0 \to -\infty}. \tag{A11}$$

Let us note that the hyperplane $\alpha(\sigma) = \alpha_0$ along which the integration is performed in (A11), is the same as it is used as an initial condition for the quasi-Heisenberg operator $\hat{\alpha}$ in (17). In a more convenient momentum representation $\hat{\pi}_2(\sigma) = k(\sigma)$, $\hat{\varphi}_2(\sigma) = i\frac{\delta}{\delta k(\sigma)}$, the wave function ψ is

$$\psi[\alpha,k] = C[k] \exp\left(-i \int \alpha(\sigma) \sqrt{k^2(x) + \pi_1^2} \, dx \right). \tag{A12}$$

Then, the mean value of an operator becomes

$$<\psi|\hat{A}[\alpha(\sigma), k(\sigma), i\frac{\delta}{\delta k(\sigma)}]|\psi> =$$
$$\int C^*[k] e^{-i \int \alpha(\sigma)\sqrt{k^2(\sigma)+\pi_1^2}\, d\sigma} \hat{A} \, e^{i \int \alpha(\sigma)\sqrt{k^2(\sigma)+\pi_1^2}\, d\sigma} C[k] \, \mathcal{D}k(\sigma) \Bigg|_{\alpha(\sigma)=\alpha_0 \to -\infty}. \tag{A13}$$

Thus, we have an exact quantization scheme consisting of the Wheeler–DeWitt equation in the vicinity of small scale-factor (A8), the operator initial conditions (A5) for the equations of motion and the expressions (A11) and (A13) for calculation of the mean values of operators.

Notes

1. See [27,28] for EP historical and philosophical overview, and [29] for compatibility of EP with QFT.
2. It should be noted that a mutual cancellation of the bosonic N_{boson} and fermionic N_{boson} degrees of freedom removes all the vacuum energy but demands exact supersymmetry, which was not observed to date [33].
3. However, when one compares $\Omega_b = \frac{8\pi G \rho_b}{3H^2}$ from nucleosynthesis with Ω_m from cosmological observations, the result could depend on the possible renormalization of the gravitational constant [56]. Then, the gravitational constant measured on the Earth or the solar system can differ from the constant used in cosmology for the uniform universe.
4. In ΛCDM, the recombination turns out to be almost instantaneous, i.e., the last scattering surface is very thin.
5. The event horizon is a region of space-time that is causality disjointed from the rest of space-time.
6. The observations revealed the phenomena such as ultra-speed star motion, accretion disks around the super-massive and extremely compact objects (e.g., see [68,69]), and gravitational waves from colliding compact objects of stellar mass [70], which fit well in the black hole concept. However, the claims about "black hole discovery" should be treated with caution because

⁷ these observations do not rule out completely the alternative theories (e.g., see [71]), which also admit the existence of extremely compact massive objects with the exterior mimicking a black hole.

⁷ Here, we obtain primitive geometrical formulas connecting the radius of a compact astrophysical object with its mass and density. To obtain nontrivial formulas expressing the radius of the object through its mass only, using the physical equation of state is needed, e.g., nucleonic matter or strange quark matter as it was done in the neutron star physics [73].

⁸ It could be compared with properties of neutron and exotic stars [75].

References

1. Schaffner, K.F. *Nineteenth-Century Aether Theories*; Pergamon: New York, NY, USA, 1972.
2. Berestetskii, V.B.; Landau, L.; Lifshitz, E.; Pitaevskii, L.P. *Quantum Electrodynamics*; Butterworth-Heinemann: Oxford, UK, 1982.
3. Klimchitskaya, G.L.; Mostepanenko, V.M. Experiment and theory in the Casimir effect. *Contemp. Phys.* **2006**, *47*, 131–144. [CrossRef]
4. Mattingly, D. Modern tests of Lorentz invariance. *Liv. Rev. Rel.* **2005**, *8*, 1–84. [CrossRef] [PubMed]
5. Dirac, P.A.M. Is there an Æther? *Nature* **1951**, *168*, 906–907. [CrossRef]
6. Collins, J.; Perez, A.; Sudarsky, D. Lorentz invariance violation and its role in quantum gravity phenomenology. *arXiv* **2006**, arXiv:hep-th/0603002.
7. Jacobson, T. Einstein-Aether gravity: A status report. *Proc. Sci.* **2008**, *QG-Ph*, 20. [CrossRef]
8. Horava, P. Quantum gravity at a Lifshitz point. *Phys. Rev. D* **2009**, *79*, 084008. [CrossRef]
9. Nilsson, N.A.; Czuchry, E. Horava-Lifshitz cosmology in light of new data. *Phys. Dark Univ.* **2019**, *23*, 100253. [CrossRef]
10. Nilsson, N.A. Aspects of Lorentz and CPT Violation in Cosmology. Ph.D. Thesis, National Centre for Nuclear Research, Otwock-Świerk, Poland, 2020.
11. Tureanu, A. CPT and Lorentz Invariance: Their Relation and Violation. *J. Phys. Conf. Ser.* **2013**, *474*, 012031. [CrossRef]
12. Cherkas, S.L.; Batrakov, K.G.; Matsukevich, D. Testing of CP, CPT, and causality violation with light propagation in vacuum in the presence of uniform electric and magnetic fields. *Phys. Rev. D* **2002**, *66*, 065011. [CrossRef]
13. Kostelecky, V.A.; Russell, N. Data tables for Lorentz and CPT violation. *Rev. Mod. Phys.* **2011**, *83*, 11–31. [CrossRef]
14. Kostelecky, V.A.; Mewes, M. Lorentz and diffeomorphism violations in linearized gravity. *Phys. Lett. B* **2018**, *779*, 136–142. [CrossRef]
15. Cherkas, S.L.; Kalashnikov, V.L. An approach to the theory of gravity with an arbitrary reference level of energy density. *Proc. Natl. Acad. Sci. Belarus Ser. Phys.-Math.* **2019**, *55*, 83–96. [CrossRef]
16. Amelino-Camelia, G.; Ellis, J.; Mavromatos, N.; Nanopoulos, D.V.; Sarkar, S. Tests of quantum gravity from observations of γ-ray bursts. *Nature* **1998**, *393*, 763–765. [CrossRef]
17. Cherkas, S.L.; Kalashnikov, V.L. An inhomogeneous toy model of the quantum gravity with the explicitly evolvable observables. *Gen. Rel. Grav.* **2012**, *44*, 3081–3102. [CrossRef]
18. Cherkas, S.L.; Kalashnikov, V.L. Quantum evolution of the Universe in the constrained quasi-Heisenberg picture: From quanta to classics? *Grav. Cosmol.* **2006**, *12*, 126–129. [CrossRef]
19. Mizner, C.W. A minisuperspace Example: The Gowdy T3 Cosmology. *Phys. Rev. D* **1973**, *8*, 3271–3285. [CrossRef]
20. Green, M.B.; Schwarz, J.; Witten, E. *Superstring Theory*; Cambridge University Press: Cambridge, UK, 1987; Volume 1.
21. Kaku, M. *Introduction to Superstrings*; Springer: New York, NY, USA, 2012.
22. Kiritsis, E. *String Theory in a Nutshell*; Princeton University Press: Princeton, NJ, USA, 2019.
23. Anischenko, S.; Cherkas, S.; Kalashnikov, V. Functional minimization method addressed to the vacuum finding for an arbitrary driven quantum oscillator. *Nonlin. Phenom. Compl. Syst.* **2009**, *12*, 16–26.
24. Unruh, W.G. Notes on black-hole evaporation. *Phys. Rev. D* **1976**, *14*, 870–892. [CrossRef]
25. Akhmedov, E.T.; Singleton, D. On the physical meaning of the Unruh effect. *JETP Lett.* **2008**, *86*, 615–619. [CrossRef]
26. Arnowitt, R.; Deser, S.; Misner, C.W. Republication of: The dynamics of general relativity. *Gen. Rel. Grav.* **2008**, *40*, 1997–2027. [CrossRef]
27. Lehmkuhl, D. The Equivalence Principle(s). 2019. Available online: http://philsci-archive.pitt.edu/17709/ (accessed on 12 August 2021).
28. Knox, E. Effective spacetime geometry. *Stud. Hist. Philos. Mod. Phys.* **2013**, *44*, 346–356. [CrossRef]
29. Shalyt-Margolin, A. The Quantum Field Theory Boundaries Applicability and Black Holes Thermodynamics. *Int. J. Theor. Phys.* **2021**, *60*, 1858–1869. [CrossRef]
30. Pauli, W. *Pauli Lectures on Physics: Vol 6, Selected Topics in Field Quantization*; MIT Press: Cambridge, MA, USA, 1971.
31. Visser, M. Lorentz Invariance and the Zero-Point Stress-Energy Tensor. *Particles* **2018**, *1*, 138–154. [CrossRef]
32. Blinnikov, S.I.; Dolgov, A.D. Cosmological acceleration. *Phys. Usp.* **2019**, *62*, 529. [CrossRef]
33. Autermann, C. Experimental status of supersymmetry after the LHC Run-I. *Progr. Part Nucl. Phys.* **2016**, *90*, 125–155. [CrossRef]
34. Cherkas, S.; Kalashnikov, V. Dark-Energy-Matter from Vacuum owing to the General Covariance Violation. *Nonlin. Phenom. Complex Syst.* **2020**, *23*, 332–337. [CrossRef]
35. Kowalski-Glikman, J.; Nowak, S. Non-commutative space-time of Doubly Special Relativity theories. *Int. J. Mod. Phys. D* **2003**, *12*, 299–315. [CrossRef]

36. Pachol, A. Short review on noncommutative spacetimes. *J. Phys. Conf. Ser.* **2013**, *442*, 012039. [CrossRef]
37. Mir-Kasimov, R.M. Noncommutative space-time and relativistic dynamics. *Phys. Part. Nucl.* **2017**, *48*, 309–318. [CrossRef]
38. Dodelson, S. *Modern Cosmology*; Elsevier: Amsterdam, The Netherlands, 2003.
39. Consoli, M.; Pluchino, A. The CMB, Preferred Reference System, and Dragging of Light in the Earth Frame. *Universe* **2021**, *7*. [CrossRef]
40. Cherkas, S.L.; Kalashnikov, V.L. Determination of the UV cut-off from the observed value of the Universe acceleration. *JCAP* **2007**, *1*, 28. [CrossRef]
41. Cherkas, S.L.; Kalashnikov, V.L. Universe driven by the vacuum of scalar field: VFD model. In Proceedings of the International Conference "Problems of Practical Cosmology", Saint Petersburg, Russia, 23–27 June 2008; pp. 135–140.
42. Haridasu, B.S.; Cherkas, S.L.; Kalashnikov, V.L. A reference level of the Universe vacuum energy density and the astrophysical data. *Fortschr. Phys.* **2020**, *68*, 2000047. [CrossRef]
43. Perlmutter, S. Supernovae, Dark Energy, and the Accelerating Universe. *Phys. Today* **2003**, *56*, 53. [CrossRef]
44. Sultana, J. The $Rh = ct$ universe and quintessence. *MNRAS* **2016**, *457*, 212–216. [CrossRef]
45. Klinkhamer, F.R.; Wang, Z.L. Instability of the big bang coordinate singularity in a Milne-like universe. *arXiv* **2019**, arXiv:gr-qc/1911.11116.
46. Wan, H.Y.; Cao, S.L.; Melia, F.; Zhang, T.J. Testing the Rh=ct universe jointly with the redshift-dependent expansion rate and angular-diameter and luminosity distances. *Phys. Dark Univ.* **2019**, *26*, 100405. [CrossRef]
47. John, M.V. $Rh = ct$ and the eternal coasting cosmological model. *MNRAS Lett.* **2019**, *484*, L35–L37. [CrossRef]
48. Manfredi, G.; Rouet, J.L.; Miller, B.N.; Chardin, G. Structure formation in a Dirac-Milne universe: Comparison with the standard cosmological model. *Phys. Rev. D* **2020**, *102*, 103518. [CrossRef]
49. Lewis, G.F.; Barnes, L.A. The one-way speed of light and the Milne universe. *Publ. Astron. Soc. Aust.* **2021**, *38*, e007. [CrossRef]
50. Chardin, G.; Dubois, Y.; Manfredi, G.; Miller, B.; Stahl, C. MOND-like behavior in the Dirac–Milne universe. *Astron. Astrophys.* **2021**, *652*, A91. [CrossRef]
51. Batra, A.; Lohiya, D.; Mahajan, S.; Mukherjee, A.; Ashtekar, A. Nucleosynthesis in a Universe with a Linearly Evolving Scale Factor. *Int. J. Mod. Phys. D* **2000**, *9*, 757–773. [CrossRef]
52. Singh, G.; Lohiya, D. Inhomogeneous nucleosynthesis in linearly coasting cosmology. *MNRAS* **2018**, *473*, 14–19. [CrossRef]
53. Lewis, G.; Barnes, L.; Kaushik, R. Primordial Nucleosynthesis in the $R_h = ct$ cosmology: Pouring cold water on the Simmering Universe. *MNRAS* **2016**, *460*, stw1003. [CrossRef]
54. Pitrou, C.; Coc, A.; Uzan, J.P.; Vangioni, E. Precision Big Bang Nucleosynthesis with the New Code PRIMAT. In Proceedings of the 15th International Symposium on Origin of Matter and Evolution of Galaxies, Kyoto University, Kyoto, Japan, 2–5 July 2019; p. 011034. [CrossRef]
55. Pitrou, C.; Coc, A.; Uzan, J.P.; Vangioni, E. A new tension in the cosmological model from primordial deuterium? *MNRAS* **2021**, *502*, 2474–2481. [CrossRef]
56. Cherkas, S.L.; Kalashnikov, V.L. Vacuum Polarization Instead of Dark Matter in a Galaxy. *Universe* **2022**, *8*, 456. [CrossRef]
57. Cherkas, S.L.; Kalashnikov, V.L. The equation of vacuum state and the structure formation in universe. *Vestnik Brest Univ. Ser. Fiz.-Mat.* **2021**, *1*, 41–59.
58. Cherkas, S.L.; Kalashnikov, V.L. Eicheons instead of Black holes. *Phys. Scr.* **2020**, *95*, 085009. [CrossRef]
59. Carr, B.; Kuhnel, F. Primordial black holes as dark matter candidates. *SciPost Phys. Lect. Notes* **2022** . [CrossRef]
60. Cassisi, S.; Castellani, V. An Evolutionary Scenario for Primeval Stellar Populations. *Astrophys. J. Suppl.* **1993**, *88*, 509. [CrossRef]
61. Coc, A.; Uzan, J.P.; Vangioni, E. Standard big bang nucleosynthesis and primordial CNO abundances after Planck. *JCAP* **2014**, *2014*, 050. [CrossRef]
62. Woosley, S. Neutrino-induced nucleosynthesis and deuterium. *Nature* **1977**, *269*, 42–44. [CrossRef]
63. Jedamzik, K. Cosmological deuterium production in non-standard scenarios. *Planet. Space Sci.* **2002**, *50*, 1239–1244. [CrossRef]
64. Cherkas, S.L.; Kalashnikov, V.L. Plasma perturbations and cosmic microwave background anisotropy in the linearly expanding Milne-like universe. In *Fractional Dynamics, Anomalous Transport and Plasma Science*; Skiadas, C.H., Ed.; Springer: Cham, Switzerland, 2018; Chapter 9. [CrossRef]
65. Fujii, H. Inconsistency of the Rh = ct Cosmology from the Viewpoint of the Redshift of the Cosmic Microwave Background Radiation. *Res. Notes AAS* **2020**, *4*, 72. [CrossRef]
66. Tutusaus, I.; Lamine, B.; Blanchard, A.; Dupays, A.; Zolnierowski, Y.; Cohen-Tanugi, J.; Ealet, A.; Escoffier, S.; Le Fèvre, O.; Ilic, S.; et al. Power law cosmology model comparison with CMB scale information. *Phys. Rev. D* **2016**, *94*, 103511. [CrossRef]
67. Lewis, A. Linear effects of perturbed recombination. *Phys. Rev. D* **2007**, *76*, 063001. [CrossRef]
68. Gillessen, S.; Eisenhauer, F.; Trippe, S.; Alexander, T.; Genzel, R.; Martins, F.; Ott, T. Monitoring stellar orbits around the Massive Black Hole in the Galactic Center. *Astrophys. J.* **2009**, *692*, 1075. [CrossRef]
69. Akiyama, K.; Algaba, J.C.; Alberdi, A.; Alef, W.; Anantua, R.; Asada, K.; Azulay, R.; Baczko, A.K.; Ball, D.; Balokovic, M.; et al. First M87 event horizon telescope results. VII. Polarization of the ring. *Astrophys. J. Lett.* **2021**, *910*, L12. [CrossRef]
70. Abbott, B.P.; Abbott, R.; Abbott, T.D.; Abernathy, M.R.; Acernese, F.; Ackley, K.; Adams, C.; Adams, T.; Addesso, P.; Adhikari, R.X.; et al. Properties of the binary black hole merger GW150914. *Phys. Rev. Lett.* **2016**, *116*, 241102. [CrossRef] [PubMed]
71. Konoplya, R.; Zhidenko, A. Detection of gravitational waves from black holes: Is there a window for alternative theories? *Phys. Lett. B* **2016**, *756*, 350–353. [CrossRef]

72. Weinberg, S. *Gravitation and Cosmology: Principles and Applications of the General Theory of Relativity*; John Wiley & Sons: New York, NY, USA, 1972.
73. Lattimer, J.M.; Prakash, M. Neutron Star Structure and the Equation of State. *Astrophys. J.* **2001**, *550*, 426–442. [CrossRef]
74. Barrau, A.; Rovelli, C. Planck star phenomenology. *Phys. Lett. B* **2014**, *739*, 405–409. [CrossRef]
75. Sandin, F. Compact stars in the standard model-and beyond. In *How and Where to Go beyond the Standard Model*; World Scientific: Singapore, 2007; pp. 411–420.
76. Bambi, C.; Cárdenas-Avendaño, A.; Dauser, T.; García, J.A.; Nampalliwar, S. Testing the Kerr Black Hole Hypothesis Using X-ray Reflection Spectroscopy. *Astrophys. J.* **2017**, *842*, 76. [CrossRef]
77. Agullo, I.; Cardoso, V.; del Rio, A.; Maggiore, M.; Pullin, J. Potential Gravitational Wave Signatures of Quantum Gravity. *Phys. Rev. Lett.* **2021**, *126*, 041302. [CrossRef] [PubMed]
78. Dong, R.; Stojkovic, D. Gravitational wave echoes from black holes in massive gravity. *Phys. Rev. D* **2021**, *103*, 024058. [CrossRef]
79. Cherkas, S.L.; Kalashnikov, V.L. Wave optics of quantum gravity for massive particles. *Phys. Scr.* **2021**, *96*, 115001. [CrossRef]
80. Petruzziello, L.; Illuminati, F. Quantum gravitational decoherence from fluctuating minimal length and deformation parameter at the Planck scale. *Nat. Commun.* **2021**, *12*, 4449. [CrossRef]
81. Belenchia, A.; Wald, R.M.; Giacomini, F.; Castro-Ruiz, E.; Brukner, C.; Aspelmeyer, M. Quantum superposition of massive objects and the quantization of gravity. *Phys. Rev. D* **2018**, *98*, 126009. [CrossRef]
82. Carney, D. Newton, entanglement, and the graviton. *Phys. Rev. D* **2022**, *105*, 024029. [CrossRef]
83. Danielson, D.L.; Satishchandran, G.; Wald, R.M. Gravitationally mediated entanglement: Newtonian field versus gravitons. *Phys. Rev. D* **2022**, *105*, 086001. [CrossRef]
84. Christodoulou, M.; Di Biagio, A.; Aspelmeyer, M.; Brukner, C.; Rovelli, C.; Howl, R. Locally mediated entanglement through gravity from first principles. *arXiv* **2022**, arXiv:quant-ph/2202.03368.
85. Krisnanda, T.; Tham, G.Y.; Paternostro, M.; Paterek, T. Observable quantum entanglement due to gravity. *NPJ Quantum Inf.* **2020**, *6*, 1–6. [CrossRef]
86. Gorbatsievich, A.K. *Quantum Mechanics in General Relativity Theory: Basic Principles and Elementary Applications*; BSU: Minsk, Belarus, 1985. Available online: https://scholar.google.ru/citations?view_op=view_citation&hl=ru&user=RTFZL9sAAAAJ&citation_for_view=RTFZL9sAAAAJ:u-x6o8ySG0sC (accessed on 3 November 2021).
87. Martinez-Huerta, H.; Lang, R.G.; de Souza, V. Lorentz Invariance Violation Tests in Astroparticle Physics. *Symmetry* **2020**, *12*, 1232. [CrossRef]
88. Wei, J.J.; Wu, X.F. Tests of Lorentz Invariance. *arXiv* **2021**, arXiv:astro-ph/2111.02029. Available online: https://arxiv.org/abs/2111.02029 (accessed on 3 November 2021).
89. Gonzalez-Diaz, P.; Garay, L. Quantum Closed Timelike Curves in General Relativity. In *Recent Developments in General Relativity*; Cianci, R., Collina, R., Francaviglia, M., Fre, P., Eds.; Springer: Genoa, Italy, 2002; pp. 459–463. [CrossRef]
90. Bini, D.; Geralico, A. On the occurrence of Closed Timelike Curves and the observer's point of view. *Eur. Phys. J. Web Conf.* **2013**, *58*, 01002. [CrossRef]
91. Faizuddin, A. Type III spacetime with closed timelike curves. *Progr. Phys.* **2016**, *12*, 329–331.
92. Novikov, I. An analysis of the operation of a time machine. *JETP* **1989**, *95*, 439–443.
93. Wuthrich, C. Time travelling in emergent spacetime. *arXiv* **2019**, arXiv:1907.11167.
94. Teo, E. Rotating traversable wormholes. *Phys. Rev. D* **1998**, *58*, 024014. [CrossRef]
95. Herrero-Valea, M.; Liberati, S.; Santos-Garcia, R. Hawking radiation from universal horizons. *JHEP* **2021**, *2021*, 1–32. [CrossRef]
96. Coogan, A.; Morrison, L.; Profumo, S. Direct Detection of Hawking Radiation from Asteroid-Mass Primordial Black Holes. *Phys. Rev. Lett.* **2021**, *126*, 171101. [CrossRef]
97. Saraswat, K.; Afshordi, N. Extracting Hawking radiation near the horizon of AdS black holes. *JHEP* **2021**, *2021*, 1–47. [CrossRef]
98. Cherkas, S.L.; Kalashnikov, V.L. Cosmological singularity as an informational seed for Everything. *Nonlin. Phenom. Complex Syst.* **2022**, *25*, 266–275. [CrossRef]
99. Burdík, Č.; Navrátil, O. Dirac formulation of free open string. *Univ. J. Phys. Appl.* **2007**, *4*, 487–506. [CrossRef]
100. Cherkas, S.L.; Kalashnikov, V.L. Quantization of the inhomogeneous Bianchi I model: Quasi-Heisenberg picture. *Nonlin. Phenom. Complex Syst.* **2015**, *18*, 1–14.
101. Cherkas, S.L.; Kalashnikov, V.L. Quantum evolution of the Universe from $\tau = 0$ in the constrained quasi-Heisenberg picture. In Proceedings of the VIIIth International School-seminar "The Actual Problems of Microworld Physics", Gomel, Belarus, 25 July–5 August 2007; JINR: Dubna, Russia, 2007; Volume 1, p. 208.

Article

Path Integral Action for a Resonant Detector of Gravitational Waves in the Generalized Uncertainty Principle Framework

Soham Sen [1,*], Sukanta Bhattacharyya [2] and Sunandan Gangopadhyay [1]

[1] S.N. Bose National Centre for Basic Sciences, Department of Astrophysics and High Energy Physics, JD Block, Sector III, Salt Lake, Kolkata 700106, India
[2] Department of Physics, West Bengal State University, Barasat, Kolkata 700126, India
* Correspondence: sensohomhary@gmail.com

Abstract: The Heisenberg uncertainty principle is modified by the introduction of an observer-independent minimal length. In this work, we have considered the resonant gravitational wave detector in the modified uncertainty principle framework, where we have used the position momentum uncertainty relation with a quadratic order correction only. We have then used the path integral approach to calculate an action for the bar detector in the presence of a gravitational wave and then derived the Lagrangian of the system, leading to the equation of motion for the configuration-space position coordinate in one dimension. We then find a perturbative solution for the coordinate of the detector for a circularly polarized gravitational wave, leading to a classical solution of the same for the given initial conditions. Using this classical form of the coordinate of the detector, we finally obtain the classical form of the on-shell action describing the harmonic oscillator–gravitational wave system. Finally, we have obtained the free particle propagator containing the quantum fluctuation term considering gravitational wave interaction.

Keywords: resonant bar detector; gravitational wave; generalized uncertainty principle; path integral

1. Introduction

Quantum mechanics and general relativity are the two most successful theories explaining the phenomena at the two most fundamental length scales of the universe. While quantum mechanics explains the intricacies of the atomic length scale, general relativity [1,2] sheds light on the large-scale structure of the universe. In order to understand the fundamental mysteries of the universe, we need a quantum theory of gravity, explaining the analytical structure of the gravitational interaction at the quantum length scale. Theories such as loop quantum gravity [3,4], string theory [5,6], and noncommutative geometry [7] have provided a convincing theoretical framework explaining the Planck-scale nature of gravity, but none of them have compelling experimental evidence to support their claim of providing an exact description of the quantum nature of gravity. Meanwhile, all of them prescribe the existence of an observer-independent minimal length, which can be incorporated by the modification of the standard Heisenberg uncertainty principle (HUP), also known as the generalized uncertainty principle (GUP). The first few attempts to improvise an integral relation between minimal length scale and gravity was shown in [8,9], followed by [10]. We also obtain strong evidence of the existence of this fundamental length scale from the various gedanken experiments in quantum gravity phenomenology as well. This GUP framework has been used to investigate several areas of theoretical physics, including black hole physics and its thermodynamics [11–21], various quantum systems, such as particle in a box and simple harmonic oscillators [22,23], optomechanical systems [24–26], and gravitational wave bar detectors [27,28]. There have been several recent studies involving the path integral formalism of a non-relativistic particle moving in an arbitrary potential in the generalized uncertainty principle framework [27,29,30]. The simplest form of the modified HUP can be written in the following form [31]:

$$\Delta q_i \Delta p_i \geq \frac{\hbar}{2}\left[1 + \gamma\left(\Delta p^2 + \langle p \rangle^2\right) + 2\gamma\left(\Delta p_i^2 + \langle p_i \rangle^2\right)\right]; \; i = 1, 2, 3 \quad (1)$$

where $p^2 = \sum_{k=1}^{3} p_k p_k$ and q_k, p_k are the phase space position and its conjugate momenta. In Equation (1), the GUP parameter γ in terms of the dimensionless parameter γ_0 can be recast as follows:

$$\gamma = \frac{\gamma_0}{m_p^2 c^2} \quad (2)$$

where m_p is the Planck mass and c is the speed of light. It is quite natural to realize that the order of magnitude of the GUP parameter will play a significant role in providing an understanding of the GUP effects. There have been several studies to find a bound on the GUP parameter itself [17,22,28,32–38].

In 1969, the first proposition to detect gravitational waves was made by J. Weber [39], which was followed by a subsequent paper in 1982 by Ferrari et al. [40]. Bar detectors currently have a sensitivity $\frac{\Delta L}{L} \sim 10^{-19}$ [41], where ΔL is the fractional variation of the length L ($\sim$1 m) of the bar detector. A historical perspective on these resonant detectors is given in [42]. The detection of gravitational waves by the LIGO [43,44] and Virgo [45] detectors has unveiled a new realm of quantum gravity phenomenology. There have been several recent investigations regarding the traces of quantum gravitational effects in these gravitational wave detectors. A great deal of investigation has been conducted to check if any signature of this fundamental Planck length, whether it is noncommutativity [46–51] or GUP [28,52], is visible in GW bar detectors. We would like to point out that, to date, there has not been a successful detection of gravitational waves in resonant bar detectors. However, there is strong hope that the sensitivity of the detectors will increase in the future, enabling the detection of these waves. The AURIGA (*Antenna Ultracriogenica Risonante per l'Indagine Gravitazionale Astronomica*) detector at INFN, Italy is probably the only functional bar detector. These bar detectors are sensitive to frequencies of the order of 1kHz, along with a strain sensitivity of the order 10^{-19} [53]. In the case of astrophysical events, collapsing and bouncing cores of supernova can be a source of huge intensities of gravitational waves having frequencies in the vicinity of 1–3 kHz. The value of the strain sensitivity can be calculated using Thorne's formula [54]. The strain sensitivity (h), according to this formula, is given by

$$h = 2.7 \times 10^{-20} \left[\frac{\Delta E_{GW}}{M_s c^2}\right]^{\frac{1}{2}} \left[\frac{1 kHz}{f}\right]^{\frac{1}{2}} \left[\frac{10 Mpc}{d}\right] \quad (3)$$

where ΔE_{GW} is the energy converted to gravitational waves, f is the characteristic frequency of the burst, M_s is the solar mass, and d is the distance of the burst source from Earth. A possible value of the fraction of energy converted to gravitational waves for supernova events is around 7×10^{-4}. Now, for $h \sim 3 \times 10^{-19}$ and $f \sim 0.9$ kHz, the distance d has a value around 25 kpc. The occurrence of such a supernova event of the required magnitude at this distance from the Earth would definitely result in the detection of gravitational waves by the bar detectors. An effort to increase the sensitivity of these detectors to $h \sim 10^{-20}$ is presently being carried out, and achieving this sensitivity would increase the distance of the supernova event from the Earth to 250 kpc, which is more likely to occur. The main motivation to work with a gravitational wave bar detector is that it is a very useful and economic alternative to the LIGO/VIRGO detectors.

In this work, we investigate the path integral formalism of a resonant gravitational wave bar detector interacting with the gravitational wave emitted from a distant source in the GUP framework. The incoming gravitational waves interact with the elastic matter in the resonant bar detector, causing tiny vibrations called phonons. Physically, we can describe these detectors as a quantum mechanical gravitational wave–harmonic oscillator (GW-HO) system, because we call these vibrations the quantum mechanical forced harmonic oscillator. To calculate the perturbative solution to the system, we use the gravitational wave and generalized uncertainty modifications as perturbations. Our study presents a path integral approach to look at such a system and is the first work using a path integral. The advantage of working with path integrals is that the effective action describing the system can be easily read off from the structure of the configuration space path integral [55].

2. The Gravitational Wave Resonant Detector Interaction Model

To begin the discussion, we need to present the Hamiltonian for the resonant bar detector in the presence of a gravitational wave in the generalized uncertainty principle framework. The modified commutation relation following from Equation (1) takes the following form [31]:

$$[\hat{q}_i, \hat{p}_j] = i\hbar \left[\delta_{ij} + \gamma \delta_{ij} \hat{p}^2 + 2\gamma \hat{p}_i \hat{p}_j \right] \quad (4)$$

where $i, j = 1, 2, 3$. The modified position and momentum operators $\hat{q}_i$ and $\hat{p}_i$ in terms of the usual variables $\hat{q}_{0i}$ and $\hat{p}_{0i}$ read

$$\hat{q}_i = \hat{q}_{0i} , \quad \hat{p}_i = \hat{p}_{0i}\left(1 + \gamma \hat{p}_0^2\right). \quad (5)$$

Here, $\hat{p}_0^2 = \sum_{k=1}^{3} \hat{p}_{0k}\hat{p}_{0k}$ and $[\hat{q}_{0i}, \hat{p}_{0j}] = i\hbar \delta_{ij}$. In order to write the Hamiltonian of the system, we start by analyzing the background metric as a superposition of a small perturbation on the flat background metric. The background metric is taken as follows:

$$g_{\mu\nu} = \eta_{\mu\nu} + h_{\mu\nu} \quad (6)$$

where $\eta_{\mu\nu} = \text{diag}\{1, -1, -1, -1\}$ and $|h_{\mu\nu}| \ll 1$. We now consider a two-dimensional harmonic oscillator with mass m and intrinsic frequency ω. The geodesic deviation equation for the aforementioned system in the proper detector frame is given as follows [56]:

$$m\ddot{q}^k = -mR^k{}_{0l0}q^l - m\omega^2 q^k$$
$$\implies \ddot{q}^k = \frac{d\Gamma^k{}_{0l}}{dt}q^l - \omega^2 q^k \, ; \, k = 1, 2 \quad (7)$$

where $R^k{}_{0l0}$ in terms of the background perturbation is given by

$$R^k{}_{0l0} = -\frac{d\Gamma^k{}_{0l}}{dt} = -\frac{\ddot{h}_{kl}}{2}. \quad (8)$$

Note that, here, we are using the transverse traceless gauge to eliminate the unphysical degrees of freedom. The Lagrangian from which Equation (8) can be obtained reads

$$L = \frac{1}{2}m\dot{q}_k^2 - m\Gamma^k{}_{0l}\dot{q}_k q^l - \frac{1}{2}m\omega^2 q_k^2. \quad (9)$$

The Hamiltonian corresponding to the Lagrangian in Equation (9) reads

$$H = \frac{1}{2m}\left(p_k + m\Gamma^k{}_{0l}q^l\right)^2 + \frac{1}{2}m\omega^2 q_l^2. \quad (10)$$

To write the Hamiltonian in Equation (10) in quantum mechanical description, we simply elevate q and p to the operator prescription. Therefore, the Hamiltonian in terms of the position and momentum operators can be expressed as follows:

$$\hat{H} = \frac{1}{2m}\left(\hat{p}_k + m\Gamma^k{}_{0l}\hat{q}^l\right)^2 + \frac{1}{2}m\omega^2 \hat{q}_l^2. \quad (11)$$

Using the representation of the position and momentum operators in Equation (5), the Hamiltonian (11) of the GW-HO system in the presence of GUP can be written as follows:

$$\hat{H} = \left(\frac{\hat{p}_{0k}^2}{2m} + \frac{1}{2}m\omega^2\hat{q}_{0k}^2\right) + \frac{\gamma}{m}\hat{p}_{0k}^2\hat{p}_0^2 + \frac{1}{2}\Gamma^k_{0l}\left(\hat{p}_{0k}\hat{q}^{0l} + \hat{q}^{0l}\hat{p}_{0k}\right) + \frac{\gamma}{2}\Gamma^k_{0l}\left(\hat{p}_{0k}\hat{p}_0^2\hat{q}^{0l} + \hat{q}^{0l}\hat{p}_{0k}\hat{p}_0^2\right). \quad (12)$$

Now, a typical bar is a cylinder of length $L \equiv 3$ m and radius $R \equiv 30$ cm [56]. Hence, in a first approximation, we can treat the GW detector in the presence of GUP as a one-dimensional HO. The Hamiltonian in Equation (12) can be recast in one dimension as follows:

$$\hat{H} = \frac{p^2}{2m} + \frac{1}{2}m\omega^2 q^2 + \gamma\frac{p^4}{m} + \frac{1}{2}\Gamma^1_{01}(pq+qp) + \frac{\gamma}{2}\Gamma^1_{01}(p^3q+qp^3) \quad (13)$$

where, for notational simplicity, we have used $\hat{p}_{01} = p$ and $\hat{q}_{01} = q$. In the next section, we will proceed to construct the path integral formalism of the GW-HO system in the presence of the GUP and calculate the propagation kernel for that system.

3. Path Integral and the Propagation Kernel

In this section, we will use the Hamiltonian in Equation (13) to calculate the propagation kernel via the path integral approach. We consider the initial and the final state of the Hamiltonian in Equation (13) at initial time t_i and final time t_f as $|q_i, t_i\rangle$ and $|q_f, t_f\rangle$, respectively. The general form of the propagation kernel can be written as follows:

$$\langle q_f, t_f | q_i, t_i \rangle = \lim_{N\to\infty} \int_{-\infty}^{+\infty} dq_{N-1}\ldots dq_1 \langle q_f, t_f | q_{N-1}, t_{N-1} \rangle \langle q_{N-1}, t_{N-1} | q_{N-2}, t_{N-2} \rangle \ldots \langle q_1, t_1 | q_i, t_i \rangle$$

$$= \lim_{N\to\infty} \int_{-\infty}^{+\infty} \prod_{\alpha=1}^{N-1} dq_\alpha \langle q_f | e^{-\frac{i\hat{H}(t_f - t_{N-1})}{\hbar}} | q_{N-1} \rangle \ldots \langle q_1 | e^{-\frac{i\hat{H}(t_1 - t_i)}{\hbar}} | q_i \rangle \quad (14)$$

$$= \lim_{N\to\infty} \int_{-\infty}^{+\infty} \prod_{\alpha=1}^{N-1} dq_\alpha \prod_{\beta=0}^{N-1} \langle q_{\beta+1} | e^{-\frac{i\hat{H}(t_{\beta+1} - t_\beta)}{\hbar}} | q_\beta \rangle$$

where $t_f = t_N$, $t_i = t_0$ and $t_N - t_{N-1} = \Delta t$. Now, we will introduce the complete set of momentum eigenstates $\left(\int_{-\infty}^{+\infty} dp |p\rangle\langle p| = 1\right)$ in the following way:

$$\langle q_f, t_f | q_i, t_i \rangle = \lim_{N\to\infty} \int \prod_{\alpha=1}^{N-1} dq_\alpha \prod_{\beta=0}^{N-1} \int dp_\beta \langle q_{\beta+1} | p_\beta \rangle \langle p_\beta | q_\beta \rangle \exp\left(-\frac{iH(q_\beta, p_\beta)(t_{\beta+1} - t_\beta)}{\hbar}\right)$$

$$= \lim_{N\to\infty} \int_{-\infty}^{+\infty} \prod_{\alpha=1}^{N-1} dq_\alpha \prod_{\beta=0}^{N-1} \int_{-\infty}^{+\infty} \frac{dp_\beta}{2\pi\hbar} \exp\left[\frac{i\Delta t}{\hbar} \sum_{\beta=0}^{N-1} \left[\frac{p_\beta(q_{\beta+1} - q_\beta)}{\Delta t} - \left(\frac{p_\beta^2}{2m} + \frac{1}{2}m\omega^2 q_\beta^2 + \frac{\gamma p_\beta^4}{m}\right)\right.\right. \quad (15)$$

$$\left.\left. + \frac{p_\beta q_\beta(h_{\beta+1} - h_\beta)}{2\Delta t} + \frac{\gamma p_\beta^3 q_\beta(h_{\beta+1} - h_\beta)}{2\Delta t}\right)\right]$$

where we have used $h_{11} = h$. The final form of Equation (15) in the $\Delta t \to 0$ limit can be recast as follows:

$$\langle q_f, t_f | q_i, t_i \rangle = \int \mathcal{D}q\mathcal{D}p \exp\left(\frac{i}{\hbar}\mathcal{S}\right) \quad (16)$$

where $\mathcal{S}$ is the phase space action. The phase space action is given as follows:

$$\mathcal{S} = \int_{t_i}^{t_f} dt \left[p\dot{q} - \left(\frac{p^2}{2m} + \frac{\dot{h}_{11}}{2}pq + \frac{1}{2}m\omega^2 q^2 + \frac{\gamma p^4}{m} + \frac{\gamma \dot{h}_{11}}{2}p^3 q\right)\right]. \quad (17)$$

To obtain the configuration space Lagrangian, we will simplify Equation (15) as follows:

$$\langle q_f, t_f | q_i, t_i \rangle \cong \lim_{N \to \infty} \int_{-\infty}^{+\infty} \prod_{\alpha=1}^{N-1} dq_\alpha \prod_{\beta=0}^{N-1} \int_{-\infty}^{+\infty} \frac{dp_\beta}{2\pi\hbar} \left[1 - \frac{i\gamma\Delta t}{m\hbar} \left(p_\beta^4 + \frac{h_{\beta+1} - h_\beta}{2\Delta t} p_\beta^3 q_\beta \right) + \mathcal{O}(\gamma^2) \right] \quad (18)$$

$$\times \exp\left[\frac{i\Delta tm}{2\hbar} \left[\left(\frac{q_{\beta+1} - q_\beta}{\Delta t} - \frac{h_{\beta+1} - h_\beta}{4\Delta t} q_\beta \right)^2 - \omega^2 q_\beta^2 \right] \right] \exp\left[-\frac{i\Delta t}{2m\hbar} \left[p_\beta - \left(\frac{m(q_{\beta+1} - q_\beta)}{\Delta t} - \frac{m(h_{\beta+1} - h_\beta)q_\beta}{4\Delta t} \right) \right]^2 \right].$$

To perform the momentum integral for each β value, we shall perform the following coordinate transformation:

$$\bar{p}_\beta = p_\beta - \left(\frac{m(q_{\beta+1} - q_\beta)}{\Delta t} - \frac{m(h_{\beta+1} - h_\beta)q_\beta}{4\Delta t} \right). \quad (19)$$

Using Equation (19) in Equation (18), the propagation kernel up to $\sim \gamma, h$ can be recast as

$$\langle q_f, t_f | q_i, t_i \rangle \cong \lim_{N \to \infty} \int_{-\infty}^{+\infty} \prod_{\alpha=1}^{N-1} dq_\alpha \prod_{\beta=0}^{N-1} \int_{-\infty}^{+\infty} \frac{d\bar{p}_\beta}{2\pi\hbar} \left[1 - \frac{i\gamma\Delta t}{m\hbar} \left[\left(\bar{p}_\beta + \left(\frac{m(q_{\beta+1} - q_\beta)}{\Delta t} - \frac{m(h_{\beta+1} - h_\beta)q_\beta}{4\Delta t} \right) \right)^4 + \right. \right.$$
$$\left. \left. \frac{h_{\beta+1} - h_\beta}{2\Delta t} \left(\bar{p}_\beta + \left(\frac{m(q_{\beta+1} - q_\beta)}{\Delta t} - \frac{m(h_{\beta+1} - h_\beta)q_\beta}{4\Delta t} \right) \right)^3 q_\beta \right] + \mathcal{O}(\gamma^2) \right] \exp\left[-\frac{i\Delta t}{2m\hbar} \bar{p}_\beta^2 \right] \quad (20)$$
$$\times \exp\left[\frac{i\Delta tm}{2\hbar} \left[\left(\frac{q_{\beta+1} - q_\beta}{\Delta t} - \frac{h_{\beta+1} - h_\beta}{4} q_\beta \right)^2 - \omega^2 q_\beta^2 \right] \right].$$

The momentum integral in Equation (20) can be obtained as follows:

$$\langle q_{\beta+1}, t_{\beta+1} | q_\beta, t_\beta \rangle \cong \sqrt{\frac{m}{2\pi i \hbar \Delta t}} \left\{ 1 - 6\gamma m^2 \left(\frac{q_{\beta+1} - q_\beta}{\Delta t} - \frac{h_{\beta+1} - h_\beta}{4\Delta t} q_\beta \right)^2 - \frac{3\gamma m^2 (h_{\beta+1} - h - \beta)}{2\Delta t} \left(\frac{q_{\beta+1} - q_\beta}{\Delta t} \right) \right.$$
$$\left. - \frac{(h_{\beta+1} - h_\beta)}{4\Delta t} q_\beta \right) q_\beta + \frac{3i\gamma m\hbar}{\Delta t} \right\} \exp\left[\frac{im\Delta t}{2\hbar} \left[\left(\frac{q_{\beta+1} - q_\beta}{\Delta t} - \frac{(h_{\beta+1} - h_\beta)q_\beta}{4} \right)^2 - 2\gamma m^2 \right. \right. \quad (21)$$
$$\left. \left. \times \left[\left(\frac{q_{\beta+1} - q_\beta}{\Delta t} - \frac{(h_{\beta+1} - h_\beta)q_\beta}{4} \right)^4 + \frac{(h_{\beta+1} - h_\beta)q_\beta}{2} \left(\frac{q_{\beta+1} - q_\beta}{\Delta t} - \frac{(h_{\beta+1} - h_\beta)q_\beta}{4} \right)^3 \right] - \omega^2 q_\beta^2 \right] \right].$$

Using Equation (21) in Equation (20), we obtain the form of the propagation kernel up to some constant factor as follows:

$$\langle q_f, t_f | q_i, t_i \rangle = \int_{-\infty}^{+\infty} \prod_{\alpha=1}^{N-1} dq_\alpha \exp\left[\sum_{\beta=0}^{N-1} \frac{im\Delta t}{2\hbar} \left\{ \left(\frac{q_{\beta+1} - q_\beta}{\Delta t} - \frac{(h_{\beta+1} - h_\beta)q_\beta}{4} \right)^2 - 2\gamma m^2 \left(\left(\frac{q_{\beta+1} - q_\beta}{\Delta t} \right) \right. \right. \right.$$
$$\left. \left. \left. - \frac{(h_{\beta+1} - h_\beta)q_\beta}{4} \right)^4 + \frac{(h_{\beta+1} - h_\beta)q_\beta}{2} \left(\frac{q_{\beta+1} - q_\beta}{\Delta t} - \frac{(h_{\beta+1} - h_\beta)q_\beta}{4} \right)^3 \right) - \omega^2 q_\beta^2 \right\} \right]. \quad (22)$$

Imposing the $\Delta t \to 0$ limit in Equation (22), the final form of the propagation kernel has the usual configuration space path integral structure as follows:

$$\langle q_f, t_f | q_i, t_i \rangle = \mathcal{N}(T, \gamma, \dot{h}) \int \mathcal{D}q e^{\frac{i}{\hbar}S}. \quad (23)$$

In the above equation, the configuration space structure of the action S is given as follows:

$$S = \int_{t_i}^{t_f} dt \left(\frac{m}{2} \left(\dot{q} - \frac{\dot{h}q}{4} \right)^2 - \frac{1}{2} m\omega^2 q^2 - \gamma m^3 \left(\dot{q} - \frac{\dot{h}q}{4} \right)^4 - \frac{\gamma m^3 \dot{h}q}{2} \left(\dot{q} - \frac{\dot{h}q}{4} \right)^3 \right)$$
$$\cong \int_{t_i}^{t_f} dt \left(\frac{m}{2} \dot{q}^2 - \frac{1}{2} m\omega^2 q^2 - \frac{m\dot{h}\dot{q}q}{4} - \gamma m^3 \dot{q}^4 + \frac{1}{2} m^3 \gamma \dot{h} \dot{q}^3 q \right). \quad (24)$$

In the last line of the above Equation (24), we have kept terms up to $\mathcal{O}(h,\gamma)$. The Lagrangian can be easily read off from Equation (24) as follows:

$$L = \frac{m}{2}\dot{q}^2 - \frac{1}{2}m\omega^2 q^2 - \frac{m\dot{h}\dot{q}q}{4} - \gamma m^3 \dot{q}^4 + \frac{1}{2}\gamma m^3 h\dot{q}^3 q. \tag{25}$$

The equation of motion following from the Lagrangian reads

$$\ddot{q} - \frac{\ddot{h}q}{4} + \omega^2 q - 12m^2\gamma\ddot{q}\dot{q}^2 + 3\gamma m^2 \ddot{h}\dot{q}\dot{q} + \frac{3}{2}\gamma m^2 \ddot{h}\dot{q}^2 q + \gamma m^2 h\dot{q}^3 = 0. \tag{26}$$

In the next section, we calculate the classical solution for the above equation of motion.

4. Obtaining the Classical Solution for a Periodic Circularly Polarized Gravitational Wave

To obtain the classical solution, we shall consider a circularly polarized gravitational wave in the transverse traceless gauge. Now, for a periodic circularly polarized gravitational wave, the perturbation term h containing the polarization information reads

$$h_{kl}(t) = 2f_0\left(\varepsilon_\times(t)\sigma^1_{kl} + \varepsilon_+(t)\sigma^3_{kl}\right); \; k,l = 1,2 \tag{27}$$

where $2f_0$ is the amplitude of the gravitational wave (here, f_0 is very small), and σ^1 and σ^3 are the Pauli spin matrices. In Equation (27), $(\varepsilon_+(t), \varepsilon_\times(t))$ are the two possible polarization states of the gravitational wave satisfying the condition $\varepsilon_+(t)^2 + \varepsilon_\times(t)^2 = 1$. In this particular scenario, the chosen functional forms of the polarization states can be written as follows:

$$\varepsilon_+(t) = \cos(\Omega t), \; \varepsilon_\times(t) = \sin(\Omega t) \tag{28}$$

with Ω being the frequency of the gravitational wave. In our case, we will consider that the only non-zero polarization state is $\varepsilon_+(t) = \cos(\Omega t)$. Therefore, in one dimension, the perturbation term can be written as $h = 2f_0\cos(\Omega t)$. The equation of motion in Equation (26) up to $\mathcal{O}(f_0, \gamma)$ takes the form as follows:

$$\ddot{q} + \omega^2 q - 12m^2\gamma\ddot{q}\dot{q}^2 = 0 \tag{29}$$

where $\omega^2 = \varpi^2 - \frac{\ddot{h}}{4}$. For the equation of motion in Equation (29), we consider a solution up to $\mathcal{O}(f_0, \gamma)$ as

$$q(t) = q_0(t) + f_0 q_{f_0}(t) + \gamma q_\gamma(t). \tag{30}$$

For the form $q(t)$ in the above equation, we obtain the solution of Equation (26) as a linear combination as $q_0(t), q_{f_0}(t)$ and $q_\gamma(t)$. The analytical forms of $q_0(t), q_{f_0}(t)$ and $q_\gamma(t)$ are given as follows:

$$q_0(t) = \mathcal{A}_1 \cos(\varpi t) + \mathcal{A}_2 \sin(\varpi t), \tag{31}$$

$$q_{f_0}(t) = \mathcal{A}_3 \cos(\varpi t) + \mathcal{A}_4 \sin(\varpi t) - \frac{\Omega}{2(4\varpi^2-\Omega^2)}[\Omega\cos(\Omega t)\{\mathcal{A}_1\cos(\varpi t) + \mathcal{A}_2\sin(\varpi t)\} \\ -2\varpi\sin(\Omega t)\{\mathcal{A}_2\cos(\varpi t) - \mathcal{A}_1\sin(\varpi t)\}], \tag{32}$$

$$q_\gamma(t) = \mathcal{A}_5\cos(\varpi t) + \mathcal{A}_6\sin(\varpi t) - \frac{3m^2\varpi^2}{2}[t\varpi\mathcal{A}_1(\mathcal{A}_1^2 + \mathcal{A}_2^2)\sin(\varpi t) - t\varpi\mathcal{A}_2(\mathcal{A}_1^2 + \mathcal{A}_2^2)\cos(\varpi t)] \\ + \frac{\mathcal{A}_1}{4}(\mathcal{A}_1^2 - 3\mathcal{A}_2^2)\cos(3\varpi t) - \frac{\mathcal{A}_2}{4}(\mathcal{A}_2^2 - 3\mathcal{A}_1^2)\sin(3\varpi t)] \tag{33}$$

where $\mathcal{A}_1, \mathcal{A}_2, \mathcal{A}_3, \mathcal{A}_4, \mathcal{A}_5$ and $\mathcal{A}_6$ are arbitrary constants, which we will calculate for the $q_{cl}(t)$. To obtain the form of the above constants, we will apply the following set of the initial conditions:

$$q(t) = \begin{cases} q_0 & \text{for } t = 0 \\ q_f & \text{for } t = T \end{cases}. \tag{34}$$

Using the initial conditions in Equation (34), the constants can be obtained as follows:

$$\mathcal{A}_1 = q_0, \quad \mathcal{A}_2 = \frac{q_f - q_0 \cos(\omega T)}{\sin(\omega T)}, \tag{35}$$

$$\mathcal{A}_3 = \frac{\mathcal{A}_1 \Omega^2}{2(4\omega^2 - \Omega^2)}, \tag{36}$$

$$\mathcal{A}_4 = \frac{\Omega\{\cos(\omega T)[\Omega\mathcal{A}_1\cos(\Omega T) - 2\omega\mathcal{A}_2\sin(\Omega T)] + \sin(\omega T)[\Omega\mathcal{A}_2\cos(\Omega T) + 2\omega\mathcal{A}_1\sin(\omega T)]\}}{2(4\omega^2 - \Omega^2)\sin(\omega T)} - \mathcal{A}_3 \cot(\omega T), \tag{37}$$

$$\mathcal{A}_5 = \tfrac{3}{8} m^2 \omega^2 \mathcal{A}_1 \left(\mathcal{A}_1^2 - 3\mathcal{A}_2^2\right), \tag{38}$$

$$\mathcal{A}_6 = \frac{3m^2\omega^2 \left[\omega T(\mathcal{A}_1 \sin(\omega T) - \mathcal{A}_2 \cos(\omega T))(\mathcal{A}_1^2 + \mathcal{A}_2^2) + \frac{\mathcal{A}_1(\mathcal{A}_1^2 - 3\mathcal{A}_2^2)\cos(3\omega T)}{4} - \frac{\mathcal{A}_2(\mathcal{A}_2^2 - 3\mathcal{A}_1^2)\sin(3\omega T)}{4}\right]}{2\sin(\omega T)} - \mathcal{A}_5 \cot(\omega T). \tag{39}$$

Using Equation (30) along with Equations (35)–(39) in Equation (24) (with h being replaced by $2 f_0 \cos(\Omega t)$), we obtain the form of the classical action up to $\mathcal{O}(\gamma, f)$ as follows:

$$S_C = S_C^{(0)} + S_C^{(\gamma)} + S_C^{(f_0)} \tag{40}$$

where $S_C^{(0)}$, $S_C^{(\gamma)}$, and $S_C^{(f_0)}$ are given by the following equations:

$$S_C^{(0)} = \frac{m\omega}{2\sin(\omega T)} \left((q_0^2 + q_f^2)\cos(\omega T) - 2 q_0 q_f \right), \tag{41}$$

$$S_C^{(\gamma)} = -\frac{\gamma m^3 \omega^3}{32\sin^4(\omega T)} \Big[12\omega T \left(q_f^4 + 4 q_f^2 q_0^2 + q_0^4\right) - 48 q_0 q_f \omega T \cos(\omega T)(q_f^2 + q_0^2) + 24 q_0^2 q_f^2 \omega T \cos(2\omega T) \\ - 44 q_0 q_f \sin(\omega T)(q_0^2 + q_f^2) + 4\sin(2\omega T)\left(2 q_0^4 + 15 q_0^2 q_f^2 + 2 q_f^4\right) - 12 q_0 q_f \sin(3\omega T)(q_0^2 + q_f^2) + \sin(4\omega T)(q_0^4 + q_f^4) \Big], \tag{42}$$

$$S_C^{(f_0)} = -\frac{f_0 m \omega \Omega}{2\sin(\omega T)(4\omega^2 - \Omega^2)} \Big[\frac{\omega \sin(\Omega T)}{\sin(\omega T)} \left(q_0^2 - 2 q_0 q_f \cos(\omega T) + q_f^2 \cos(2\omega T) \right) + 2 q_0 q_f \Omega \cos^2\left(\tfrac{\Omega T}{2}\right) \\ - \Omega \cos(\omega T)\left(q_0^2 + q_f^2 \cos(\Omega T)\right) \Big]. \tag{43}$$

Therefore, we now have the final form of the propagator for the resonant bar detector interacting with a gravitational wave as follows:

$$\langle q_f, T | q_0, 0 \rangle = \sqrt{\frac{m\omega}{2\pi i \hbar \sin(\omega T)}} \tilde{\mathcal{N}}(T, \gamma, f_0) e^{\frac{i}{\hbar} S_{cl}}. \tag{44}$$

To obtain an overall structure of the fluctuation parameter in the above equation, we consider the free particle structure involving gravitational wave (GW) interaction only. In this case, the infinitesimal propagator considering the particle GW interaction from Equation (15) can be extracted as follows (in the $\omega \to 0$ limit):

$$\langle q_1, \Delta t | q_0, 0 \rangle = \int_{-\infty}^{\infty} \frac{dp_0}{2\pi\hbar} \exp\left[\frac{i\Delta t}{\hbar}\left(p_0 \frac{(q_1 - q_0)}{\Delta t} - \left(\frac{p_0^2}{2m} + \frac{\gamma p_0^4}{m} + \frac{p_0 q_0 f_0}{\Delta t}(\cos(\Omega\Delta t) - 1)\right)\right)\right] \\ \simeq \sqrt{\frac{m}{2\pi i \hbar \Delta t}} e^{\frac{im}{2\hbar\Delta t}(q_1 - q_0)^2} \left[1 + \frac{3 i m \gamma \hbar}{\Delta t} - 6\gamma m^2 \left(\frac{q_1 - q_0}{\Delta t}\right)^2 - \frac{i \gamma m^3 (q_1 - q_0)^4}{\hbar \Delta t^3} \\ - \frac{i f_0 q_0}{\hbar}\left(\frac{m(q_1 - q_0)}{\Delta t}\right)(\cos(\Omega\Delta t) - 1)\right]. \tag{45}$$

Now, the total propagator can be written using the set of infinitesimal propagators as follows:

$$\langle q_f, T|q_0, 0\rangle \simeq \left(\frac{m}{2\pi i\hbar\Delta t}\right)^{\frac{N}{2}} \int dq_1 dq_2 \cdots dq_{N-1} e^{\frac{im}{2\hbar\Delta t}[(q_1-q_0)^2+(q_2-q_1)^2+\cdots+(q_f-q_{N-1})^2]} \left[1 + \frac{3i\gamma mh N}{\Delta t} - \frac{6\gamma m^2}{\Delta t^2}((q_1-q_0)^2 + (q_2-q_1)^2 + \cdots + (q_f-q_{N-1})^2) - \frac{i\gamma m^3}{\hbar \Delta t^3}((q_1-q_0)^4 + (q_2-q_1)^4 + \cdots + (q_f-q_{N-1})^4) - \frac{if_0 m}{\hbar \Delta t^2}\left[q_0(q_1-q_0)(\cos(\Omega\Delta t)-1) + \cdots + q_{N-1}(q_f-q_{N-1})(\cos(N\Omega\Delta t) - \cos((N-1)\Omega\Delta t))\right]\right]. \quad (46)$$

In the absence of the gravitational wave [29], the form of the propagator in Equation (46) reads

$$\langle q_f, T|q_0, 0\rangle = \sqrt{\frac{m}{2\pi i\hbar T}} e^{\frac{im}{2\hbar T}(q_f-q_0)^2} \left(1 + \frac{3i\gamma m\hbar}{T} - 6\gamma m^2 \left(\frac{q_f-q_0}{T}\right)^2 - \frac{i\gamma m^3}{\hbar T^3}(q_f-q_0)^4\right). \quad (47)$$

In the presence of the gravitational wave, the propagator has the form given as

$$\langle q_f, T|q_0, 0\rangle \simeq \sqrt{\frac{m}{2\pi i\hbar T}} e^{\frac{im}{2\hbar T}(q_f-q_0)^2} \left(1 + \frac{3i\gamma m\hbar}{T} - 6\gamma m^2\left(\frac{q_f-q_0}{T}\right)^2 - \frac{i\gamma m^3}{\hbar T^3}(q_f-q_0)^4 \right.$$
$$\left. + \frac{if_0 mT}{\hbar}\left(\frac{(q_f-q_0)}{T}\right)^2 [\cos(\Omega T) - 1] - \frac{if_0 q_f}{\hbar}\left(\frac{m(q_f-q_0)}{T}\right)[\cos(\Omega T) - 1]\right) \quad (48)$$
$$\simeq \sqrt{\frac{m}{2\pi i\hbar T}} \cdot \tilde{\mathcal{N}}(T, \gamma, f_0) e^{\frac{i}{\hbar} S_{cl}^{(f)}}$$

where $S_{cl}^{(f)}$ is the classical action involving free particles and gravitational waves given by

$$S_{cl}^{(f)} = \frac{m}{2T}(q_f-q_0)^2 - \frac{\gamma m^3}{T^3}(q_f-q_0)^4 - \frac{mf_0}{2T}(q_f-q_0)\left[(q_f\cos[\Omega T] - q_0) - (q_f-q_0)\frac{\sin[\Omega T]}{\Omega T}\right] \quad (49)$$

and the form of the fluctuation term is given as follows:

$$\tilde{\mathcal{N}}(T, \gamma, f_0) \simeq 1 + \frac{3i\gamma m\hbar}{T} - 6\gamma m^2\left[\frac{q_f-q_0}{T}\right]^2 + \frac{if_0 mT}{\hbar}\left[\frac{q_f-q_0}{T}\right]^2[\cos(\Omega T) - 1] - \frac{if_0 mq_f}{\hbar}\left[\frac{q_f-q_0}{T}\right][\cos(\Omega T) - 1]$$
$$- \frac{imf_0(q_f-q_0)}{2\hbar T}\left[\frac{(q_f-q_0)\sin(\Omega T)}{\Omega T} - (q_f\cos(\Omega T) - q_0)\right]. \quad (50)$$

5. Summary

In this work, we have constructed the path integral formalism of the propagation kernel for a resonant bar detector in the presence of a gravitational wave in the generalized uncertainty principle framework. In this framework, we have considered only quadratic-order correction in the momentum. We have obtained the configuration space action for this system using the path integral formalism. With the action in hand, we have then obtained the equation of motion of the system. From the equation of motion, we observe that the overall frequency of the resonant detector shifts due to interaction with the gravitational wave. Next, we have used the form of the perturbation term for a circularly polarized gravitational wave to calculate the classical solution of the detector coordinate $q(t)$. Using this form of $q(t)$, we have finally obtained the classical action for a resonant bar detector interacting with a gravitational wave in the generalized uncertainty principle framework. We have then investigated the quantum fluctuation parameter of the bar detector in the presence of a circularly polarized gravitational wave. In order to obtain the final form of the fluctuation, we have considered a free particle interacting with the gravitational wave. The final form of the fluctuation picks up correction terms due to both GUP correction and gravitational wave interaction. In this process, we have neglected cross terms considering both GUP and GW interactions as it would result in a much smaller correction to the fluctuation factor than the other corrections present in the analytical form of the quantum fluctuation. It would also be important to carry out the above analysis in a linear

GUP framework. However, we would like to report this in future. From an observational point of view, the importance of our work lies in the fact that resonant bar detectors have the potential for detecting gravitational waves with their present sensitivity at distances of the order of 10^2 kpc from the Earth. The propagator captures the quantum effects also. Hence, detectability of such quantum effects in resonant bar detectors is also a possibility in the near future. Knowledge of the propagator of the detector coordinates is therefore necessary, if not absolutely essential.

Author Contributions: Writing—original draft, S.S., S.B. and S.G.; Writing—review & editing, S.S., S.B. and S.G. All the authors have equally contributed towards the conceptualization and the preparation of the manuscript. All authors have read and agreed to the published version of the manuscript.

Funding: Full APC funding via the discount voucher "0d7d8275791061d8".

Institutional Review Board Statement: The study did not require ethical approval.

Informed Consent Statement: Not applicable.

Data Availability Statement: Not applicable.

Conflicts of Interest: The authors declare no conflicts of interest.

References

1. Einstein, A. Die feldgleichungen der gravitation. *Sitzungsber Preuss Akad Wiss* **1915**, *25*, 844–847.
2. Einstein, A.Die Grundlage der allgemeinen Relativitätstheorie.*Ann. Physik* **1916**, *49*, 769. [CrossRef]
3. Rovelli, C. Loop Quantum Gravity. *Living Rev. Relativ.* **1998**, *1*, 1–69.
4. Carlip, S. Quantum Gravity: A progress report. *Rep. Prog. Phys.* **2001**, *64*, 885. [CrossRef]
5. Amati, D.; Ciafaloni, M.; Veneziano, G. Can spacetime be probed below the string size? *Phys. Lett. B* **1989**, *216*, 41–47. [CrossRef]
6. Konishi, K.; Paffuti, G.; Provero, P. Minimum physical length and the generalized uncertainty principle in string theory. *Phys. Lett. B* **1990**, *234*, 276–284. [CrossRef]
7. Girelli, F.; Livine, E.R.; Oriti, D. Deformed special relativity as an effective flat limit of quantum gravity. *Nucl. Phys. B* **2005**, *708*, 411–433. [CrossRef]
8. Bronstein, M.P. Kvantovanie gravitatsionnykh voln (Quantization of gravitational waves). *Zh. Eksp. Teor. Fiz.* **1936**, *6*, 195.
9. Bronstein, M.P. Quantentheorie schwacher gravitationsfelder. *Phys. Z. Sowjetunion* **1936**, *9*, 140–157.
10. Mead, C.A. Possible Connection Between Gravitation and Fundamental Length. *Phys. Rev. B* **1964**, *135*, B849. [CrossRef]
11. Maggiore, M. The algebraic structure of the generalized uncertainty principle. *Phys. Lett. B* **1993**, *319*, 83–86. [CrossRef]
12. Scardigli, F. Generalized uncertainty principle in quantum gravity from micro-black hole gedanken experiment. *Phys. Lett. B* **1999**, *452*, 39–44. [CrossRef]
13. Adler, R.J.; Santiago, D.I. On gravity and the uncertainty principle. *Mod. Phys. Lett. A* **1999**, *14*, 1371–1381. [CrossRef]
14. Adler, R.J.; Chen, P.; Santiago, D.I. The Generalized Uncertainty Principle and Black Hole Remnants. *Gen. Relativ. Gravit.* **2001**, *33*, 2101–2108. [CrossRef]
15. Banerjee, R.; Ghosh, S. Generalised uncertainty principle, remnant mass and singularity problem in black hole thermodynamics. *Phys. Lett. B* **2010**, *688*, 224–229. [CrossRef]
16. Gangopadhyay, S.; Dutta, A.; Saha, A. Generalized uncertainty principle and black hole thermodynamics. *Gen. Relativ. Gravit.* **2014**, *46*, 1661. [CrossRef]
17. Scardigli, F.; Casadio, R. Gravitational tests of the generalized uncertainty principle. *Eur. Phys. J. C* **2015**, *75*, 425. [CrossRef]
18. Mandal, R.; Bhattacharyya, S.; Gangopadhyay, S. Rainbow black hole thermodynamics and the generalized uncertainty principle. *Gen. Relativ. Gravit.* **2018**, *50*, 143. [CrossRef]
19. Ong, Y.C. Generalized uncertainty principle, black holes, and white dwarfs: A tale of two infinities. *J. Cosmol. Astropart. Phys.* **2018**, *2018*, 015. [CrossRef]
20. Buoninfante, L.; Luciano, G.G.; Petruzziello, L. Generalized uncertainty principle and corpuscular gravity. *Eur. Phys. J. C* **2019**, *79*, 663. [CrossRef]
21. Majumder, B. Quantum black hole and the modified uncertainty principle. *Phys. Lett. B* **2011**, *701*, 384–387. [CrossRef]
22. Das, S.; Vagenas, E.C. Universality of Quantum Gravity Corrections. *Phys. Rev. Lett.* **2008**, *101*, 221301. [CrossRef] [PubMed]
23. Das, S.; Vagenas, E.C. Phenomenological implications of the generalized uncertainty principle. *Can. J. Phys.* **2009**, *87*, 233–240. [CrossRef]
24. Pikovski, I.; Vanner, M.R.; Aspelmeyer, M.; Kim, M.S.; Brukner, Č. Probing Planck-scale physics with quantum optics. *Nat. Phys.* **2012**, *8*, 393–397. [CrossRef]
25. Bosso, P.; Das, S.; Pikovski, I.; Vanner, M.R. Amplified transduction of Planck-scale effects using quantum optics. *Phys. Rev. A* **2017**, *96*, 023849. [CrossRef]
26. Kumar, S.P.; Plenio, M.B. Quantum-optical tests of Planck-scale physics. *Phys. Rev. A* **2018**, *97*, 063855. [CrossRef]

27. Gangopadhyay, S.; Bhattacharyya, S. Path-integral action of a particle with the generalized uncertainty principle and correspondence with noncommutativity. *Phys. Rev. D* **2019**, *99*, 104010. [CrossRef]
28. Bhattacharyya, S.; Gangopadhyay, S.; Saha, A. Generalized uncertainty principle in resonant detectors of gravitational waves. *Class. Quant. Grav.* **2020**, *37*, 195006. [CrossRef]
29. Das, S.; Pramanik, S. Path integral for nonrelativistic generalized uncertainty principle corrected Hamiltonian. *Phys. Rev. D* **2012**, *86*, 085004. [CrossRef]
30. Gangopadhyay, S.; Bhattacharyya, S. Path integral action in the generalized uncertainty principle framework. *Phys. Rev. D* **2021**, *104*, 026003.
31. Kempf, A.; Mangano, G.; Mann, R.B. Hilbert space representation of the minimal length uncertainty relation. *Phys. Rev. D* **1995**, *52*, 1108. [CrossRef] [PubMed]
32. Bawaj, M.; Biancofiore, C.; Bonaldi, M.; Bonfigli, F.; Borrielli, A.; Di Giuseppe, G.; Marin, F. Probing deformed commutators with macroscopic harmonic oscillators. *Nat. Commun.* **2015**, *6*, 7503. [CrossRef] [PubMed]
33. Feng, Z.W.; Yang, S.Z.; Li, H.L.; Zu, X.T. Constraining the generalized uncertainty principle with the gravitational wave event GW150914. *Phys. Rev. B* **2017**, *768*, 81–85. [CrossRef]
34. Bushev, P.A.; Bourhill, J.; Goryachev, M.; Kukharchyk, N.; Ivanov, E.; Galliou, S.; Tobar, M.E.; Danilishin, S. Testing the generalized uncertainty principle with macroscopic mechanical oscillator and pendulums. *Phys. Rev. D* **2019**, *100*, 066020. [CrossRef]
35. Scardigli, F. The deformation parameter of the generalized uncertainty principle. *J. Phys. Conf. Ser.* **2019**, *1275*, 012004. [CrossRef]
36. Girdhar, P.; Doherty, A.C. Testing generalized uncertainty principles through quantum noise. *New J. Phys.* **2020**, *22*, 093073. [CrossRef]
37. Chatterjee, R.; Gangopadhyay, S. Violation of equivalence in an accelerating atom-mirror system in the generalized uncertainty principle framework. *Phys. Rev. D* **2021**, *104*, 124001. [CrossRef]
38. Sen, S.; Bhattacharyya, S.; Gangopadhyay, S. Probing the generalized uncertainty principle through quantum noises in optomechanical systems. *Class. Quant. Grav.* **2022**, *39*, 075020. [CrossRef]
39. Weber, J. Evidence for Discovery of Gravitational Radiation. *Phys. Rev. Lett.* **1969**, *22*, 1320. [CrossRef]
40. Ferrari, V.; Pizzella, G.; Lee, M.; Weber, J. Search for correlations between the University of Maryland and the University of Rome gravitational radiation antennas. *Phys. Rev. D* **1982**, *24*, 2471. [CrossRef]
41. Giazotto, A. Status of Gravitational Wave Detection. In *General Relativity and John Archibald Wheeler*; Astrophysics and Space Science Library 367; Ciufolini, I., Matzner, R.A., Eds.; Springer: Berlin/Heielberg, Germany, 2010.
42. Aguiar, O.D. Past, present and future of the Resonant-Mass gravitational wave detectors. *Res. Astron. Astrophys.* **2010**, *11*, 1. [CrossRef]
43. Aasi, J.; Abbott, B.P.; Abbott, R.; Abbott, T.; Abernathy, M.R.; Ackley, K.; DeSalvo, R. Advanced LIGO. *Class. Quant. Grav.* **2015**, *32*, 074001.
44. Abott, B.P. LIGO Scientific Collaboration and Virgo Collaboration. GW170817: Observation of Gravitational Waves from a Binary Neutron Star Inspiral. *Phys. Rev. Lett.* **2017**, *119*, 161101.
45. Acernese, F.A.; Agathos, M.; Agatsuma, K.; Aisa, D.; Allemandou, N.; Allocca, A.; Meidam, J. Advanced Virgo: A second-generation interferometric gravitational wave detector. *Class. Quant. Grav.* **2015**, *32*, 024001. [CrossRef]
46. Saha, A.; Gangopadhyay, S. Noncommutative quantum mechanics of a test particle under linearized gravitational waves. *Phys. Lett. B* **2009**, *681*, 96–99. [CrossRef]
47. Saha, A.; Gangopadhyay, S.; Saha, S. Noncommutative quantum mechanics of a harmonic oscillator under linearized gravitational waves. *Phys. Rev. D* **2011**, *83*, 025004. [CrossRef]
48. Saha, A.; Gangopadhyay, S. Resonant detectors of gravitational wave as a possible probe of the noncommutative structure of space. *Class. Quant. Grav.* **2016**, *33*, 205006. [CrossRef]
49. Saha, A.; Gangopadhyay, S.; Saha, S. Quantum mechanical systems interacting with different polarizations of gravitational waves in noncommutative phase space. *Phys. Rev. D* **2018**, *97*, 044015. [CrossRef]
50. Bhattacharyya, S.; Gangopadhyay, S.; Saha, A. Footprint of spatial noncommutativity in resonant detectors of gravitational wave. *Class. Quant. Grav.* **2018**, *36*, 055006. [CrossRef]
51. Gangopadhyay, S.; Bhattacharyya, S.; Saha, A. Signatures of Noncommutativity in Bar Detectors of Gravitational Waves. *Ukr. J. Phys.* **2019**, *64*, 1029. [CrossRef]
52. Bosso, P.; Das, S.; Mann, R.B. Potential tests of the generalized uncertainty principle in the advanced LIGO experiment. *Phys. Rev. B* **2018**, *785*, 498–505. [CrossRef]
53. Pizzella, G. Search for Gravitational Waves with Resonant Detectors. In *General Relativity and John Archibald Wheeler*; Astrophysics and Space Science Library, vol 367; Ciufolini, I., Matzner, R.A., Eds.; Springer: Dordrecht, Germany, 2010.
54. Thorne, K.S. *300 Years of Gravitation*; Hawking, S.W., Israel, W., Eds.; Cambridge University Press: Cambridge, UK, 1987; p. 330.
55. Gangopadhyay, S.; Scholtz, F.G. Path-Integral Action of a Particle in the Noncommutative Plane. *Phys. Rev. Lett.* **2009**, *102*, 241602. [CrossRef] [PubMed]
56. Maggiore, M. *Gravitational Waves. Vol. 1: Theory and Experiments*; Oxford Master Series in Physics; Oxford University Press: London, UK, 2007.

Communication

Gauss's Law and a Gravitational Wave

Olamide Odutola [1,2,†] and Arundhati Dasgupta [2,*,†]

1. Department of Physics, University of Durham, Durham DH1 3LE, UK; olamide.odutola@durham.ac.uk
2. Department of Physics and Astronomy, University of Lethbridge, Lethbridge, AB T1K 3M4, Canada
* Correspondence: arundhati.dasgupta@uleth.ca
† These authors contributed equally to this work.

Abstract: In this paper, we discuss the semi-classical gravitational wave corrections to Gauss's law and obtain an explicit solution for the electromagnetic potential. The gravitational wave perturbs the Coulomb potential with a function that propagates it to the asymptotics.

Keywords: gravitationalwave; semi-classical gravity; loop quantum gravity

1. Introduction

The discovery of gravitational waves (GWs) has not only opened a window to astrophysical events, but it has also given us instruments that are sensitive enough to test very weak gravitational phenomena [1,2]. Therefore, new theoretical work acquires meaning and some of the results can be tested, thereby providing evidence for the correctness of the physical theories. In particular, quantum gravity, which has no experimental confirmation as of yet, needs to be tested. Our entire understanding of the visible matter universe is based on the standard model of particle physics, which is quantized. The quantum of the GW—the graviton—is yet to be detected, and theoretical predictions regarding it have non-renormalizable quantum interactions. What, therefore, is the story of gravity at tiny length scales? In [3], we explored a coherent state for the GW, which would help to predict semi-classical phenomena at higher length scales than the 10^{-33} cm Planck length. Verification of the predictions from the coherent states would provide evidence for an underlying quantum world, which we hope to probe at a later time with more sophisticated instruments and understanding. On this note, we will briefly discuss a modified GW metric that was obtained in [3] and has a semi-classical correction to it. A similar computation of generalized uncertainty principle correction to a GW detector has appeared in this volume [4]. We will then solve Gauss's law and find that there are interesting results with the GW metric when used by itself. What we will find could be interpreted as the charge density receiving a correction that is measurable. We will consider a configuration with a point charge at the origin, which thus places us in the realm of electrostatics. Coulomb's law is valid and gives the electric field but no magnetic field. We found that if the background of this is not flat spacetime but a GW, then there is a non-zero 'current' generated. An interesting discussion of a similar phenomenon and its applications can be found in [5]. Note our work is also different from the example of an oscillatory electron, which is discussed in [6]. As the change in source is proportional to the GW amplitude, we studied a 'perturbation' of Coulomb's law that is time-dependent and gives rise to a magnetic field. The time-dependent scalar potential does not fall off at infinity but rises with distance. The electric field's radial component runs to zero at infinity, but the angular components rise as they have the same radial behavior as the potential; this can be measured and we will provide some numerical estimates. We also show that the magnetic potential is generated in a similar way as the electric potential. A magnetic field will be obtained from this as non-zero, though one that is very weak. In the conclusion, we will discuss the results in detail.

2. Gauss's Law and the Gravitational Wave

We solved for Maxwell's equation when investigating the background of a gravitational wave metric, which was corrected using semi-classical coherent states [3]. For the Maxwell field, the Lagrangian is:

$$\mathcal{L} = -\frac{\sqrt{-g}}{4} F^{\mu\nu} F_{\mu\nu} = -\frac{\sqrt{-g}}{4} F_{\sigma\rho} F_{\mu\nu} g^{\sigma\mu} g^{\rho\nu}$$

$$= -\frac{\sqrt{-g}}{4} (\partial_\mu A_\nu - \partial_\nu A_\mu)(\partial_\sigma A_\rho - \partial_\rho A_\sigma) g^{\sigma\mu} g^{\rho\nu},$$

where we assumed a non-trivial metric.

From the Euler–Lagrange equations, we obtained the following EoM in the presence of a four-source current j^ν:

$$\frac{1}{\sqrt{-g}} \partial_\mu(\sqrt{-g}\, F^{\mu\nu}) = \frac{1}{\sqrt{-g}} \partial_\mu(\sqrt{-g}\, g^{\mu\rho} g^{\nu\sigma} F_{\rho\sigma}) = j^\nu. \quad (1)$$

In [3], which appeared in this volume, we found semi-classical corrections to a GW metric. We used the coherent states in a system of loop quantum gravity (LQG) [7,8], which was defined on the phase space of the LQG canonical variables, i.e., holonomies $h_{e_a}(A)$ and conjugate momenta $P_{e_a}^I(\mathcal{E})$. The holonomy of the gauge connection A_a was obtained from the exponential of a path-ordered integral of a gauge connection over a one-dimensional 'edge' e_a, which formed the links of a graph; meanwhile, the momentum (built from the densitized triads $\mathcal{E}_a$) was obtained by smearing the triads $\mathcal{E}_a$ over surfaces S_{e_a} which the edges intersected. In this calculation, we used only the momentum variables,

$$P_{e_a}^I = \int_{S_{e_a}}^{*} \mathcal{E}^I; \quad P_{e_a} = \sqrt{P_{e_a}^I P_{e_a}^I}. \quad (2)$$

and the following relation:

$$\mathcal{E}_I^a \mathcal{E}_I^b = q q^{ab}, \quad (3)$$

where $\mathcal{E}_I^a$ are the density triads; $a, I = 1, 2, 3$ represent the space and internal SU(2) indices respectively; q_{ab} is the three-space metric of the background; and q is its determinant. The coherent states were also characterized using a semi-classical parameter $\tilde{t} \sim l_p^2/\lambda^2$, which is a ratio of the Planck length to the length scale of a system (here λ is the GW wavelength) and has a range of $0 < \tilde{t} < 1$. For these purposes, we considered a measurable $\tilde{t} \sim 10^{-16}$ for a GW with a frequency of 10^{35} Hz. This, however, was too high for the observed waves (which had a frequency of 100 Hz) as their $\tilde{t}$ was far smaller. For the next generation of detectors which will detect higher frequency waves, see [9] for a review.

The momenta were generated by smearing the triads over the faces of a cube, which were perpendicular to the edges e_a which were straight lines along the three axes. This type of discretization is not unique; however, with respect to the continuum limit, it serves the purpose of helping to find a semi-classical correction to the metric, as defined from the operator expectation values of the momentum (a detailed discussion on this topic can be found in [3]). The LQG-corrected metric of a gravitational wave with the polarizations of $h_+ = A_+ \cos(\omega(t-z))$, $h_\times = A_\times \cos(\omega(t-z))$ (as derived in [3]) is as follows:

$$g_{\mu\nu} = \begin{pmatrix} -1 & 0 & 0 & 0 \\ 0 & (1+h_+)(1+2\tilde{t}f_x) & h_\times(1+\tilde{t}f_x+\tilde{t}f_y) & 0 \\ 0 & h_\times(1+\tilde{t}f_x+\tilde{t}f_y) & (1-h_+)(1+2\tilde{t}f_y) & 0 \\ 0 & 0 & 0 & 1+2\tilde{t}f_z \end{pmatrix}. \quad (4)$$

The determinant of the metric was simplified to a first order in $\tilde{t}, h_{\times,+}$, which yielded the following:

$$g \approx (1 + 2\tilde{t}f_x + 2\tilde{t}f_y + 2\tilde{t}f_z)(h_\times^2 + h_+^2 - 1), \quad (5)$$

where the semi-classical correction functions in the metric were

$$f_i = f(P_{e_i}), \ f(P) = \frac{1}{P}\left(\frac{1}{P} - \coth(P)\right),$$

$$P_{e_x} = \frac{\epsilon^2}{\kappa}\left(1 + \frac{1}{2}h_+\right)$$

$$P_{e_y} = \frac{\epsilon^2}{\kappa}\left(1 - \frac{1}{2}h_+\right)$$

$$P_{e_z} = \frac{\epsilon^2}{\kappa},$$

where e_i refers to the straight edges along the x, y, z directions of the three spatial slices of the system [3]; ϵ represents the graph edge lengths; $\epsilon \to 0$ gives the continuum geometry; and κ is the dimensional gravitational constant, which is expressed in natural units as the Planck length squared. We then found the 0th component of the Maxwell's equations in a vacuum, i.e., in the presence of no sources. In flat geometry, this gives us Gauss's law, but in the background of the new metric, one instead obtains the following:

$$-\frac{1}{\sqrt{-g}}\partial_i\left(\sqrt{-g}g^{ij}F_{j0}\right) = 0$$

$$\Longrightarrow g^{xx}\frac{\partial E_x}{\partial x} + g^{yy}\frac{\partial E_y}{\partial y} + g^{zz}\frac{\partial E_z}{\partial z} + g^{xy}\left(\frac{\partial E_y}{\partial x} + \frac{\partial E_x}{\partial y}\right) + g_{zz}E_z\frac{1}{\sqrt{-g}}\frac{\partial\sqrt{-g}}{\partial z} = 0.$$

As the metric semi-classical corrections were proportional to the GW, these corrections were found to be functions of t, z (which has been found as such only in [3]). However, the derivative terms were proportional to $\tilde{t}A_+$, which is a product of small quantities; therefore, we could neglect them in the first approximation. Thus, we obtained

$$\vec{\nabla}\cdot\vec{E} = 2\tilde{t}(f_x\frac{\partial E_x}{\partial x} + f_y\frac{\partial E_y}{\partial y} + f_z\frac{\partial E_z}{\partial z}) + h_+\left(\frac{\partial E_x}{\partial x} - \frac{\partial E_y}{\partial y}\right) + h_\times\left(\frac{\partial E_y}{\partial x} + \frac{\partial E_x}{\partial y}\right). \quad (6)$$

In the approximation, we wrote the electric field as a zero-eth order field plus a small perturbation, and the RHS of the above equation could be interpreted as a source for the perturbation. The zeroeth order field was a static EM field, which was generated by a point source at the origin. Hence, we obtained

$$\vec{E} = \frac{1}{4\pi\epsilon_0}\frac{\hat{r}}{r^2} + \vec{\tilde{E}}, \quad (7)$$

where we assumed a point source charge at the origin, or at least a charge of 1 Coulomb within a small radius ϵ (which is where our considerations were outside the radius). As the source was time-dependent, we took the perturbation to be composed of the potentials

$$\vec{\tilde{E}} = -\vec{\nabla}\Phi + \frac{\partial\vec{A}}{\partial t}. \quad (8)$$

In the Coulomb gauge $\vec{\nabla}\cdot\vec{A} = 0$, the following was yielded:

$$\nabla^2\Phi(x,y,z,t) = 6h_\times\frac{xy}{r^5} + 3h_+\frac{(x^2 - y^2)}{r^5}, \quad (9)$$

which is clearly Poisson's equation with a time-dependent source. Seeing as the divergence of the electric field was zero and the first order in the corrections, all of the $f(P_{e_i})$ were found to be equal; as such, we can ignore the semi-classical term ($=2\tilde{t}f\vec{\nabla}\cdot\vec{E} = 0$). A way through which to understand the GW-generated oscillation of the source was to observe that the charge density fluctuated with time as the volume changed.

To simplify the system, at $\theta = \pi/2$, we solved for the equations. As such, we obtained, as the particular solution, the following:

$$\Phi(r,t) = \left(-\frac{3A_+}{4r}\right)\cos(2\phi)\cos(\omega t). \tag{10}$$

Clearly, this potential is different in behavior to the regular $1/r$ spherical potential of the point-charge source at the origin. Here, the ϕ dependence makes the potential acquire different signs as it approaches the x and y axes. If we write the above equation in spherical coordinates, in which we assume a form of the potential in spherical harmonics with the same frequency as that of the GW in its time dependence, we obtain

$$\Phi(r,\theta,\phi,t) = \sum_{lm}\Phi_{lm}(r,t)Y_l^m(\theta,\phi), \tag{11}$$

which gives, from Gauss's law, the following:

$$\sum_{l,m}\left[\frac{d}{dr}\left(r^2\frac{d\Phi_{lm}}{dr}\right) - l(l+1)\Phi_{lm}\right]Y_{lm}(\theta,\phi) = \frac{3A_+ e^{i\omega(t-z)}}{r}\sin^2\theta\cos(2\phi). \tag{12}$$

We then assumed that $\Phi_{lm}(r,t) = e^{i\omega t}\Phi_{lm}(r)$. If we keep the plane wave e^{ikz} in the source ($k = \omega$), then we have to use the spherical wave expansion of the function $e^{ikr\cos\theta}$, where we obtain the following:

$$e^{ikr\cos\theta} = \sum_{l=0}^{\infty}i^l(2l+1)j_l(kr)P_l(\cos\theta). \tag{13}$$

Using the partial wave analysis of the above RHS (with the assumption that the EM potential has the same frequency as the GW), a propagating mode was generated in the case of the oscillating sources. We also wrote the equation $\cos(2\phi) = 1/2(\exp(2i\phi) + \exp(-2i\phi))$. We found that the ODE for $\Phi_{l2}(r)$ was the same as the ODE for $\Phi_{l-2}(r)$; therefore, we dropped the second index and solved for the following equation:

$$\sum_l\left[\frac{d}{dr}\left(r^2\frac{d\Phi_l}{dr}\right) - l(l+1)\Phi_l\right]\sqrt{\frac{(2l+1)}{4\pi}\frac{(l-2)!}{(l+2)!}}P_l^2(\cos\theta)$$
$$= \frac{3A_+}{2}\sum_{l'}i^{l'}(2l'+1)\frac{j_{l'}(kr)}{r}P_{l'}(\cos\theta)\sin^2\theta. \tag{14}$$

The associated Legendre function $P_l^2(\cos(\theta))$ is on the left and the usual Legendre function $P_l(\cos\theta)$ is on the right. If we take the orthonormality property of the associated Legendre functions by first multiplying with $P_n^2(\cos\theta)d(\cos\theta)$ and then integrating both sides of the Equation for $-1 \leq \cos\theta \leq 1$, we obtain

$$\sum_l\left[\frac{d}{dr}\left(r^2\frac{d\Phi_l}{dr}\right) - l(l+1)\Phi_l\right]\lambda_l\int_{-1}^{1}P_l^2(x)P_n^2(x)dx$$
$$= \frac{3A_+}{2}\sum_{l'}i^{l'}(2l'+1)\frac{j_{l'}(kr)}{r}\int_{-1}^{1}P_{l'}(x)(1-x^2)P_n^2(x)dx, \tag{15}$$

where λ_l represents the normalization constant from $Y_{lm}(\theta,\phi)$. Furthermore, we replaced $\cos\theta$ with x for brevity. The LHS uses the orthogonality condition; but, on the RHS, the integral was difficult to compute. Given the Legendre function recursion equations [10] and integrals [11], we obtained non-zero values for $l = n - 2$ and $n, n + 2$. Therefore, we found

$$\left[\frac{d}{dr}\left(r^2\frac{d\Phi_n(r)}{dr}\right) - n(n+1)\Phi_n(r)\right]\frac{2(n+2)!\lambda_n}{(2n+1)(n-2)!} =$$
$$\frac{1}{r}(\Lambda_{n-2}j_{n-2}(kr) + \Lambda_n j_n(kr) + \Lambda_{n+2}j_{n+2}(kr)], \quad (16)$$

where there were also the following constants:

$$\Lambda_{n-2} = \frac{3A_+}{2}i^{n-2}\left[\frac{2n(n^2-1)(n+2)}{(2n+1)(2n-1)}\right], \quad (17)$$

$$\Lambda_n = -\frac{3A_+}{2}i^n\left[\frac{4n(n+1)(n-1)(n+2)}{(2n-1)(2n+3)}\right], \quad (18)$$

$$\Lambda_{n+2} = \frac{3A_+}{2}i^{n+2}\frac{2n(n+1)(n+2)(n-1)}{(2n+1)(2n+3)}. \quad (19)$$

There were, therefore, three independent $l = n - 2$, n, $n + 2$ partial waves, which gave non-zero values for the RHS of the equation and generated the 'source' for the EM potential of the nth angular mode. We used MAPLE to generate the solution to the above ODE, and we found a very elongated formula that contained LommelS1 and Hypergeometric functions, which, nevertheless, gave the RHS particular solution. It must be noted that, if we keep the $\tilde{t}$ term detailed in the above equation, the particular solution will become corrected with static functions as there are no-time dependent contributions of the first order in $\tilde{t}$. As mentioned earlier, we ignored the $\tilde{t}A_+$ product terms, which are equivalent to the second-order infinitesimal corrections to Gauss's law.

The general solution is as follows:

$$\Phi_n(r) = A_0 r^n + \frac{B_0}{r^{n+1}} - \frac{k^{3/2}}{\Gamma\left(\frac{7}{2}+n\right)2^{n-1/2}}\left(A\frac{(rk)^{n+1}}{32(n+1)(n+1/2)}H([n+1],[2+n,\frac{7}{2}+n],-\frac{r^2k^2}{4})\right.$$
$$+ B\frac{(kr)^{n-1}(n+\frac{5}{2})(n+\frac{3}{2})}{8n(n+\frac{1}{2})}H([n],[n+1,n+\frac{3}{2}],-\frac{r^2k^2}{4})$$
$$+ C\frac{(n-\frac{1}{2})(n+\frac{3}{2})(n+\frac{5}{2})(rk)^{n-3}}{2(n-1)}H([n-1],[n,n-\frac{1}{2}],-\frac{r^2k^2}{4})\right)$$
$$+ \frac{(rk)^n k^{3/2}}{96n(n+1)(n+1/2)(2+n)}\left[\left(-\frac{1}{8}(rk)J_{n-1/2}(kr) + \frac{1}{4}(n+\frac{1}{2})J_{n+\frac{1}{2}}(kr)\right)W(A,B,C)S_{3/2-n,n+1/2}(kr)\right.$$
$$+ -n(kr)W(A,B,C)J_{n+1/2}(kr)S_{1/2-n,3/2+n}(kr)] + \frac{1}{96(n+\frac{1}{2})n(n+1)(2+n)}\left[\left(-\frac{1}{4}(kr)^{1/2}W(A,B,C)\right.\right.$$
$$- 2An(n+1)(n+\frac{3}{2})(n+\frac{1}{2})(kr)^{-7/2} + \frac{1}{4}(kr)^{-3/2}nW(A,B,C)\right)J_{n+1/2}(kr)$$
$$+ \left.\left(\frac{1}{4}(kr)^{-1/2}nW(A,B,C) + (kr)^{3/2}\frac{1}{8}W(A,B,C) + (kr)^{-5/2}V(A,B,C)\right)J_{n-1/2}(kr)\right]. \quad (20)$$

In the above, we have $J_n(x)$ as the Bessel function of the first kind, $S_{n,m}(x)$ as the LommelS1 functions, and $H(a,b;c,d,e,x)$ and $H(a;b,c,x)$ as the Hypergeometric functions of the (2, 3) and (1, 2) type, respectively. In addition, Φ_n was set to have the usual partial wave potentials of the form r^n and r^{-n-1}, which were also the solutions of the homogeneous equation. The particular solutions represent the functions generated by the GW-induced oscillations and are propagating EM potentials. There were singularities hidden in the LommelS1 functions for the integer values of n, which we regulated. Note that we can trust only the solutions for $r \neq 0$, and this is justified as we have a semi-classical parameter

$\tilde{t} \approx 0$ and the discretization ϵ length scale, which provide a minimum length to which the geometry can be probed. The constants were

$$A = \frac{(2n+1)(n-2)!}{2(n+2)!\lambda_n}\Lambda_{n+2}, \qquad (21)$$

$$B = \frac{(2n+1)(n-2)!}{2(n+2)!\lambda_n}\Lambda_n, \qquad (22)$$

$$C = \frac{(2n+1)(n-2)!}{2(n+2)!\lambda_n}\Lambda_{n-2}, \qquad (23)$$

$$W(A,B,C) = \frac{4}{3}(Cn^2) + (-2B+4C)n + A - 4B + \frac{8}{3}(C), \qquad (24)$$

$$V(A,B,C) = -\frac{2}{3}(Cn^2) + (A-C)n + \frac{3}{2}A + \frac{2}{3}(C). \qquad (25)$$

Note that the above results were true only for $n = 2$ and higher. As the behavior of the functions for general n were difficult to plot, we simply took one representative partial wave and observed the difference from a regular solution. We took $n = 3$ and observed the behavior of $\Phi_3(r)$ as $r \to \infty$. The $\Phi_3(r)$ function had a real component that fell of as the r^{-4} was obtained from the homogeneous equation solution, and an imaginary component (which was evident from the coefficients on the RHS) was the particular solution for $n = 3$. Additionally, as our ansatz for the potential was of the form $\Phi(r)e^{i\omega t}$, it was not surprising that the solution was complex. We then plotted the function $|\Phi_3(r)|^2$ to examine its asymptotic behavior. We found that, despite putting the particular solution strength as 10^{-10} of the r^{-4} term, the function started increasing after a certain interval. We know that $r^{-4} \to 0$ as $r \to \infty$, but the presence of GWs reverses the fall off. This behavior persists for a higher n, thus confirming our claim that the electric potential now extends to the asymptotic region.

In general, the solutions will be of the form

$$\Phi(r,\theta,\phi,t) = \sum_l \Phi_l(r)\left(P_l^2(\cos\theta)e^{2i\phi} + P_l^{-2}(\cos\theta)e^{-2i\phi}\right)e^{i\omega t}. \qquad (26)$$

To obtain the observable function, one must take the real part of the summed solution. As shown above in Equation (20), $\Phi_l(r)$ is composed of solutions to the homogeneous equations of the form $A_l r^l + B_l r^{-(l+1)}$. In addition, for each l, there is a particular solution. It is plausible that the sum over l for the particular solution has a finite convergent answer. We tried finding a convergent answer but the summation was not simple; thus, work is still in progress. We instead used a numerical method of summing up the partial waves to some finite number. We then plotted the particular solution and summed up to $l = 3, \ldots, m$, where m is some large number. This evidently represents a truncated GW wave contribution that is up to the $m + 2$ mode in the source, but it is a good-enough approximation to what might be the real system. Therefore, we—in the following—plotted the plane wave that was summed up to $m = 50, 100$, as well as showed the corresponding Coulomb potential that was generated by the system.

We investigated the analytic formula in Equation (20) and the partial wave summation of the spherical wave solution. We found that the potential started growing as had been observed for the $l = 3$ solution of the potential, as shown in Figure 1. We then plotted the potential in 3d and for $\phi = 0$. This showed that the GW effect on the Coulomb potential was non-trivial and was, in principle, detectable using an electrometer, which is sensitive to the electric potential. This approach will aid in the detection of a GW in a very isolated environment.

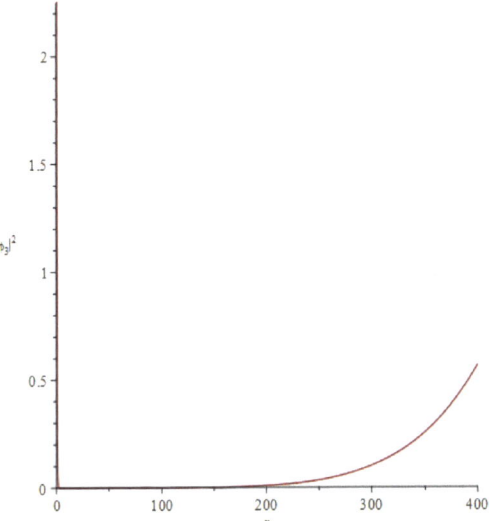

Figure 1. The modulus square of the potential for $l = 3$.

As is evident from the above plots, i.e., in Figure 2a,b (one for $m = 50$ and another for $m = 100$), the potential increased as a function of r, and the image on the $x = \cos\theta$ axis showed oscillations due to the Legendre function. If one plots the sum over a small interval, then these features are also evident, as shown in Figure 3a. If one plots the potential on the sphere, the oscillations would of course appear as 'petals' in a spherical coordinates plot, as shown in Figure 3b.

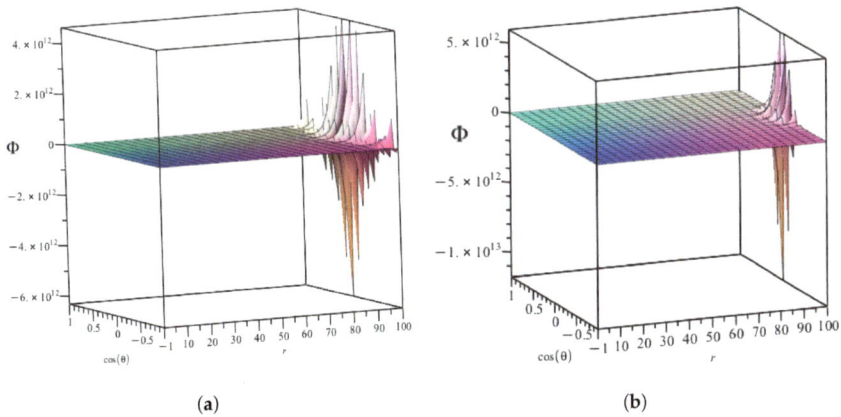

Figure 2. Real parts of $\Phi(r, \theta, 0)$. (**a**) Potential with partial modes summed from $l = 3, \ldots, 50$, $\phi = 0$. (**b**) Potential with the partial modes summed for $l = 3, \ldots, 100$, $\phi = 0$.

The electric field defined from the above potential was expressed simply as

$$\vec{E} = -\vec{\nabla}\Phi(r,\theta,\phi) = -\left(\frac{\partial \Phi}{\partial r}\hat{r} + \frac{1}{r}\frac{\partial \Phi}{\partial \theta}\hat{\theta} + \frac{1}{r\sin\theta}\frac{\partial \Phi}{\partial \phi}\hat{\phi}\right). \tag{27}$$

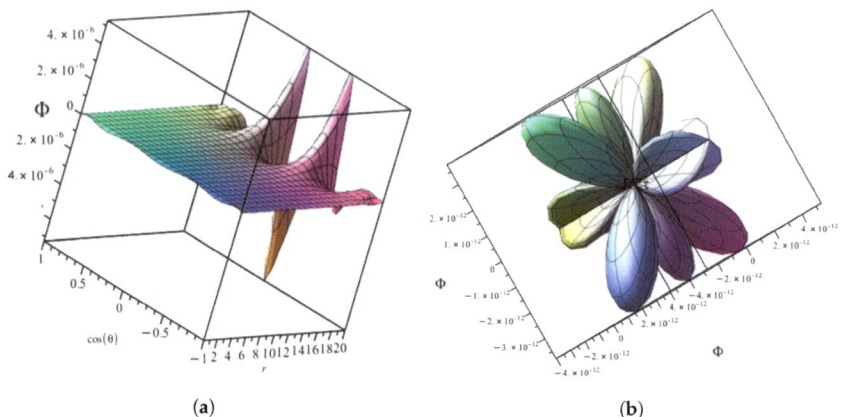

Figure 3. Real parts of $\Phi(r,\theta,0)$. (a) Potential with partial modes summed from $l = 3,\ldots,50$ and plotted for $k = 1$, $r = [0, 20]$, $\phi = 0$. (b) Potential with partial modes summed for $l = 3,\ldots,50$ and plotted in θ, ϕ, $k = 1$, $r = 1$.

The electric field in the $\hat{r}$ direction had a non-trivial derivative in the radial direction. The derivatives of θ and ϕ acted on the $P_l^2(\cos\theta)$ and the $\cos(2\phi)$ functions. We found that the E_r function was the derivative of the potential function that is given in Equation (20); it was also found to be very lengthy and involved SturveH functions. Instead of quoting that, we show a graphical representation of the functions in the following Figure 4a,b for the $l = 3$ partial wave only.

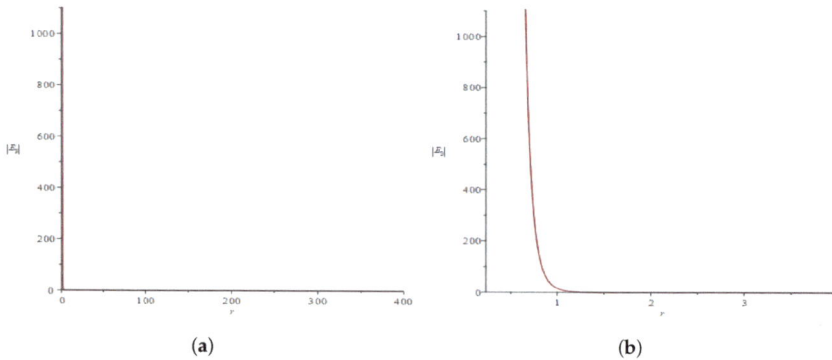

Figure 4. Magnitude of the radial electric field solution for $l = 3$. (a) Magnitude of $E_r \propto -\partial_r \Phi_3$ for $\phi = 0$, $\theta = \pi/4$; $r = [0, 400]$. (b) The E_r field for $\phi = 0$, $\theta = \pi/4$, $r = [0,4]$.

As evident from the above, the radial component decreased with distance. However, it must be mentioned that the particular part of the solution did show an increase as a function of r. As in the potential, we took the ratio of the Coulomb term and GW-induced term as 10^{-10}. In the event that this ratio was different, the nature of the electric field's radial component would again change. As shown in Figure 5, the contribution from the GW-induced electric field increased with r. It also remained that there were angular components of the electric field, which were generated due to the GW, and these should be detectable in an electrometer.

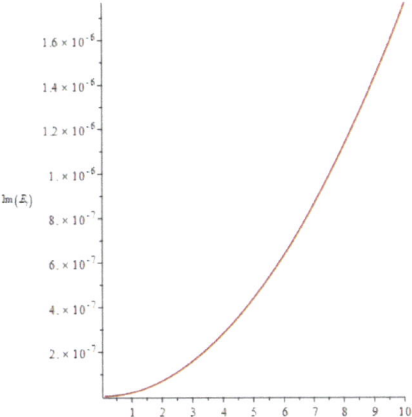

Figure 5. The GW-induced electric field radial component for $l = 3, k = 1$.

Next, we also found that the electric field's radial component for the summed potential was $E_r(r, \theta) = -\partial_r \left(\sum_{l=3}^{50} \Phi_l(r) P_l^2(\theta) \right)$. This showed behavior that was almost similar to the electric field for $l = 3$, where the function shows a fall off as a function of r. We plotted the particular solution of the GW-induced electric field, which is non-trivial, for $k = 1$, as shown in Figure 6.

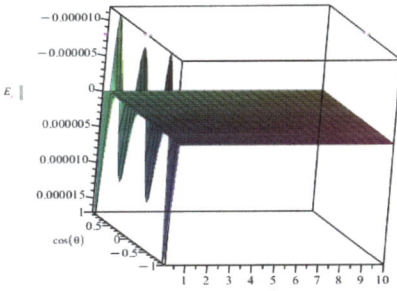

Figure 6. The GW-induced electric field radial component for the partial wave summed as $l = 3, \ldots, 50, k = 1$.

Before we end this discussion, the obvious question is whether a calibrated electrometer will detect the above-generated fluctuating electric field, and the answer is yes. If we find the potential function at a distance of 10 m from the origin where a 10^{-9} Coulomb charge has been placed ($q/4\pi\epsilon_0 \sim 1$) and where the GW has a frequency of 10 Hz with an amplitude of 10^{-21}, then the E_θ component at a fixed angle being proportional to the potential is almost of a 0.1 N/C order. Small changes in the magnetic fields were detected by SQUIDS [12], we therefore needed to discuss the magnetic field generated by the GW.

In the above, we showed how a GW can modify Gauss's law but where our electric field perturbation was time dependent. Therefore, the discussion is incomplete without discussing the magnetic field and studying the vector potential. To obtain the magnetic field, we studied Maxwell's equations for $\nu = i$, where i is a space component and the current density is $j^i = 0$, as we are only studying Coulomb's law for a static source in this discussion. We found that Maxwell's equation is as follows:

$$-\frac{1}{\sqrt{-g}} \partial_0 \left(\sqrt{-g} g^{ij} F_{0j} \right) + \frac{1}{\sqrt{-g}} \partial_k \left(\sqrt{-g} g^{kl} g^{ij} F_{lj} \right) = 0. \tag{28}$$

As the magnetic field was initially zero, the contribution to a non-zero magnetic field $\vec{B}$ at a first order in the GW amplitude was

$$(\partial_z B_y - \partial_y B_z) = -\partial_0 h_+ E_x^0 + \partial_0 \tilde{E}_x, \qquad (29)$$
$$(\partial_x B_z - \partial_z B_x) = \partial_0 h_+ E_y^0 + \partial_0 \tilde{E}_y, \qquad (30)$$
$$(\partial_x B_y - \partial_y B_x) = \partial_0 \tilde{E}_z. \qquad (31)$$

In the above, E_i^0 is the components of the Coulomb field and $\tilde{E}_i$ is the perturbations that were computed due to the GW. If we use the Lorenz gauge and write the magnetic field in terms of a Gauge potential $\vec{B} = \vec{\nabla} \times \vec{A}$, such that $\vec{\nabla} \cdot \vec{A} = 0$, one obtains

$$\nabla^2 A_x = -\partial_0 h_+ E_x^0 + \partial_0 \tilde{E}_x, \qquad (32)$$
$$\nabla^2 A_y = \partial_0 h_+ E_y^0 + \partial_0 \tilde{E}_y, \qquad (33)$$
$$\nabla^2 A_z = \partial_0 \tilde{E}_z. \qquad (34)$$

The above equations can be solved using the same method as the scalar potential solution for Gauss's law. Thus, apart from modifying Gauss's law, the GW also induces a magnetic field, and this can be calculated. We hope to discuss this in a future work. The fact that a tiny magnetic field was generated is important for detection purposes as small changes in magnetic fields can be found using SQUIDS [12].

3. Conclusions

In this short article, we have shown that the GW generates a source for a perturbation of the EM potential, which is time-dependent. The solution is complicated in form but was exactly obtained. As GWs were detected, we predicted the corrections to the Coulomb potential being of a point source charge, and we hope to find an experimental verification of our results. The semi-classical corrections to the metric described in the paper will also correct Gauss's law in a slightly similar functional form but will also be of a next order in the perturbation. Previously, and in recent years, GW wave-induced corrections to Maxwell's equations have been studied [12–16], but our results specifically discussed corrections to a static electric Coulomb potential using partial wave analysis. We also showed how a magnetic field is generated by the GW. We found that, when using numerical values, the GW-induced electric fields propagated and can be almost of an order 1. The question then is, have we already seen the GW-induced correction to Gauss's law in some detector? To attribute the EM detection to a GW would therefore be the next task.

Funding: O.O. was funded by a MITACS summer fellowship.

Data Availability Statement: This paper did not report any experimental data.

Acknowledgments: We are grateful to Narasimha Reddy Gosala for their useful discussions.

Conflicts of Interest: The authors declare no conflict of interest.

References

1. Abbott, B.P.; Abbott, R.; Abbott, T.D.; Abraham, S.; Acernese, F.; Ackley, K.; Bulik, T.; LIGO–Virgo–KAGRA Collaboration. Prospects for Observing and Localizing Gravitational-Wave Transients with Advanced LIGO, Advanced Virgo and KAGRA. *Living Rev. Rel.* **2020**, *23*, 3. [CrossRef] [PubMed]
2. Saulson, P.R. *Fundamentals of Interferometric Gravitational Wave Detectors*; World Scientific: Singapore, 1994.
3. Dasgupta, A.; Montenegro, J.L.F. Aspects of Quantum Gravity phenomenology and Astrophysics. *Universe* **2023**, *9*, 128. [CrossRef]
4. Sen, S.; Bhattacharya, S.; Gangopadhay, S. Path Integral Action for a Resonant Detector of Gravitational Waves in the Generalized Uncertainty Principle Framework. *Universe* **2022**, *8*, 450. [CrossRef]
5. Bruschi, D.E. Gravity-induced electric currents. *arXiv* **2023**, arXiv:2306.03742.
6. Audagnotto, G.; Keitel, C.H.; Piazza, A.D. Proportionality of gravitational and electromagnetic radiation by an electron in an intense plane wave. *Phys. Rev. D* **2022**, *106*, 076009. [CrossRef]

7. Thiemann, T.; Winkler, O. Gauge field theory coherent states (GCS): II. Peakedness properties. *Class. Quantum Gravity* **2001**, *14*, 2561. [CrossRef]
8. Thiemann, T. Introduction to Modern Canonical General Relativity. *arXiv* **2001**, arXiv:0110034.
9. Aggarwal, N.; Aguiar, O.D.; Bauswein, A.; Cella, G.; Clesse, S.; Cruise, A.M.; White, G. Challenges and Opportunities of Gravitational Wave Searches at MHz to GHz frequencies. *Liv. Rev. Rel.* **2021**, *24*, 4. [CrossRef]
10. Gradshteyn, I.S.; Ryzhik, I.M. *Tables of Integrals, Series, Products*; Elsevier: Amsterdam, The Netherlands, 2007.
11. Samaddar, S.N. *Some Integrals Involving Associated Legendre Functions*. Math. Comp. **1974**, *128*, 257.
12. Cabral, F.; Lobo, F.S.N. Gravitational Waves and Electrodynamics: New perspectives. *Eur. Phys. J. C* **2017**, *77*, 237. [CrossRef] [PubMed]
13. Patel, A.; Dasgupta, A. Interaction of Electromagnetic field with a Gravitational wave in Minkowski and de-Sitter space-time. In *General Relativity and Quantum Cosmology*; Cornell University: Ithaca, NY, USA, 2021.
14. Kim, D.; Park, C. Detection of gravitational waves by light perturbation. *Eur. Phys. J. C* **2021**, *81*, 563. [CrossRef]
15. Ganjali, M.A.; Sedaghatmanesh, Z. Laser interferometer in presence of scalar field on gravitational wave background. *Class. Quant.Grav.* **2021**, *38*, 105010. [CrossRef]
16. Calura, M.; Montinari, E. Exact Solution to the homogeneous Maxwell Equations in the Field of a Gravitational Wave in Linearized Theory. *Class. Quant. Grav.* **1999**, *16*, 643–652. [CrossRef]

Disclaimer/Publisher's Note: The statements, opinions and data contained in all publications are solely those of the individual author(s) and contributor(s) and not of MDPI and/or the editor(s). MDPI and/or the editor(s) disclaim responsibility for any injury to people or property resulting from any ideas, methods, instructions or products referred to in the content.

Review

Space–Time Physics in Background-Independent Theories of Quantum Gravity

Martin Bojowald

Institute for Gravitation and the Cosmos, The Pennsylvania State University, 104 Davey Lab, University Park, PA 16802, USA; bojowald@gravity.psu.edu

Abstract: Background independence is often emphasized as an important property of a quantum theory of gravity that takes seriously the geometrical nature of general relativity. In a background-independent formulation, quantum gravity should determine not only the dynamics of space–time but also its geometry, which may have equally important implications for claims of potential physical observations. One of the leading candidates for background-independent quantum gravity is loop quantum gravity. By combining and interpreting several recent results, it is shown here how the canonical nature of this theory makes it possible to perform a complete space–time analysis in various models that have been proposed in this setting. In spite of the background-independent starting point, all these models turned out to be non-geometrical and even inconsistent to varying degrees, unless strong modifications of Riemannian geometry are taken into account. This outcome leads to several implications for potential observations as well as lessons for other background-independent approaches.

Keywords: background independence; space–time physics; geometry; loop quantum gravity; covariance

Citation: Bojowald, M. Space–Time Physics in Background-Independent Theories of Quantum Gravity. *Universe* **2021**, *7*, 251. https://doi.org/10.3390/universe7070251

Academic Editors: Alfredo Iorio and Arundhati Dasgupta

Received: 29 June 2021
Accepted: 17 July 2021
Published: 20 July 2021

Publisher's Note: MDPI stays neutral with regard to jurisdictional claims in published maps and institutional affiliations.

Copyright: © 2021 by the author. Licensee MDPI, Basel, Switzerland. This article is an open access article distributed under the terms and conditions of the Creative Commons Attribution (CC BY) license (https://creativecommons.org/licenses/by/4.0/).

1. Introduction

A key feature of general relativity is its ability to determine both the dynamics and the structure of space–time. A complete quantum theory of gravity should therefore refrain from presupposing space–time structure; only then can it be considered a proper quantization of the theory. As a conclusion, space–time structure must be derived after quantization for a subsequent physical analysis, and the result may be modified compared with the familiar Riemannian structure. Depending on the quantization procedure, it may even happen that no consistent space–time structure exists for its solutions. A detailed analysis is then required to see whether the theory can be considered a valid candidate for quantum gravity, even if it is formally consistent, judged by non-geometrical standards such as conditions commonly imposed on quantizations of gauge theories. These questions are highly non-trivial in any approach. A detailed analysis is now available in models of loop quantum gravity, but it remains preliminary owing to the tentative nature of physical models of space–time in this theory.

Loop quantum gravity is often advertised as a background-independent approach to quantum gravity. This characterization suggests that the theory might indeed be free of pre-supposed space–time structures. In practice, however, the rather involved nature of methods suitable for derivations of space–time structures, combined with the canonical treatment used in the more successful realizations of loop quantum gravity, has for some time obscured the role and nature of space–time in this theory. In fact, several long-standing doubts exist as to the possibility of covariance in models of loop quantum gravity. For instance, the "bounce" idea, used in a majority of cosmological and black-hole models in this setting, is largely based on calculations available for the dynamics in homogeneous cosmological models, introducing formal properties of discreteness or boundedness seen in the kinematics of the full theory of loop quantum gravity. Since it remains unknown whether there is space–time dynamics consistent with the kinematics of the full theory, there is no guarantee that kinematical ingredients exported to homogeneous models of quantum cosmology can give rise to a meaningful structure of space–time and some sense of general covariance.

More specifically, kinematical features that apply spatial discreteness work as a cut-off which, if it is a fixed scale, is hard to reconcile with the transformations required for covariance. If one accepts the possibility that quantum gravity may well lead to non-classical space–time structures that require a modified and perhaps weakened version of general covariance, consistency requires a detailed demonstration of how one can avoid various low-energy problems that may then trickle down from the Planck regime, as pointed out in [1,2]. Moreover, in such a situation, it is important to determine how a modified space–time structure can be described in meaningful terms, for instance by addressing the question of whether such a theory can still be considered geometrical and whether there is an extended range of parameters (such as $\hbar$) in which effective line elements may still be available.

A consideration of space–time structure in bounce models also raises the question of how exactly singularity theorems are evaded. In models of loop quantum cosmology, bounce solutions are obtained without modifying matter Hamiltonians. The standard energy conditions therefore remain satisfied, obscuring the possibility of avoided singularities often claimed in this setting. Since singularity theorems make statements about boundaries of space–time and use the general properties of Riemannian geometry such as the Ricci curvature and the geodesic deviation equation, they depend on and require a consistent form of space–time structure. Unfortunately, however, bounce models of loop quantum gravity are often accompanied by poorly justified and contradictory statements about space–time. For instance, standard line elements are commonly used to express modified gravitational dynamics in tractable form, implicitly presupposing that space–time remains Riemannian. However, then, singularity theorems should be applicable to the resulting modified solutions since the behavior of matter energy is assumed to remain unchanged, making it impossible to evade singularities by a bounce. (The behavior of singularities may depend on a possibly modified relationship between stress–energy and Ricci curvature even if one maintains positive-energy conditions. However, simple bounce models based on modified Friedmann equations do not provide such a relationship because their space–time structure remains unclear.) The fact that this contradiction has gone unnoticed for several years in this field serves to highlight the challenging nature of questions about space–time in loop quantum gravity.

Independently of bounce claims, results about space–time structure in models of loop quantum gravity have been accumulating in recent years. This review presents a summary, highlighting the similarities between different ways in which covariance can be and often is violated. By now, all the high-profile claims made in the last decade in the context of loop quantum gravity, including [3–6], have been shown to rest upon inconsistent assumptions about space–time structure and covariance. It is therefore of interest to combine and compare the various ways in which covariance can be violated in order to arrive at a general perspective. (Some of these models have already been presented in an overview form in [7]. The focus of this previous review was on implications for models of black holes, while the present one emphasizes the role of these results for general aspects of background independence and the viability of quantum gravity. Moreover, it presents further comparisons between the different results).

A discrete fundamental theory is not expected to respect all the properties that we are used to from classical space–time. Some violation of classical covariance may therefore be allowed. Nevertheless, because covariance does not only describe a property of classical space–time but also implies that all consistency conditions are met for gravity as a gauge theory, the requirement of general covariance cannot just be abandoned without suitable replacements. One task to be completed for a consistent theory of quantum gravity is to find suitable middle ground between completely broken covariance and the strictly classical notion of general covariance. Considerations of covariance therefore remain important even if one believes that quantum gravity may completely change the structure of space–time in its fundamental formulation.

The examples of violations of covariance discussed here do not directly apply to fundamental quantum gravity but rather to models used for phenomenological studies of cosmology or black holes. In this context, the question of covariance is even more pressing because a general (but often implicit) strategy in this context is to use well-understood Riemannian geometry to analyze potential modifications in the dynamical equations of quantum gravity. Since these modifications may easily affect space–time structure as well, any implicit assumptions about space–time must be uncovered and analyzed before an analysis can be considered meaningful. In this phenomenological context, the question of space–time structure is not as challenging as it is at the fundamental level, but it is still relevant. The task is to show that a certain geometrical structure applies to solutions of an effective description of quantum gravity not only in the strict classical limit where $\hbar = 0$ but also within some finite range of the expansion parameter, given for instance by $\rho/\rho_{\rm P}$ in a cosmological model with energy density ρ relative to the Planck density.

The studies [3–6] of interest here implicitly assume that space–time structure remains unmodified even in the presence of modified dynamics, and sometimes even all the way to the Planck scale [3,6]. This strong assumption is implemented by inserting solutions of modified equations in a standard line element, without checking whether the modified solutions obey gauge transformations compatible with coordinate transformations such that an invariant line element results. Such a line element is crucial in these studies because it enables the formulation of new claims of potential physical effects that make these studies interesting and publishable in high-profile journals. The same ingredient makes these studies vulnerable to violations of covariance, as reviewed in detail in the following sections.

The concluding section of this review points out general properties of covariance in models of loop quantum gravity that may be useful for other approaches. It is generally expected that quantum gravity leads to new geometrical features at large curvature that can no longer be described by a classical form of space–time with its common sense of covariance. Loop quantum gravity is only one approach in which a specific example of discreteness or other non-classical geometrical effects is being explored. The general question to be addressed is then whether quantum gravity at large curvature remains a geometrical theory in the sense that its solutions can still be described in terms of space–time with a certain generalized meaning compared with our classical notion.

2. Models of Loop Quantum Gravity

In order to set up our analysis, we should first introduce the general form of modifications implemented in models of loop quantum gravity (see [8] for more details). It is sufficient to illustrate these modifications by recalling the basics of loop quantum cosmology for spatially flat, isotropic models.

2.1. Holonomy Modifications and Space–Time Structure

The classical dynamics of the scale factor a can be expressed by a canonical pair (q, p) where $q = \dot{a}$ (a proper-time derivative) and $|p| = a^2$, subject to the Friedmann constraint:

$$-\frac{q^2}{|p|} + \frac{8\pi G}{3}\rho = 0 \tag{1}$$

with the energy density ρ. Kinematical aspects of loop quantization suggest the replacement, or "holonomy modification":

$$\frac{q^2}{|p|} \mapsto \frac{\sin(\ell q/\sqrt{|p|})^2}{\ell^2} \tag{2}$$

where ℓ is a suitable, possibly running length scale, such as the Planck length ℓ_P in simple cases.

Taken in isolation, holonomy modifications imply non-singular behavior in isotropic models with a modified Friedmann constraint:

$$\frac{\sin(\ell q/\sqrt{|p|})^2}{\ell^2} = \frac{8\pi G}{3}\rho \qquad (3)$$

because the energy density of any solution to this equation must be bounded (assuming that ℓ is constant, as commonly done in this context). However, this equation includes only one type of expected quantum corrections. In addition, a complete effective description of some underlying dynamics of quantum gravity (of any kind) should also include the remnants of higher-curvature terms in an isotropic model. Higher-curvature terms, just like holonomy modifications, require a given length scale, which we may assume to equal ℓ if holonomy modifications and higher-curvature terms are derived from a single quantum theory of gravity. It is easy to see that higher-curvature terms are not described by (3) because they generically imply higher time derivatives and therefore extend the phase space by additional momenta.

The Equation (3) is therefore incomplete from the viewpoint of effective theory. Nevertheless, it may be useful because it determines at least one type of quantum corrections. However, knowing that there are additional terms not included in (3) that also depend on ℓ, we cannot trust the full function $\sin^2(\ell q/\sqrt{|p|})/\ell^2$ but should rather expand:

$$\frac{\sin(\ell q/\sqrt{|p|})^2}{\ell^2} \sim \frac{q^2}{|p|}\left(1 - \frac{1}{3}\ell^2\frac{q^2}{|p|} + \cdots\right) \qquad (4)$$

and only include leading-order terms. If $\ell \sim \ell_P$, these leading corrections are of the order $\ell_P^2 q^2/|p| \sim \rho/\rho_P$, which is the same as the order expected for higher-curvature terms. Even the leading corrections in (3) should therefore not be considered to be definitely certain and considered with caution. Interpreting the full series expansion or its sum to the sine function as an indication of bounded densities is unjustified in the absence of information about higher-curvature terms.

Higher-curvature terms are also of interest from the point of view of space–time structure. We already used the fact that they generically include higher time derivatives, but the specific appearance of such terms is not arbitrary and is instead guided by requirements of general covariance. In loop quantum cosmology, the form of quantum corrections that may appear in addition to holonomy modifications can therefore be determined only if there is good control on space–time structure in this setting.

Isotropic and homogeneous models are not sufficient for an analysis of space–time structure and covariance because these questions rely on how spatial and temporal dependencies are related in differential equations and their solutions. At least one spatial direction of inhomogeneity should then be included in suitable models, in addition to the non-trivial time dependence already described by models such as (3). While such (midisuperspace) models have been considered in loop quantum gravity for some time, their application to the question of covariance is rather new and has led to several surprising results.

2.2. Three Examples and One Theorem

We will review three examples of the proposed methods to describe inhomogeneity in models of loop quantum gravity and the reasons why they turn out to violate covariance in ways that render them inconsistent. The first example, the dressed-metric approach for cosmological inhomogeneity [9], has been used several times as a crucial ingredient in cosmological model building, leading to claims of observational testability that, given the underlying problems with space–time structure, turn out to be unfounded. (Similar arguments regarding violations of covariance apply to the hybrid approach to inhomogeneity in loop quantum cosmology [10]). The remaining two examples, given by partial

Abelianizations of constraints in spherically symmetric models [4] as well as a misleadingly named "covariant polymerization" [11] in related studies apply to proposed scenarios for quantum black holes. (The proposal of [11] was intended to justify modified equations used for a study of critical collapse in [5]).

In addition, we will describe a detailed no-go theorem based on a minisuperspace description of the static Schwarzschild exterior by a homogeneous time-like slicing, as originally proposed for a different purpose in [6].

3. Dressed-Metric Approach

In classical gravity, as is well known, it is possible to describe cosmological inhomogeneity in the early universe as a coupled system of two independent sets of degrees of freedom, given by inhomogeneous perturbations evolving on a homogeneous background with THE choice of a time coordinate (such as proper time or conformal time). In a discussion of possibly modified dynamics and space–time structure, it is important to remember that these two ingredients, background and perturbations, have rather different properties related to covariance.

3.1. Background and Perturbations

The dynamics of any homogeneous background can be modified without violating covariance because there is a single constraint, (3), which is always consistent with itself in any modified form: because $\{C,C\} = 0$ for any Poisson bracket, Hamilton's equations generated by a constraint C are guaranteed to preserve the constraint equation $C = 0$ imposed on initial values.

Applied to the Friedmann constraint C, we generate equations of motion:

$$\frac{df}{dt} = \{f, NC\} \tag{5}$$

for any phase-space function f, with respect to a time coordinate t indirectly determined by the lapse function $N > 0$. The generic time derivative, applied to solutions of the constraint $C = 0$, can be rewritten as

$$\frac{1}{N}\frac{df}{dt} = \{f, C\} = \frac{df}{d\tau} \tag{6}$$

introducing proper time τ in the last step by the usual definition $d\tau = Ndt$.

All allowed choices of time coordinates (monotonically related to τ) can therefore be described by a single line element:

$$ds^2 = -d\tau^2 + \tilde{a}(\tau)^2 d\sigma^2 \tag{7}$$

where $\tilde{a}(\tau)$ denotes the scale factor subject to potentially modified dynamics, and $d\sigma^2$ is a standard isotropic spatial line element. Because the definition of τ implies that the line element is correctly transformed to:

$$ds^2 = -N^2 dt^2 + \tilde{a}(t)^2 d\sigma^2 \tag{8}$$

for any other time coordinate t, there is a suitable way to describe any modified homogeneous dynamics, subject to a single constraint, by a space–time geometry that is invariant with respect to the full coordinate changes allowed by the symmetry, given by reparameterizations of time.

Coordinate changes are more involved in the case of spatial inhomogeneity because several independent coordinates may be related by transformations. In the canomical language of constraints, the presence of a multitude of independent ones, one Hamiltonian constraint per spatial point as well as diffeomorphism constraints, which implies that a modification of one or more constraints no longer implies the consistency of their Hamiltonian flows with respect to the other constraints. Since the relevant constraints implement space–time transformations, a dedicated space–time analysis then becomes important.

For small, perturbative inhomogeneity, there is a standard way to describe curvature perturbations in terms of combinations of metric and matter fields that are invariant with respect to small coordinate changes [12]. However, compared with the reparameterizations of time relevant for the background, it is much harder to derive a suitable invariant line element extending (8) in a way that is consistent with Hamilton's equations generated by modified constraints for perturbative inhomogeneity. In fact, the standard derivations of curvature perturbations [12,13] as well as the canonical version given in [14] assume that space–time is of its classical form, for instance by directly working with the coordinate substitutions in a line element. A modified or quantum treatment then cannot take it for granted that the form of these curvature perturbations remains unchanged, because the space–time structure itself may be modified in quantum gravity.

The dressed-metric approach proceeds by quantizing standard curvature perturbations on a modified background, leading to wave equations for perturbations on a modified background line element $ds^2 = \tilde{g}_{\alpha\beta} dx^\alpha dx^\beta$ of the form (8). The approach therefore implicitly assumes that space–time structure remains classical even while the dynamics of at least the background are modified. Upon closer inspection, this assumption turns out to be unjustified.

3.2. The Metric's New Clothes

As already pointed out in [15], Bardeen variables or curvature perturbations are "gauge invariant" under small coordinate changes, but not necessarily under all coordinate changes relevant for a given cosmological situation. In particular, in cosmological models of perturbative inhomogeneity, we also need invariance under potentially large background transformations of time, such as transforming from proper time to conformal time.

Small coordinate changes of perturbations and large reparameterizations of background time are not independent of each other. Algebraically, they form a semidirect product rather than a direct one, as shown in [16]. The non-trivial interplay between these transformations can be deduced from vector-field commutators such as:

$$\left[f(t) \frac{\partial}{\partial t}, \zeta^\alpha \frac{\partial}{\partial x^\alpha} \right] = f \dot{\zeta}^\alpha \frac{\partial}{\partial x^\alpha} - \dot{f} \zeta^0 \frac{\partial}{\partial t} \tag{9}$$

which in general are not zero (in contrast to what a direct product would imply) but rather form a small inhomogeneous transformation. This interplay is a general property of perturbations in Riemannian geometry, as encoded in line elements suitable for perturbative inhomogeneity.

The applicability of standard line elements requires the precise algebra of coordinate transformations to be modeled by gauge transformations in a canonical formulation of any gravity theory. However, while the dressed-metric approach assumes the availability of standard line elements with the usual coordinate dependence (but possibly modified metric coefficients), it violates the algebraic condition by its independent treatment of background and perturbations: quantizing the background separately from the perturbations evolving on it implicitly assumes a direct product of coordinate changes. Writing a line element $ds^2 = \tilde{g}_{\alpha\beta} dx^\alpha dx^\beta$ based on modified metric components $\tilde{g}_{\alpha\beta}$ in a dressed-metric model is therefore meaningless.

3.3. Effective Line Element

Because a line element $ds^2 = g_{\alpha\beta} dx^\alpha dx^\beta$ is defined as the square of an infinitesimal distance, it can be meaningful as a description of geometry only if it is independent of coordinate choices that affect dx^α as well as $g_{\alpha\beta}$. For ds^2 to be invariant, the metric coefficients $g_{\alpha\beta}$ must be subject to the standard tensor-transformation law:

$$g_{\alpha'\beta'} = \frac{\partial x^\alpha}{\partial x^{\alpha'}} \frac{\partial x^\beta}{\partial x^{\beta'}} g_{\alpha\beta} \tag{10}$$

if coordinates x^α are transformed to $x^{\alpha'}$.

Canonical quantization in its usual form, as applied in models of loop quantum gravity, does not modify space–time coordinates x^α and their transformations, but it may alter the equations of motion (with respect to these coordinates) for the spatial metric q_{ij} in the generic canonical line element:

$$ds^2 = -N^2 dt^2 + q_{ij}(dx^i + M^i dt)(dx^j + M^j dt). \qquad (11)$$

Modifications of the remaining components, the lapse function N and shift vector M^i, are also determined by canonical equations, although more indirectly because N and M^i do not have unconstrained momenta. In the presence of modifications, altered equations for q_{ij}, N and M^i must remain consistent with coordinate transformations if an effective line element ds^2 is to be meaningful.

A crucial ingredient in a canonical analysis of covariance is therefore given by the transformations of N and M^i, in addition to the more obvious transformations of q_{ij}. The full set of canonical transformations makes use of the specific properties of the constraints of the theory. At this point, the analysis of geometrical properties relevant for effective line elements benefits from a discussion of hypersurface deformations in space–time, which are generated from the constraints. While properties of hypersurface deformations constitute some of the classic results in canonical general relativity [17–21], they do not appear to be widely known. What follows is a construction of hypersurface deformations based on elementary properties of special relativity.

3.3.1. Hypersurface Deformations

In special relativity, an observer moving at speed v assigns new coordinates to events in space–time according to a Lorentz transformation:

$$x' = \frac{x - vt}{\sqrt{1 - v^2}} \quad , \quad t' = \frac{t - vx}{\sqrt{1 - v^2}}. \qquad (12)$$

Interpreting this transformation as a linear deformation of axes in a space–time diagram, as shown in Figure 1, the set of all Poincaré transformations can be geometrically represented by linear hypersurface deformations with respect to lapse functions $N(\mathbf{x}) = \Delta t + \mathbf{v} \cdot \mathbf{x}$ (deformations in the normal direction of a spatial slice) and shift vector fields $\mathbf{M}(\mathbf{x}) = \Delta \mathbf{x} + \mathbf{R}\mathbf{x}$ (tangential deformations within a spatial slice). The parameters in these expressions for linear lapse functions and shift vector fields determine a time translation Δt, a boost velocity $\mathbf{v}$, a spatial shift $\Delta \mathbf{x}$ and a spatial rotation matrix $\mathbf{R}$.

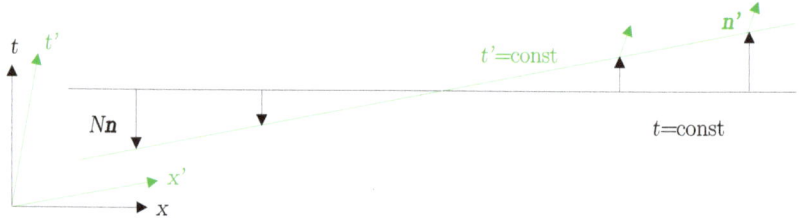

Figure 1. A Lorentz transformation in Minkowski space–time, shown in the traditional way by means of axes as well as in terms of linear normal deformations of a spatial slice. A slice $t = $ const in the original coordinate system was transformed to a new spatial slice $t' = $ const by a linear deformation with position-dependent displacement $N(x) = N_0 + vx$ along the unit normal vector field $\mathbf{n}$.

We extend these considerations to general relativity by replacing the restricted set of translations, rotations and Lorentz boosts with arbitrary non-linear coordinate changes. Correspondingly, hypersurfaces are subject to non-linear deformations [17]. Infinitesimal hypersurface deformations in Riemannian space–time, split into "temporal" deformations

$T(N)$ in a normal direction and "spatial" deformations $S(\mathbf{M})$ in tangential directions, can be shown to obey the commutators:

$$[S(\mathbf{M_1}), S(\mathbf{M_2})] = S((\mathbf{M_1} \cdot \nabla)\mathbf{M_2} - (\mathbf{M_2} \cdot \nabla)\mathbf{M_1}) \qquad (13)$$
$$[T(N), S(\mathbf{M})] = -T(\mathbf{M} \cdot \nabla N) \qquad (14)$$
$$[T(N_1), T(N_2)] = S(N_1 \nabla N_2 - N_2 \nabla N_1) \qquad (15)$$

when they are applied in two alternative orderings. A visualization is shown in Figure 2. The brackets (13)–(15) represent general covariance in canonical form. While specific expressions for S and T can vary depending on the gravitational theory, such as different higher-curvature actions [22], the brackets remain the same as long as the underlying geometry of space–time is Riemannian. Conversely, deviations of the brackets from their Riemannian form can be used to detect non-classical space–time structures in modified canonical gravity. The algebraic nature of the brackets makes it possible to analyze gravitational theories without presupposing specific geometrical formulations of space–time, constituting a major strength of the canonical approach.

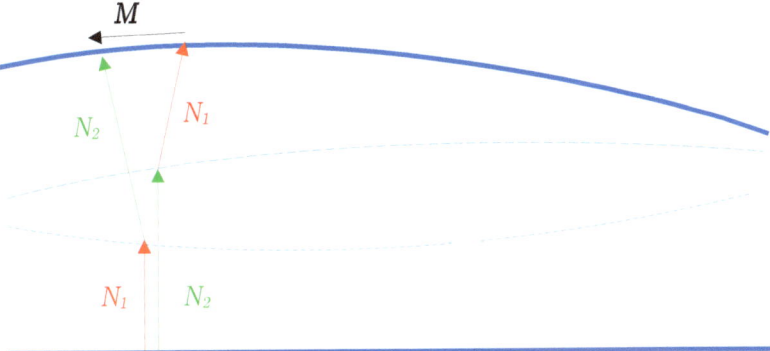

Figure 2. Two non-linear normal deformations, one with a lapse function N_1 and one with a lapse function N_2, applied in two different orderings, show the commutator (15) given by a spatial displacement **M**.

Figure 3 represents the commutator of an infinitesimal time translation and an infinitesimal normal deformation. This picture can be interpreted as a version of the vector-field commutator (9) of a background transformation and a small perturbative transformation. The non-zero result of (9) corresponds to the presence of a spatial shift on the right-hand side of Figure 3. Even though there is no immediate time dependence of the canonical data on which a background vector field as in (9) would act, the semidirect product of background and perturbative transformations is clear. In canonical language, the failure of the dressed-metric approach to realize the correct semidirect product means that there is no common $T(N)$ for background and perturbations in this setting. The non-existence of consistent temporal deformations signals the break-down of space–time and covariance.

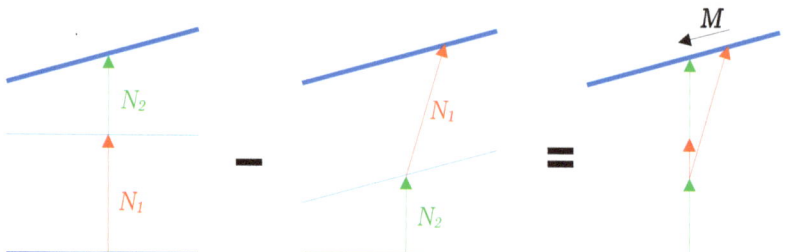

Figure 3. Semidirect product of time reparameterizations and inhomogeneous transformations as in (9), represented in the picture of hypersurface deformations: The commutator of two such normal deformations produces a non-zero spatial shift **M**.

3.3.2. Structure Functions

The brackets of hypersurface deformations have structure functions because the gradient in (15) requires the use of the spatial metric, and therefore depend on the geometry described by these brackets. A canonical realization of these brackets is given by the Hamiltonian and diffeomorphism constraints, $H[N]$ and $D[M^i]$, of a given gravity theory. Written in the form:

$$\{D[M_1^i], D[M_2^j]\} = D[[M_1, M_2]^i] \tag{16}$$
$$\{H[N], D[M^i]\} = -H[M_1^i \nabla_i N] \tag{17}$$
$$\{H[N_1], H[N_2]\} = D[q^{ij}(N_1 \nabla_j N_2 - N_2 \nabla_j N_1)], \tag{18}$$

they make the appearance of structure functions explicit, depending on the inverse spatial metric q^{ij}. Formally, we may write the constraint brackets as $\{C_A, C_B\} = F_{AB}^D C_D$ with indices A, B and D that combine spatial positions with the type of constraint (Hamiltonian or a component of the diffeomorphism constraint). The coefficients F_{AB}^D are not constants but phase-space functions.

The presence of structure functions causes long-standing problems in the quantization of canonical gravity [23,24]: upon quantization, q^{ij} as well as D and H are turned into operators. Maintaining closed brackets therefore requires specific ordering, regularization, or other choices. Even if the brackets can remain closed under certain conditions, quantized structure functions may be quantum corrected. A question relevant for covariance is then that of whether a meaningful interpretation of the generators as hypersurface deformations in space–time still exists.

As shown in [25], a meaningful space–time interpretation does exist at least in some cases of modified structure functions. To see this, it is necessary to construct a space–time line element that is consistent with the modified gauge transformations generated by (18) with the quantum-corrected structure functions. If these functions are modified, so are the versions of hypersurface deformations they represent, and therefore the objects q_{ij}, N and M^i in which the brackets are formulated, do not directly define the components of a meaningful line element because this notion is based on classical space–time with standard hypersurface deformations. However, in some cases, suitable redefinitions of the canonical fields are available that can serve this purpose.

A derivation of proper effective line elements is based on the general property of Hamiltonian and diffeomorphism constraints as generators of evolution equations, giving the time derivative:

$$\dot{f} = \mathcal{L}_t f = \{f, H[N] + D[M^i]\} \tag{19}$$

of any phase-space function f with respect to the time-evolution vector field $t^\alpha = Nn^\alpha + M^\alpha$. (The space–time vector field M^α is the push-forward of the spatial vector field M^i by the embedding map of a spatial slice in space–time). In addition, the constraints generate gauge transformations:

$$\delta_\epsilon f = \{f, H[\epsilon] + D[\epsilon^i]\} \tag{20}$$

which would correspond to coordinate changes generated by the vector field $\zeta^\alpha = \epsilon n^\alpha + \epsilon^\alpha$ if structure functions were unmodified.

In all cases—modified and unmodified structure functions—evolution equations and gauge transformations must be consistent with each other: a gauge-transformed f must evolve according to the general Equation (19) with the same generators H and D as the original f, but possibly with a new time-evolution vector field. Since the direction of the time-evolution vector field within a given theory is determined by lapse and shift, this consistency condition can be used to derive gauge transformations for N and M^i. Together with the gauge transformations of q_{ij}, directly determined by (20) because q_{ij} are phase-space functions, all components of a candidate space–time line element can therefore be unambiguously transformed.

For generic structure functions F^D_{AB}, evolution and gauge transformations are consistent with each other, provided the multipliers $(N^A) = (N, M^i)$ gauge transform according to [26]:

$$\delta_\epsilon N^A = \dot{\epsilon}^A + N^B \epsilon^C F^A_{BC}. \tag{21}$$

Unlike in the case of $\delta_\epsilon q_{ij} = \{q_{ij}, H[\epsilon] + D[\epsilon^i]\}$, the structure functions appear explicitly in (21). Structure functions, and their possible modifications, are therefore directly relevant for space–time structure and the existence of meaningful effective line elements:

$$ds^2 = -\tilde{N}^2 dt^2 + \tilde{q}_{ij}(dx^i + \tilde{M}^i dt)(dx^j + \tilde{M}^j dt) \tag{22}$$

which may require field redefinitions of $\tilde{N}$, $\tilde{M}^i$ as well as $\tilde{q}_{ij}$ if the structure functions F^A_{BC} are modified.

3.4. Lessons from Hypersurface Deformations

In canonical models of modified gravity, control on space–time structure requires full expressions for the Hamiltonian constraint $H[N]$ and the diffeomorphism constraint $D[M^i]$ with closed brackets. This condition is violated in the dressed-metric approach (as well as in hybrid loop quantum cosmology) because the independent treatment of remnant coordinate freedom in background and perturbations, the former through deparameterization and the latter by using curvature perturbations, precludes the construction of joint constraints for both sets of degrees of freedom. The common assumption that space–time in this setting can still be described by a line element, presupposing a Riemannian structure of space–time, is therefore unjustified. Detailed discussions of the underlying modifications of contributions to the Hamiltonian constraint from background and perturbations show that the implicit assumption of unmodified brackets, and thus Riemannian structures, is inconsistent [16].

For a consistent space–time structure, the gauge behavior of the classical theory must remain intact, even while it may be modified and subject to quantum effects. In general, this condition requires anomaly freedom, such that the same number of physical degrees of freedom as in classical gravity is realized in a modified version. If this condition is violated, the modified theory cannot have the correct classical limit owing to a discontinuity in the number of degrees of freedom. An anomalous modification or quantization of gravity does not permit a semiclassical or effective treatment by line elements in any form because it is incompatible with the gauge structure of space–time.

A formal statement of the condition that the gauge behavior remains intact is the existence of closed Poisson brackets of $H[N]$ and $D[M^i]$ for all relevant N and M^i, depending on whether one considers the full theory or a restricted version such as a midisuperspace model. This condition allows for possible quantum corrections in the structure functions of the gauge algebra, given in the case of gravity by the inverse spatial metric q^{ij} as it appears in:

$$\{H[N_1], H[N_2]\} = D[\beta(q,p)q^{ij}(N_1\nabla_j N_2 - N_2\nabla_j N_1)] \tag{23}$$

with a possible modification function $\beta(q,p)$ on phase space. We have the classical space–time structure if $\beta = \pm 1$, giving two possible choices of the signature of a classical four-dimensional metric, where $\beta = 1$ for Lorentzian-signature space–time and $\beta = -1$ for 4-dimensional Euclidean-signature space. (In each case, the name only refers to the signature and does not imply flatness).

We have a consistent non-classical space–time structure if the brackets are closed such that $\beta \neq \pm 1$. The modification function β determines the structure functions of hypersurface-deformation brackets in the modified theory. Modified structure functions, in turn, show via (21) how lapse and shift transform and whether it is possible to find suitable field redefinitions of these fields that can be used in a proper effective line element as discussed in detail in [25].

As we saw in the present section, suitable transformations of lapse and shift as components of the space–time metric require knowledge of the structure functions of $H[N]$ and $D[M^i]$. If the brackets do not close, as in the dressed-metric approach, there are no meaningful transformations of lapse and shift and it is impossible to construct a valid structure of space–time. Such a structure exists only in anomaly-free modifications of the constraints. However, the condition of anomaly-freedom is not sufficient if it does not imply a clear modification of the structure function of hypersurface-deformation brackets, for instance in cases in which the constrained system is reformulated before it is modified or quantized. An example for such an approach is given by a partial Abelianization of the constraints [4], to which we turn next.

4. Spherical Symmetry

An instructive set of examples is given by spherically symmetric space–time geometries with the line element:

$$ds^2 = -N^2 dt^2 + L^2(dx + Mdt)^2 + S^2(d\vartheta^2 + \sin^2\vartheta d\varphi^2) \tag{24}$$

where N, L, M and S are functions of t and x. Together with the momenta p_L and p_S of L and S, respectively, the components L and S of the spatial metric in classical general relativity are subject to the Hamiltonian constraint:

$$H[N] = \int_{-\infty}^{\infty} N\left(-\frac{p_L p_S}{S} + \frac{L p_L^2}{2S^2} + \frac{(S')^2}{2L} + \frac{SS''}{L} - \frac{SS'L'}{L^2} - \frac{L}{4}\right) dx \tag{25}$$

and the diffeomorphism constraint:

$$D[\epsilon] = \int_{-\infty}^{\infty} \epsilon(p_S S' - L p_L') dx. \tag{26}$$

The relevant bracket with a stucture function is given by

$$\{H[N_1], H[N_2]\} = D[L^{-2}(N_1 N_2' - N_2 N_1')]. \tag{27}$$

4.1. Reformulating the Constrained System

In [4], a reformulation of the constraints has been suggested that can remove the structure function and even partially Abelianize the brackets. Instead of $H[N]$, this reformulation uses the linear combination:

$$H[2PS'/L] + D[2Pp_L/(SL)] = \int_{-\infty}^{\infty} P\frac{d}{dx}\left(-\frac{p_L^2}{S} + \frac{S(S')^2}{L^2} - S\right) dx \tag{28}$$

of Hamiltonian and diffeomorphism constraints. Specifically, the combination replaces $H[N]$ with a new constraint whose integrand (except for the multiplier P) is a complete derivative. Imposing (28) as a constraint therefore requires that the parenthesis in this expression equals a constant, C_0. The same condition can be expressed by the alternative constraint:

$$C[Q] = \int_{-\infty}^{\infty} Q \left(-\frac{p_L^2}{S} + \frac{S(S')^2}{L^2} - S - C_0 \right) dx. \tag{29}$$

(The constant can be related to boundary values). Because $C[Q]$ depends neither on p_S nor on spatial derivatives of L, it is easy to see that two such constraints always have a vanishing Poisson bracket, unlike two Hamiltonian constraints. Together with the original diffeomorphism constraint, we have the brackets:

$$\{C[Q], D[\epsilon]\} = -C[(\epsilon Q)'] \quad , \quad \{C[Q_1], C[Q_2]\} = 0 \tag{30}$$

free of structure functions. Therefore, it may be expected that using the reformulated constraints greatly simplifies the quantization procedure or the derivation of viable modifications.

However, the reformulation has made use of metric-dependent coefficients S'/L and $p_L/(SL)$ in (28). In general, it is not clear whether these coefficients will be subject to quantum corrections, in which case it may be difficult or impossible to reconstruct valid hypersurface-deformation brackets with the correct classical limit from a quantization or modification of the system (30). The non-trivial nature of this question has been shown in [27] and the related [28], where examples were presented in which (30) can easily be modified while no hypersurface-deformation brackets can be reconstructed at all or only in modified form.

For instance, the modification:

$$C_f[Q] = \int_{-\infty}^{\infty} Q \left(-\frac{f(p_L)^2}{S} + \frac{S(S')^2}{L^2} - S - C_0 \right) dx \tag{31}$$

with a free function $f(p_L)$, such as $\sin(\ell p_L)/\ell$ where ℓ is a suitable length scale analogous to (3), and an unchanged $D[\epsilon]$ maintains the brackets (30) and is therefore anomaly-free in the reformulated system. By reverting the steps undertaken in (28), it can be seen that (31) corresponds to the modified Hamiltonian constraint:

$$H_f[N] = \int_{-\infty}^{\infty} N \left(-\frac{p_S}{S} \frac{df(p_L)}{dp_L} + \frac{Lf(p_L)}{2S^2} + \frac{(S')^2}{2L} + \frac{SS''}{L} - \frac{SS'L'}{L^2} - \frac{L}{4} \right) dx. \tag{32}$$

This modification of the Hamiltonian constraint, which has already been found in [29], also turns out to be anomaly-free, but with a modified bracket:

$$\{H_f[N_1], H_f[N_2]\} = D[\beta(p_L)L^{-2}(N_1 N_2' - N_2 N_1')] \tag{33}$$

where:

$$\beta(p_L) = \frac{1}{2} \frac{d^2 f}{dp_L^2}. \tag{34}$$

The modified structure function is an example of signature change because β is negative around any local maximum of f.

If spherically symmetric gravity is coupled to a scalar field, the partial Abelianization of [4] is still available and can be modified as in (31). However, in this case, there is no consistent set of hypersurface-deformation generators [27]. Therefore, the modified theory is formally consistent but not geometrical: its solutions cannot be described by Riemannian geometry or effective line elements, even after a field redefinition. This problem poses a significant challenge to loop quantization because an application to vacuum models would only be too restrictive. Moreover, the problem is broader because polarized Gowdy models, which can also be partially Abelianized, do not admit a consistent set of modified hypersurface-deformation brackets [28]. To date, therefore, midisuperspace models with local physical degrees of freedom cannot be geometrically described in the presence of loop modifications.

4.2. Non-Bijective Canonical Transformation

To circumvent this problem, ref. [11] proposed a modification of spherically symmetric gravity based on a non-bijective canonical transformation:

$$p_L = \frac{\sin(\ell \tilde{p}_L)}{\ell} \quad , \quad L = \frac{\tilde{L}}{\cos(\ell \tilde{p}_L)} . \tag{35}$$

The transformation can be applied to the Abelianized constraint $C[Q]$ or to the Hamiltonian constraint by inserting $p_L(\tilde{p}_L)$ and $L(\tilde{L}, \tilde{p}_L)$ in their classical expressions. (The diffeomorphism constraint is not modified by this transformation.) Terms depending on p_L in $C[Q]$ are then modified as before in (31) with a specific version of $f(p_L)$, and there are new modifications in the L-term. As postulated in [11], this procedure, based on a canonical transformation, might be able to preserve the covariance of the classical theory even in the presence of a scalar field, and yet allow room for new quantum effects because of the non-bijective nature of the canonical transformation.

Unfortunately, this hope remains unfulfilled precisely because the transformation is not bijective [30]. In particular, the bijective nature breaks down at hypersurfaces defined by $\ell p_L = \pm 1$ or $\ell \tilde{p}_L = (n + 1/2)\pi$, and p_L as well as $\tilde{p}_L$ are spatial scalars but not space–time scalars. Therefore, while the transformation preserves symmetries of the classical theory when it can be restricted to regions of phase space in which it is bijective, these regions themselves are defined in terms that are not space–time covariant. The resulting theory is not covariant.

For the same reason, $\tilde{p}_L$ not being a space–time scalar, the variable $\tilde{L}$ introduced by the canonical transformation does not have the same behavior as $L = \tilde{L}/\cos(\ell \tilde{p}_L)$ under space–time transformations. As a consequence, $\tilde{L}$ cannot be used in a space–time line element based on $\tilde{L}^2 \mathrm{d}x^2$. A meaningful effective line element is obtained only after a suitable field redefinition that leads to a function of $\tilde{L}$ with the correct transformation properties. Since we already know that $\tilde{L}$ was derived from such a function, L, the field redefinition simply sends us back from $\tilde{L}$ to L in regions in which the canonical transformation is invertible, undoing the modification of the theory in such regions. (More systematically, such a field redefinition can be derived using the methods introduced in [31].) In these regions, exact classical solutions without any modifications are produced, but different regions are connected along hypersurfaces (again, given by $\ell p_L = \pm 1$ or $\ell \tilde{p}_L = (n + 1/2)\pi$) that are not covariantly defined. Since these hypersurfaces refer to fixed values of certain components of extrinsic curvature, their positions in space–time depend on choices of coordinates and spatial slicings.

In particular, slicings with large $p_L \sim 1/\ell$ exist even in flat space–time, and therefore violations of covariance in this model cannot be considered a "large-curvature effect". These violations can occur at a low space–time curvature (in an invariant meaning), and therefore the model cannot be considered a permissible model of quantum gravity that would have non-standard geometrical features only at the Planck scale. The model could be permissible only if it were combined with a mechanism that somehow prevents one from choosing slicings that lead to large extrinsic curvature p_L. However, preventing such slicings (or any slicing) from being allowed requires violations of covariance that are hard to reconcile with the application of line elements, even if they were only used in low-curvature regions.

4.3. Bijective Canonical Transformation

As discussed in more detail in [30], the application of canonical transformations makes an analysis of space–time structure rather non-trivial even if the transformation is bijective. A bijective canonical transformation from (L, p_L) to some $(\tilde{L}, \tilde{p}_L)$ may well be such that all possible values of p_L are mapped to a finite range of $\tilde{p}_L$. One could then conclude that the transformed theory resolves singularities if $\tilde{p}_L$, interpreted as some curvature expression in the new theory, remains bounded. However, the new theory was obtained by applying a bijective canonical transformation that cannot modify the physics of classical spherically symmetric models.

The answer to this conundrum relies on effective line elements. For a transformation with a significantly modified $\tilde{p}_L$ to be canonical, $\tilde{L}$ must also be modified compared with L. Then, the structure function in (27) is modified when expressed in terms of $\tilde{L}$ instead of L, and solutions of the transformed theory cannot be directly interpreted in terms of a line element where $\tilde{L}$ directly takes the place of L. An effective line element, derived again as in [31]), requires the undoing of the canonical transformation for a valid coefficient of dx^2, sending us back to the classical theory in its geometrical interpretation.

Models of loop quantum gravity are not obtained by bijective canonical transformations and could lead to new physics. However, the example of a bijective canonical transformation demonstrates that predictions can only be reliable if a proper effective line element is derived. Unfortunately, this task is rarely performed in phenomenological studies of models of loop quantum gravity. In several proposals, as in the dressed-metric approach, it is even impossible to construct an effective line element because they do not amount to consistent modifications of the crucial bracket (27) that determines the structure of space–time.

5. Homogeneity in Schwarzschild Space–Time

It is well known that a spatially homogeneous geometry of Kantowski–Sachs type [32], with the line element:

$$ds^2 = -N(t)^2 dt^2 + a(t)^2 dx^2 + b(t)^2 \left(d\vartheta^2 + \sin^2\vartheta d\varphi^2 \right) \tag{36}$$

is realized in the Schwarzschild interior—in the (almost) standard version:

$$ds^2 = -(1 - 2M/r) d\tilde{t}^2 + \frac{dr^2}{2M/r - 1} + r^2 \left(d\vartheta^2 + \sin^2\vartheta d\varphi^2 \right) \tag{37}$$

of the Schwarzschild line element, $\tilde{t}$ is a time coordinate only for $r > 2M$, outside of the horizon. For $r < 2M$, the coordinate r may be used as time while $\tilde{t}$ contributes to a positive, space-like part of the line element. Indicating the modified roles of the coordinates in the notation, we define $t = r$ and $x = \tilde{t}$ for $r < 2M$, such that the line element turns into:

$$ds^2 = -\frac{dt^2}{2M/t - 1} + (2M/t - 1) dx^2 + t^2 \left(d\vartheta^2 + \sin^2\vartheta d\varphi^2 \right) \tag{38}$$

for $t < 2M$. A suitable identification of $N(t)$, $a(t)$ and $b(t)$ shows that this line element is of the general form (36).

The coordinates t and x determine a homogeneous space-like slicing in the interior of Schwarzschild space–time. It is therefore possible to apply minisuperspace quantizations to the interior region. However, such models do not show how a modified quantum interior may be connected to an inhomogeneous exterior, and they do not reveal properties of space–time structure (let alone physical processes such as occasionally hypothesized explosions of black holes).

5.1. Time-Like Homogeneity of Exterior Static Solutions

A complex canonical transformation $A = ia$ and $p_A = -ip_a$ together with $n = iN$ in (36) implies a Kantowski–Sachs line element of the form:

$$ds^2 = n(t)^2 dt^2 - A(t)^2 dx^2 + b(t)^2 \left(d\vartheta^2 + \sin^2\vartheta d\varphi^2 \right). \tag{39}$$

The complex transformation has the same effect as crossing the horizon in the Schwarzschild geometry: it flips the roles of t and x as time and space coordinates. Defining $X = t$ and $T = x$, the transformed line element (40) takes the form:

$$ds^2 = -A(X)^2 dT^2 + n(X)^2 dX^2 + b(X)^2 \left(d\vartheta^2 + \sin^2 \vartheta d\varphi^2 \right). \tag{40}$$

The exterior Schwarzschild line element:

$$ds^2 = -(1 - 2M/X) dT^2 + \frac{dX^2}{1 - 2M/X} + X^2 \left(d\vartheta^2 + \sin^2 \vartheta d\varphi^2 \right) \tag{41}$$

with $X > 2M$ is now of this general form. In particular, the coordinates T and X determines a homogeneous time-like slicing in the exterior. Methods of minisuperspace quantization can therefore be applied even to inhomogeneous geometries [6], possibly leading to modified space–time structures.

Symmetries of individual space–time solutions such as homogeneity, as opposed to general covariance which relates different solutions of the underlying partial differential equations, are built into the setup of the model. Therefore, they are preserved by minisuperspace quantization. Time-like homogeneity then remains intact for any modified dynamics in this setting. As shown in Figure 4, time-like homogeneity with the given number of degrees of freedom, in turn, implies the existence of a static spherically symmetric configuration if the resulting theory is covariant and slicing-independent (described by a meaningful line element). Since the black-hole analysis of [6] is based on line elements and refers to notions of Riemannian geometry, such as horizons, curvature scalars or Penrose diagrams, slicing independence is one of the ingredients of the construction and does not need to be assumed independently. It must therefore be possible to formulate the same physics claimed in [6] for a homogeneous time-like slicing also within a covariant spherically symmetric theory, restricted to static solutions.

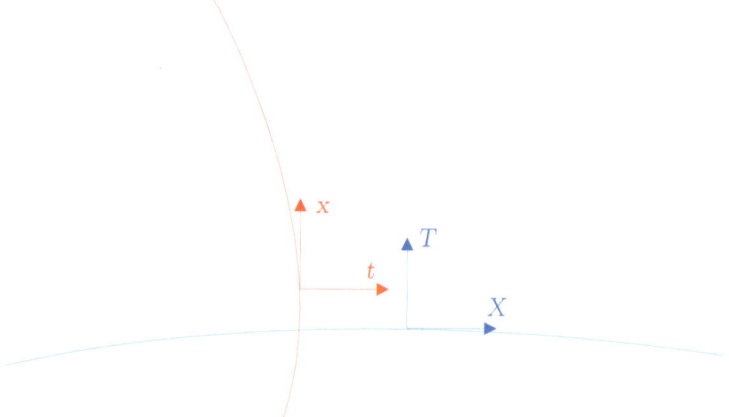

Figure 4. A homogeneous time-like slicing with coordinates (t, x) and an inhomogeneous space-like slicing with coordinates (T, X), both in the same static spherically symmetric space–time.

Covariant versions of spherically symmetric gravity models and their static solutions are under good control, thanks to work on dilaton gravity [33,34] and its generalizations [35,36]. It is therefore possible to check whether a proposed modification of the homogeneous time-like slicing has a chance of corresponding to a covariant theory.

5.2. Line Elements

Time-like homogeneity with modified dynamics leads to a formal line element:

$$ds^2 = \tilde{n}(t)^2 dt^2 - \tilde{A}(t)^2 dx^2 + \tilde{b}(t)^2 \left(d\vartheta^2 + \sin^2\vartheta d\varphi^2 \right) \tag{42}$$

if solutions $\tilde{n}$, $\tilde{A}$ and $\tilde{b}$ are simply inserted in the classical line element. Since properties of space–time transformations have not been checked at this point, there is no guarantee that (42) presents a proper effective line element.

Assuming that the Kantowski–Sachs-like (42) is a proper line element that describes a slicing-independent theory, it is equivalent to the Schwarzschild-like:

$$ds^2 = -K(X)^2 dT^2 + L(X)^2 dX^2 + S(X)^2 \left(d\vartheta^2 + \sin^2\vartheta d\varphi^2 \right) \tag{43}$$

where $X = t$, $T = x$ and:

$$\tilde{A} = K \quad, \quad \tilde{b} = S \quad, \quad \tilde{n} = L. \tag{44}$$

By construction, the coefficients in (43) depend on X but not on T. The line element therefore presents a static solution in a spherically symmetric model, subject to some modified dynamics because K, L and S are only Schwarzschild-like but not exactly of Schwarzschild form if the dynamics of the underlying homogeneous model is modified.

If the assumption of covariance, made implicitly in [6] is justified, (43) must be a solution of a $1+1$-dimensional gravity model in terms of time and space coordinates (T, X). Such theories are under strong control: all local covariant theories of this midisuperspace form are known as generalized dilaton gravity models [35]. (Their equivalence to Horndeski theories in $1+1$ dimensions has been shown in [36]). While several-free functions exist in this general setting to specify the dynamics, for instance through an action, as they can only depend on the variable analogous to our field S. Loop quantum cosmology applied to the homogeneous time-like slicing, however, implies modifications that do not fulfill this condition: such minisuperspace modifications depend non-linearly on momenta $p_{\tilde{A}}$ and $p_{\tilde{b}}$, which are linear combinations of $d\tilde{A}/dt$ and $d\tilde{b}/dt$ that, according to (44), are translated to $\partial K/\partial X$ and $\partial S/\partial X$ in the spherically symmetric slicing. Therefore, no holonomy modified dynamics of Kantowski–Sachs-style models can be part of a covariant space–time theory [37].

6. Conclusions

We discussed the main constructions that were supposed to circumvent difficulties in earlier applications of loop quantization to inhomogeneous models. Instead of solving older problems, however, these constructions led to no-go results for covariance in models of loop quantum gravity. A complete understanding of covariance in any given model is important not only to demonstrate its consistency, but also to evaluate possible observational implications of the underlying theory. For instance, if one neglects the identification of suitable space–time structures for a model of modified or quantum gravity, one could be led to posing initial conditions at an inadmissible place where there is, in fact, no meaningful version of time. A detailed space–time analysis may well show other regions in which initial values could reliably be posed, however, the altered location, perhaps at a different range of curvature values, would affect implied phenomenological effects. Addressing such questions requires an understanding of different ways in which covariance can be violated, which we compare in the next subsection. The final two subsections will discuss general implications for loop quantum gravity and a brief outlook on covariance in other approaches.

6.1. Comparison of Different Violations of Covariance

The examples reviewed in the preceding sections show different ways in which covariance can be violated in models of loop quantum gravity. The dressed-metric approach, just as hybrid loop quantum cosmology, is based on the incorrect assumption that background and perturbations can be quantized or modified independently in an inhomogeneous model. This assumption ignores a crucial feature of space–time and covariance, according to which background and perturbative transformations form a semidirect product but not a direct one as an independent treatment would require. The fundamental nature of this property implies that covariance is completely broken in these models, which are therefore inconsistent as a description of (quantum) space–time.

As usual, one may expect that space–time is non-classical at large curvature and may exhibit properties different from classical space–time. However, this expectation does not redeem quantum models that violate covariance unless they can demonstrate that the classical properties are recovered in a suitable classical limit. Moreover, the dressed-metric and the hybrid approach both refer to features of classical space–time, such as line elements or curvature perturbations, even close to the Planck curvature.

The inconsistency of these approaches is rooted not so much in possible modifications of classical space–time properties near the Planck curvature, but rather in the unquestioned (and often implicit) application of classical space–time ingredients for an analysis in this regime. For a model to be consistent, such an assumption must be justified, but this crucial step has not been attempted in the dressed-metric and hybrid approaches. There is therefore reason to doubt the validity of these constructions and their implications.

The technical observation that a key property of classical space–time is violated, given by the semidirect-product nature of transformation, serves as a concrete property that turns this doubt into a proof that the models are inconsistent, not only in the Planck regime but to any order in a semiclassical expansion by $\hbar$ or ℓ_P. Consistency is recovered only in the strict limit of $\hbar \to 0$, just because we happen to know that the classical theory is covariant and has solutions that can be described by line elements. In such modifications, there is a strong discontinuity at $\hbar = 0$ in geometrical structures, seen as an $\hbar$-dependent family of modifications. In practice, this discontinuity translates into low-curvature physical problems, as discovered in the case of black-hole models of loop quantum gravity in [38,39].

Similarly, the original attempt in [6] to describe the inhomogeneous Schwarzschild exterior by homogeneous models, using time-like slicings in a static geometry, was based on an untested assumption that is true in classical space–time but may be violated in the presence of quantum modifications. The description of inhomogeneity in this case is different from the preceding example because it is non-perturbative in a space-like slicing. Here, homogeneous and inhomogeneous configurations do not appear as background and perturbations, but rather as models of a single space–time geometry using two different slicings.

Classically, any slicing gives an equivalent description of the full geometry, but this does not need to be the case once equations have been modified, in contrast to what has implicitly been assumed in [6]. The good control on covariant local theories for spherically symmetric dynamics makes it possible to test and invalidate this assumption. Again, it is the application of line elements in [6] even in the presence of quantum modifications that makes it possible to demonstrate inconsistency. It is not necessary to assume additional classical features in the inconsistency proof, beyond properties that have already been used in [6], explicitly or implicitly.

Models that work directly with spherically symmetric inhomogeneity usually tread more carefully because the appearance of first-class constraints is explicit. A consistent quantization or modification then requires that the first-class nature be preserved, i.e., that there are no anomalies, in order to prevent spurious degrees of freedom or over-constraining the theory. However, even in an anomaly-free modification, the structure of space–time and geometry may remain unclear without further analysis. Here, our remaining two examples are relevant, given by different modifications implemented for reformulated, partially Abelianized constraints and modification through a non-bijective canonical transformation, respectively. These modifications are anomaly-free and therefore consistent in a formal sense used for general constrained systems. Nevertheless, they turn out to violate covariance in different ways, even though the papers in which they have been proposed go on and analyze their solutions by standard line elements.

6.2. Covariance Crisis of Loop Quantum Gravity

As we just saw, a crucial ingredient of proofs of inconsistency and non-covariance in models of loop quantum gravity focuses on the application of line elements used routinely to evaluate solutions of modified equations in canonical gravity. Since modifications of canonical equations need not preserve covariance, even if they may remain formally consistent and anomaly-free, line elements are rendered meaningless. It might therefore be possible to evade some of the no-go results by foregoing line elements or related and more advanced methods, such as Penrose diagrams. In principle, a physical analysis would still be possible, at least in the anomaly-free case, by expressing solutions of anomaly-free modified equations in terms of suitable canonical observables.

However, this option is rarely exercised in interesting models because of the complicated nature of deriving strict observables, compared with the simple procedure of modifying coefficients in a formal line element. Furthermore, if such an analysis could be performed, it would not be clear in which sense solutions of the modified theory could still be considered geometrical, even when quantum modifications are very small, or more practically, how one would define the horizon of a black hole or curvature perturbations for cosmology in the absence of geometry. The important covariant form of general relativity and its geometrical nature would be a mere accident of the classical theory, rather than a fundamental property of gravity that could be extended to even the tiniest of corrections. While requiring a geometrical nature for quantum gravity may be largely a matter of taste, it also has practical implications because most of the gravitational methods and definitions that we know and understand are based on geometry.

A few additional ways might remain to solve these deep problems. First, in the context of Section 5, non-local effects might help because they would evade the strong control on possible covariant theories with spherical symmetry. However, the underlying analysis of minisuperspace dynamics in [6] implicitly assumes locality because there is a single momentum for each classical metric or triad component. If one were to try non-locality in order to solve the covariance problem in models of loop quantum gravity, the entire formalism used until now would have to change, even in minisuperspace models. Moreover, non-locality is often pathological and there is no indication that loop quantization could lead to more controlled situations.

Secondly, one may try to understand non-Riemannian space–time structures as they would be implied by modified hypersurface-deformation brackets ($\beta \neq \pm 1$). In some (but not all) cases, these modified geometries can be described by an effective Riemannian line element after suitable field redefinitions. At present, such models, recently analyzed in [25,40–42], are the only well-defined descriptions of geometries that may incorporate quantum modifications. If suitable field redefinitions exist, strict effective line elements are available, but in the presence of holonomy modifications they generically imply a signature change at high curvature.

There has been progress in constructing anomaly-free versions of the Hamiltonian constraint directly at the operator level in various versions of loop quantum gravity [43–48]. These constructions do not directly refer to symmetry-reduced models but, for now, implement restrictions of general ingredients such as the spatial dimension, the local gauge group, or the signature of gravity. In this approach, progress is usually made by reformulating the constraints, simplifying their brackets in a way that is conceptually similar to partial Abelianizations discussed in Section 4. As in this case, the successful construction of anomaly-free reformulated constraints does not immediately reveal whether they describe a consistent structure of space–time or covariance.

6.3. Lessons for Other Approaches

Background independence implies that space–time structure must be derived in some way and cannot be presupposed. We should not simply assume that inserting modified solutions in classical-type line elements is consistent. As a consequence, quantum gravity may not be "geometrical" as we understand it from general relativity. In the main body of this paper, we discussed how the canonical nature of loop quantum gravity gives access to powerful space–time methods, based on algebra, that can be used to rule out many models that might otherwise look reasonable.

It is not easy to see whether there may be possible analogs of our results in alternative approaches to quantum gravity if they are not canonical. Nevertheless, we are able to draw several lessons of general form. First, non-canonical theories do not directly aim to quantize generators of hypersurface deformations, but it should still be of interest to construct them and consider their properties in order to facilitate a space–time analysis. Instead of using these generators, covariance is often expressed in terms of coordinate choices or embeddings of discrete structures, but these ingredients do not directly refer to the actual degrees of freedom of gravity. Moreover, the explicit application of these space–time ingredients reduces the freedom in formulating suitable modifications of space–time structures if they are called for by modified dynamics.

Secondly, the no-go results we encountered are very general. In particular, they do not require a specific form of modifications but only qualitative features related to discreteness, such as bounded modification functions with local maxima. They should therefore be expected to be largely independent of the specific approach. Even though they were derived for canonical quantum gravity, the no-go results can be applied to any modified cosmological dynamics that can be presented in canonical form, even if it has been derived from a non-canonical approach. It would be interesting to see how other approaches might be able to circumvent our no-go theorems, for instance by requiring new quantum degrees of freedom or specific non-local behaviors. (For an example of non-local effects derived for effective actions, see [49]).

Finally, hypersurface-deformation generators make it possible to analyze different space–time structures because they express geometrical properties through algebra. It is easier to control possible modifications or deformations of algebras (or algebroids), compared with geometrical structures. The strong algebraic background of canonical gravity is therefore the main reason why it is possible to analyze space–time structures in detail with canonical methods. Non-canonical approaches are often viewed as preferable because they can provide a direct four-dimensional space–time picture, at least heuristically. However, this proximity to the standard four-dimensional formulation of classical gravity also implies that hidden assumptions about the underlying geometry may easily and unwittingly be incorporated in a specific approach. As shown in the present paper, even canonical approaches are not immune to such hidden assumptions, but they also provide strong methods to spot and test unjustified assumptions.

Funding: This research was funded by NSF grant number PHY-1912168.

Institutional Review Board Statement: Not applicable.

Informed Consent Statement: Not applicable.

Acknowledgments: This paper was based on a talk given to the quantum gravity group at Radboud University, Nijmegen. The author is grateful to Renate Loll for an invitation and insightful questions, as well as to Jan Ambjørn, Suddhasattwa Brahma and Timothy Budd for discussions.

Conflicts of Interest: The author declares no conflict of interest.

References

1. Collins, J.; Perez, A.; Sudarsky, D.; Urrutia, L.; Vucetich, H. Lorentz invariance and quantum gravity: An additional fine-tuning problem? *Phys. Rev. Lett.* **2004**, *93*, 191301. [CrossRef] [PubMed]
2. Polchinski, J. Small Lorentz violations in quantum gravity: Do they lead to unacceptably large effects? *arXiv* **2011**, arXiv:1106.6346.

3. Agulló, I.; Ashtekar, A.; Nelson, W. A Quantum Gravity Extension of the Inflationary Scenario. *Phys. Rev. Lett.* **2012**, *109*, 251301. [CrossRef] [PubMed]
4. Gambini, R.; Pullin, J. Loop quantization of the Schwarzschild black hole. *Phys. Rev. Lett.* **2013**, *110*, 211301. [CrossRef]
5. Benítez, F.; Gambini, R.; Lehner, L.; Liebling, S.; Pullin, J. Critical collapse of a scalar field in semiclassical loop quantum gravity. *Phys. Rev. Lett.* **2020**, *124*, 071301. [CrossRef] [PubMed]
6. Ashtekar, A.; Olmedo, J.; Singh, P. Quantum Transfiguration of Kruskal Black Holes. *Phys. Rev. Lett.* **2018**, *121*, 241301. [CrossRef]
7. Bojowald, M. Black-hole models in loop quantum gravity. *Universe* **2020**, *6*, 125. [CrossRef]
8. Bojowald, M. Quantum cosmology: A review. *Rep. Prog. Phys.* **2015**, *78*, 023901. [CrossRef]
9. Agulló, I.; Ashtekar, A.; Nelson, W. An Extension of the Quantum Theory of Cosmological Perturbations to the Planck Era. *Phys. Rev. D* **2013**, *87*, 043507. [CrossRef]
10. Martín-Benito, M.; Garay, L.J.; Mena Marugán, G.A. Hybrid Quantum Gowdy Cosmology: Combining Loop and Fock Quantizations. *Phys. Rev. D* **2008**, *78*, 083516. [CrossRef]
11. Benítez, F.; Gambini, R.; Pullin, J. A covariant polymerized scalar field in loop quantum gravity. *arXiv* **2021**, arXiv:2102.09501.
12. Bardeen, J.M. Gauge-invariant cosmological perturbations. *Phys. Rev. D* **1980**, *22*, 1882–1905. [CrossRef]
13. Mukhanov, V.F.; Feldman, H.A.; Brandenberger, R.H. Theory of cosmological perturbations. *Phys. Rep.* **1992**, *215*, 203–333. [CrossRef]
14. Langlois, D. Hamiltonian formalism and gauge invariance for linear perturbations in inflation. *Class. Quantum Gravity* **1994**, *11*, 389–407. [CrossRef]
15. Stewart, J.M. Perturbations of Friedmann—Robertson—Walker cosmological models. *Class. Quantum Gravity* **1990**, *7*, 1169–1180. [CrossRef]
16. Bojowald, M. Non-covariance of the dressed-metric approach in loop quantum cosmology. *Phys. Rev. D* **2020**, *102*, 023532. [CrossRef]
17. Hojman, S.A.; Kuchař, K.; Teitelboim, C. Geometrodynamics Regained. *Ann. Phys. (N. Y.)* **1976**, *96*, 88–135. [CrossRef]
18. Kuchař, K.V. Geometrodynamics regained: A Lagrangian approach. *J. Math. Phys.* **1974**, *15*, 708–715. [CrossRef]
19. Kuchař, K.V. Geometry of hypersurfaces. I. *J. Math. Phys.* **1976**, *17*, 777–791. [CrossRef]
20. Kuchař, K.V. Kinematics of tensor fields in hyperspace. II. *J. Math. Phys.* **1976**, *17*, 792–800. [CrossRef]
21. Kuchař, K.V. Dynamics of tensor fields in hyperspace. III. *J. Math. Phys.* **1976**, *17*, 801–820. [CrossRef]
22. Deruelle, N.; Sasaki, M.; Sendouda, Y.; Yamauchi, D. Hamiltonian formulation of f(Riemann) theories of gravity. *Prog. Theor. Phys.* **2010**, *123*, 169–185. [CrossRef]
23. Komar, A. Constraints, Hermiticity, and Correspondence. *Phys. Rev. D* **1979**, *19*, 2908–2912. [CrossRef]
24. Komar, A. Consistent Factor Ordering Of General Relativistic Constraints. *Phys. Rev. D* **1979**, *20*, 830–833. [CrossRef]
25. Bojowald, M.; Brahma, S.; Yeom, D.H. Effective line elements and black-hole models in canonical (loop) quantum gravity. *Phys. Rev. D* **2018**, *98*, 046015. [CrossRef]
26. Bojowald, M. *Canonical Gravity and Applications: Cosmology, Black Holes, and Quantum Gravity*; Cambridge University Press: Cambridge, UK, 2010.
27. Bojowald, M.; Brahma, S.; Reyes, J.D. Covariance in models of loop quantum gravity: Spherical symmetry. *Phys. Rev. D* **2015**, *92*, 045043. [CrossRef]
28. Bojowald, M.; Brahma, S. Covariance in models of loop quantum gravity: Gowdy systems. *Phys. Rev. D* **2015**, *92*, 065002. [CrossRef]
29. Reyes, J.D. Spherically Symmetric Loop Quantum Gravity: Connections to 2-Dimensional Models and Applications to Gravitational Collapse. Ph.D. Thesis, The Pennsylvania State University, University Park, PA, USA, 2009.
30. Bojowald, M. Non-covariance of "covariant polymerization" in models of loop quantum gravity. *Phys. Rev. D* **2021**, *103*, 126025. [CrossRef]
31. Tibrewala, R. Inhomogeneities, loop quantum gravity corrections, constraint algebra and general covariance. *Class. Quantum Grav.* **2014**, *31*, 055010. [CrossRef]
32. Kantowski, R.; Sachs, R.K. Some spatially inhomogeneous dust models. *J. Math. Phys.* **1966**, *7*, 443. [CrossRef]
33. Strobl, T. Gravity in Two Spacetime Dimensions. *arXiv* **2000**, arXiv:hep-th/0011240.
34. Grumiller, D.; Kummer, W.; Vassilevich, D.V. Dilaton Gravity in Two Dimensions. *Phys. Rep.* **2002**, *369*, 327–430. [CrossRef]
35. Kunstatter, G.; Maeda, H.; Taves, T. New 2D dilaton gravity for nonsingular black holes. *Class. Quantum Gravity* **2016**, *33*, 105005. [CrossRef]
36. Takahashi, K.; Kobayashi, T. Generalized 2D dilaton gravity and KGB. *Class. Quantum Gravity* **2019**, *36*, 095003. [CrossRef]
37. Bojowald, M. No-go result for covariance in models of loop quantum gravity. *Phys. Rev. D* **2020**, *102*, 046006. [CrossRef]
38. Bouhmadi-López, M.; Brahma, S.; Chen, C.Y.; Chen, P.; Yeom, D.H. Asymptotic non-flatness of an effective black hole model based on loop quantum gravity. *Phys. Dark Univ.* **2020**, *30*, 100701. [CrossRef]
39. Faraoni, V.; Giusti, A. Unsettling physics in the quantum-corrected Schwarzschild black hole. *Symmetry* **2020**, *12*, 1264. [CrossRef]
40. Ben Achour, J.; Lamy, F.; Liu, H.; Noui, K. Polymer Schwarzschild black hole: An effective metric. *EPL Europhys. Lett.* **2018**, *123*, 20006. [CrossRef]
41. Ben Achour, J.; Lamy, F.; Liu, H.; Noui, K. Non-singular black holes and the limiting curvature mechanism: A Hamiltonian perspective. *JCAP* **2018**, *2018*, 072. [CrossRef]

42. Aruga, D.; Ben Achour, J.; Noui, K. Deformed General Relativity and Quantum Black Holes Interior. *Universe* **2020**, *6*, 39. [CrossRef]
43. Henderson, A.; Laddha, A.; Tomlin, C. Constraint algebra in LQG reloaded: Toy model of a $U(1)^3$ Gauge Theory I. *Phys. Rev. D* **2013**, *88*, 044028. [CrossRef]
44. Henderson, A.; Laddha, A.; Tomlin, C. Constraint algebra in LQG reloaded: Toy model of an Abelian gauge theory—II Spatial Diffeomorphisms. *Phys. Rev. D* **2013**, *88*, 044029. [CrossRef]
45. Tomlin, C.; Varadarajan, M. Towards an Anomaly-Free Quantum Dynamics for a Weak Coupling Limit of Euclidean Gravity. *Phys. Rev. D* **2013**, *87*, 044039. [CrossRef]
46. Varadarajan, M. Towards an Anomaly-Free Quantum Dynamics for a Weak Coupling Limit of Euclidean Gravity: Diffeomorphism Covariance. *Phys. Rev. D* **2013**, *87*, 044040. [CrossRef]
47. Laddha, A. Hamiltonian constraint in Euclidean LQG revisited: First hints of off-shell Closure. *arXiv* **2014**, arXiv:1401.0931.
48. Varadarajan, M. The constraint algebra in Smolins' $G \to 0$ limit of 4d Euclidean Gravity. *Phys. Rev. D* **2018**, *97*, 106007. [CrossRef]
49. Knorr, B.; Saueressig, F. Towards reconstructing the quantum effective action of gravity. *Phys. Rev. Lett.* **2018**, *121*, 161304. [CrossRef]

Review

WKB Approaches to Restore Time in Quantum Cosmology: Predictions and Shortcomings

Giulia Maniccia [1,2,*], Mariaveronica De Angelis [3] and Giovanni Montani [1,4]

1. Physics Department, "La Sapienza" University of Rome, P.le A. Moro 5, 00185 Roma, Italy
2. INFN Section of Rome, "La Sapienza" University of Rome, P.le A. Moro 5, 00185 Roma, Italy
3. School of Mathematics and Statistics, University of Sheffield, Hounsfield Road, Sheffield S3 7RH, UK
4. FNS Department, ENEA, C.R. Frascati, Via E. Fermi 45, 00044 Frascati, Italy
* Correspondence: giulia.maniccia@roma1.infn.it

Abstract: In this review, we analyse different aspects concerning the possibility to separate a gravity-matter system into a part which lives close to a quasi-classical state and a "small" quantum subset. The considered approaches are all relying on a WKB expansion of the dynamics by an order parameter and the natural arena consists of the Bianchi universe minisuperspace. We first discuss how, limiting the WKB expansion to the first order of approximation, it is possible to recover for the quantum subsystem a Schrödinger equation, as written on the classical gravitational background. Then, after having tested the validity of the approximation scheme for the Bianchi I model, we give some applications for the quantum subsystem in the so-called "corner" configuration of the Bianchi IX model. We individualize the quantum variable in the small one of the two anisotropy degrees of freedom. The most surprising result is the possibility to obtain a non-singular Bianchi IX cosmology when the scenario is extrapolated backwards in time. In this respect, we provide some basic hints on the extension of this result to the generic cosmological solution. In the last part of the review, we consider the same scheme to the next order of approximation identifying the quantum subset as made of matter variables only. This way, we are considering the very fundamental problem of non-unitary morphology of the quantum gravity corrections to quantum field theory discussing some proposed reformulations. Instead of constructing the time dependence via that one of the classical gravitational variables on the label time as in previous works, we analyse a recent proposal to construct time by fixing a reference frame. This scheme can be reached both introducing the so-called "kinematical action", as well as by the well-known Kuchar–Torre formulation. In both cases, the Schrödinger equation, amended for quantum gravity corrections, has the same morphology and we provide a cosmological implementation of the model, to elucidate its possible predictions.

Keywords: quantum cosmology; WKB approximation; minisuperspace dynamics; canonical methods of quantization; Born–Oppenheimer separation

Citation: Maniccia, G.; De Angelis, M.; Montani, G. WKB Approaches to Restore Time in Quantum Cosmology: Predictions and Shortcomings. *Universe* 2022, 8, 556. https://doi.org/10.3390/universe8110556

Academic Editors: Arundhati Dasgupta and Alfredo Iorio

Received: 9 September 2022
Accepted: 19 October 2022
Published: 25 October 2022

Publisher's Note: MDPI stays neutral with regard to jurisdictional claims in published maps and institutional affiliations.

Copyright: © 2022 by the authors. Licensee MDPI, Basel, Switzerland. This article is an open access article distributed under the terms and conditions of the Creative Commons Attribution (CC BY) license (https://creativecommons.org/licenses/by/4.0/).

1. Introduction

All the canonical formulations for quantum gravity [1–6] have to deal with two fundamental questions: one concerning the construction of a suitable time variable for the dynamics, and the other one on the determination of a correct classical limit coinciding with General Relativity (GR).

The first question has been widely addressed in the literature [7–14] and the most commonly accepted idea is that the dynamics must be described via the introduction of a "relational time" [7]. The second point on the classical limit is a rather natural question for the metric approach, related to the Wheeler–DeWitt (WDW) equation, but it becomes a puzzling question in Loop Quantum Gravity [15,16]. However, these two points are unavoidably related to the possibility to reconstruct an evolutionary quantum field theory when, starting from a purely quantum gravity approach in the presence of matter, we consider the classical limit on the geometrical component only. This theme contains also the challenging perspective to determine quantum gravity corrections to quantum field theory.

The first well-known attempt to reconstruct a Schrödinger functional theory for quantum matter from Canonical Quantum Gravity was performed in [17], limiting attention to a minisuperspace model, in which a "small" quantum subset (not necessarily restricted to matter) is recovered on a quasi-classical geometrodynamics. For previous approaches where gravity was treated on a classical background level, see [18,19]. This line of research was then expanded in [20], where a full development of the gravity-matter dynamics has been performed in terms of a single order parameter, combining the Planck length and the Newton constant (see also [21] where such expansion was considered but not fully developed). As shown in [22], the latter analysis is very similar to that in [17], but it is extrapolated to the next order where quantum gravity corrections to quantum field theory must appear. However, constructing a time variable via the time dependence of the classical limit of the metric has been recognized as affected by a non-trivial shortcoming, i.e., the emergence of non-unitarity features in the Schrödinger equation.

Here, we review the original formulations giving some sample of the implementation to the idea in [17] to specific cosmological situations, with particular reference to the "corner configuration" of a Bianchi IX dynamics. We also discuss the comparison of the Wentzel–Kramer–Brillouin (WKB) analysis to the standard Arnowitt–Deser–Misner (ADM) reduction [23] of the dynamics for the Bianchi I model, confirming that a basic assumption of the proposed formulation is the "smallness" of the quantum phase space available to the subsystem.

Then, in the second part of the review, the question concerning the non-unitarity problem is addressed in more detail and proposals for its solutions [22,24–26] are presented and discussed. In particular, we focus our attention to a cosmological implementation of the idea developed in [26] that the time variable can be constructed *a la* Kuchar–Torre [8], i.e., fixing a Gaussian reference frame, which is "materialized" as a fluid in the dynamics. In this framework, both the violation of the so-called strong energy condition and the non-unitarity of the quantum corrections to quantum field theory are simultaneously overcome.

The main aim of the present review is to focus attention to a theory that lives between quantum gravity and quantum field theory on curved space-time, i.e., the co-existence of quantum field theory for matter with weak quantum features of the background gravitational field. A convincing solution to the non-unitarity problem is therefore a central theme in this perspective and we provide a valuable picture on both the existing problems and the most promising formulations.

The review is structured as follows. In Section 2, we illustrate the general formalism of the minisuperspace reduction of a gravity-matter system, addressing the problem of time concerning the cosmological wave function. In Section 3, we present the work [17] that proposes a solution by a semiclassical separation of the system, with a brief discussion on the boundary conditions for such wave function in Section 3.1. In Section 4 we show the implications of such model for a Bianchi I universe. Section 5 presents instead the results of the model for the Bianchi IX universe, considering the vacuum case (Section 5.1), the presence of a cosmological constant and scalar field (Section 5.2), the Taub model (Section 5.3) and the generic inhomogenenous extension (Section 5.4). In Section 6 we present the Wentzel–Kramer–Brillouin expansion, whose special case is [17], discussing in Section 6.1 the proposal [20] that uses such a method to compute quantum gravity corrections to the matter sector dynamics and the following non-unitarity issue, while in Section 6.2 we review the Born–Oppenheimer scheme proposed in [25] discussing its shortcomings. Section 7 contains the recent proposals to solve both the problems of time and non-unitarity by implementing as a clock either the kinematical action (Section 7.1) or the reference frame fixing procedure (Section 7.2), presenting a cosmological implementation of the latter in Section 7.3. Further discussion and conclusions are provided in Section 8.

2. General Formalism: The Minisuperspace Analysis

Let us preliminarily fix the general context in which we will develop our analysis. In this respect, we consider a minisuperspace cosmological model [27–29], namely a reduction of the Wheeler superspace in presence of symmetries, with the line element in the Arnowitt–Deser–Misner (ADM) formulation [30] such as:

$$ds^2 = N^2(t)dt^2 - h_{ab}\sigma^a\sigma^b, \tag{1}$$

where h_{ab} ($a,b = 1,2,3$) is a function of n time dependent variables g^a, the 1-forms σ^a define the specific isometry of the considered model, e.g., the Bianchi universes [31,32] and N is the lapse function, whose specification determines the adopted time variable. An application of the above considerations can be developed for $f(R)$ gravity as analysed in [33,34].

In the Hamiltonian representation [27], the action for the minisuperspace takes the general form

$$S_{MSS} = \int dt \{p_a \dot{g}^a - NH_{MSS}\}, \tag{2}$$

p_a being the conjugate momenta to the configurational variables g^a ($a = 1, 2, ..., n$), and the superHamiltonian reads

$$H_{MSS}(g^a, p_a) = G^{ab} p_a p_b + V(g^a). \tag{3}$$

Here G^{ab} denotes the minisupermetric, encoding metric properties in the minisuperspace and in general having a pseudo-Riemmannian character, while $V(g^a)$ is a potential term due to the spatial curvature of the considered cosmological model. Additional contributions to both of them can come from the introduction of matter in the dynamics. Of particular reference is, in this respect, the presence of a self-interacting scalar field ϕ, interpretable as the inflaton and responsible for the inflationary phase of the Universe. In such a case, the superHamiltonian becomes

$$H_{MSS}(g^a, p_a) + \frac{1}{2\sqrt{h}} p_\phi^2 + \sqrt{h}\, U(\phi), \tag{4}$$

p_ϕ being the conjugate momentum to the scalar field, $U(\phi)$ its self-interaction potential and $h \equiv \det h_{ij}$.

Clearly, by varying the action with respect to N, we get that the (total) superHamiltonian (4) identically vanishes and this fact reflects the time diffeomorphism invariance of the theory. Thus, implementing the Dirac prescription [6] for the canonical quantization of a constrained system, we naturally arrive to the following Wheeler–DeWitt (WDW) equation

$$\left[-\hbar^2 G^{ab}(g^a)\frac{\partial}{\partial g^a}\frac{\partial}{\partial g^b} - \frac{\hbar^2}{2\sqrt{h}}\frac{\partial^2}{\partial \phi^2} + V(g^a) + \sqrt{h}\,U(\phi)\right]\psi = 0, \tag{5}$$

where the Universe wave function $\psi(g^a, \phi)$ is intrinsically taken over 3-geometries [27] since the spatial diffeomorphisms leave the 1-forms σ^a invariant. Above, we have chosen the so-called natural operator ordering, i.e., the functions of g^a are taken always on the left of the corresponding partial differentiations in constructing the quantum operator constraint. In this case, the minisupermetric is often redefined by a global scaling as $G^{ab} \to \sqrt{h}\,G^{ab}$, when the whole constraint is multiplied by $\sqrt{h} \neq 0$. Other operator orderings are available and classes of equivalence can be established [4]; in particular, we mention the choice of a symmetric superHamiltonian operator (for a justification see [35]).

Equation (5) is affected by the so-called "frozen formalism" problem, i.e., no time evolution emerges in terms of the wave function dependence on an external time parameter, as will be discussed in Section 2.1. However, it is a well-known result [1–3] that the WDW equation has a Klein–Gordon-like structure due to the pseudo-Riemannian nature of

G^{ab}. In fact, taking $h^{1/4}$ as a generalized coordinate, we easily see that it has a different signature with respect to the remaining ones, including also the scalar field. In the spirit of the relational approach proposed in [7], see also [28], the scalar field can be taken as a matter clock, even though it has the same signature of the "space-like" variables in G^{ab}.

Thus, the quantization of a minisuperspace model corresponding to a Bianchi Universe is reduced to the quantum dynamics of a relativistic particle [36] which is affected by a subtle question concerning the construction of a Hilbert space. In particular, the presence of the two potential terms in Equation (5) prevents, in many situations, the possibility for a frequency separation, which can be achieved under specific assumptions or in suitable asymptotic limits. In this respect, it is worth stressing that we consider here the WDW equation as a single particle dynamics [37], see also [38], without considering the so-called "third quantization approach" [39,40], that was first introduced as production of "baby Universes" in relation to the cosmological constant problem [41–46]. Finally, we observe that in Quantum Gravity, according to a very general prescription [47], the choices of $h^{1/4}$ or of ϕ as internal time coordinates can be performed after or before the quantization procedure. In the former case, the quantization is covariantly performed, without specifying any explicit expression for the lapse function. Instead, in the latter case, the choice is performed on a classical level by fixing the temporal gauge which naturally leads to the ADM-reduction [23] of the classical variational principle and therefore to a Schrödinger-like quantum dynamics for the Universe wave function.

2.1. The Wave Function and the Problem of Time in Quantum Cosmology

An important approach to quantum cosmology and its many applications regards the semiclassical approximation of the Universe. Indeed, in the full quantum picture, there is still some discussion regarding the probabilistic interpretation of the Universe wave function. This aspect is not straightforward since the wave function itself does not evolve in "time" due to the vanishing of the WDW Equation (5) [6,48–50]. In the canonical quantum picture [51], this is equivalent to a timeless Schrödinger equation with null eigenvalues describing a trivial evolution leading to the so-called *problem of time* [11,14, 47,52,53]. This issue has been long discussed in the literature since the formulation of the DeWitt theory [54–60]: for example, in [56] some time choices (scalar field, cosmological constant conjugate, and proper time) models are discussed via a semiclassical expansion in $\hbar$. Indeed, the time coordinate could in principle be regarded together with the gravitational degrees of freedom and integrated over [17], such that there is no clear choice for the definition of another time parameter; subsequently, the definition of a conserved and well-defined probability distribution is troublesome, unless one imposes further conditions, e.g., hermicity of the Hamiltonian [61] or finiteness of the probability density [62]. One of the most followed approaches is the definition of a relational time [5,7–9,37,54,63,64] to recover a time parameter leading to a Schrödinger dynamics; such "emergence of time" has been discussed not only for quantum gravity but also in the context of non-relativistic quantum mechanics, for example in [65].

This identification of a proper time-like variable avoiding the frozen formalism leads to different results whether it is tackled before or after quantization [66]. Hence, to give a meaningful probabilistic interpretation to the wave function of the Universe, one can pursue two different approaches. In the first, the super-Hamiltonian constraint is classically solved and then the resulting Schrödinger equation is quantized [67,68], i.e., the reduced phase space quantization (RPSQ) [69,70]. The RPSQ is the most straightforward method because it is an exact procedure requiring no WKB approximation based on the wave function of the Universe, even if its mass-like term is time-dependent and the Hamiltonian density is non-local. While in the second case, one implements both the WKB and Born–Oppenheimer (BO) approximation (see Sections 6 and 6.2), that is essentially Vilenkin's approach (see also the discussion and application in Section 4).

DeWitt himself observed that (5) is equivalent to a n-dimensional Klein–Gordon equation with variable mass term [1–3] given by $-\sqrt{h}\,R^{(3)}$, being $R^{(3)}$ the scalar curvature associated to the induced metric h_{ij}, and so one could implement a Klein–Gordon-like inner product. However, being the mass term not necessarily positive, and the Hamiltonian containing second derivatives in the metric coordinates, such definition could give negative probabilities for the wave functional, i.e., negative frequency components. This feature can be avoided in some special cases [37] with *ad-hoc* conditions, but it remains standing in the general case, leaving some concerns on how to interpret the wave functional itself.

3. Semiclassical and Quantum Universes: Vilenkin's Approach

Vilenkin's proposal [17] will be the starting point of our analysis, that aims to reconcile the WDW equation with a functional field theory formalism for gravity in the minisuperspace via a semiclassical expansion. To better show the feasibility of this model in the context of quantum cosmology, some key implementations to Bianchi universes are then examined in Sections 4 and 5.

Starting from the interpretation of DeWitt and following the path of a relational time study, Vilenkin's work [17] suggested to separate the Universe variables into semiclassical and quantum components. This separation is indeed valid at some point, since gravity has a full quantum behaviour only near the Planck scale, and many physical phenomena relevant in cosmology happen at lower energies.

Vilenkin first considered the case in which the whole Universe behaves semiclassically. In the homogeneous minisuperspace setting, such a system can be described by a wave function of the form

$$\Psi(h) = A(h)e^{\frac{i}{\hbar}S(h)}, \qquad (6)$$

where we label by h^a all the semiclassical superspace variables (both gravity and matter fields) and $S(h^a)$ the classical action that must be a real function, while $A(h)$ encodes the semiclassical features. The WDW equation reads as

$$\left(-\hbar^2 \nabla^{(c)2} + U^{(c)}\right)\Psi = 0, \qquad (7)$$

being $U^{(c)} = \sqrt{h}\,U(\phi) + V(g^a)$ the potential associated to all semiclassical variables and $\nabla_a^{(c)}$ the derivative with respect to h^a (we are using the superscript $^{(c)}$ to identify the semiclassical components). Such a writing is superfluous since at this stage the whole Universe behaves semiclassically, but it will come into play later by considering the more general case. A perturbative expansion of S, and so of ψ, can be implemented in powers of the Planck constant due to the semiclassical feature of the Universe (in the original work, the expansion was performed in a parameter proportional to $\hbar$ and the $\hbar$ in (6) was absorbed inside the function S; here, for clarity, it is collected in front). This allows to study the dynamics going from the lowest order corresponding to the classical limit $\hbar \to 0$, to higher orders in such parameter. The procedure is clearly linked to the Wentzel–Kramer–Brillouin (WKB) approximation [71] explained in Section 6, that uses an ansatz very similar to (6) but with a complex exponential and without the explicit separation of a semiclassical amplitude. The expansion of (7) brings at $\mathcal{O}(\hbar^0)$

$$(\nabla^{(c)}S)^2 + U^{(c)} = 0, \qquad (8)$$

that is the Hamilton-Jacobi (HJ) equation for S, ensuring the classical limit of the model. The next order $\mathcal{O}(\hbar)$ gives

$$2\nabla^{(c)}A \cdot \nabla^{(c)}S + A\nabla^{(c)2}S = 0, \qquad (9)$$

where the supermetric G_{ab} is implicitly assumed by the scalar product symbol $(\cdot)$; this is equivalent to the conservation of the following current

$$j^{(c)a} = |A|^2 \nabla^{(c)a} S, \qquad (10)$$

whose interpretation can now be understood together with the associated semiclassical probability distribution $\rho^{(c)}$. Indeed, the action S defines a congruence of classical trajectories, as follows from (8); each point h^a in a classically allowed region in the superspace belongs to a trajectory with associated momenta $p_b = \nabla_b^{(c)} S$ and velocity

$$\dot{h}^a = 2N \nabla^{(c)\,a} S, \tag{11}$$

that depends on the choice of $N(t)$ from the foliation. Here, we can infer the form of the time derivative

$$\frac{\partial}{\partial \tau} = 2N \nabla^{(c)} S \cdot \nabla^{(c)}, \tag{12}$$

which will come into play later. The points that satisfy $\nabla^{(c)} S = 0$ separate the classically allowed and forbidden regions, breaking down the semiclassical approximation. By requiring that each hypersurface is crossed only once by the congruence of trajectories, i.e.,

$$\dot{h}^a \, d\Sigma_a^{(c)} > 0, \tag{13}$$

then the probability density

$$dP = j^{(c)\,a} \, d\Sigma_a^{(c)} \tag{14}$$

is positive semi-definite, thus the Universe wave function can be properly normalized. The same can be implemented for a wave function that is a superposition $\sum_k \Psi_k$ of terms defined as in (6) when the condition (13) is satisfied for each k, such that the total probability is conserved.

One could then wonder if a similar implementation is possible in the more general case, when only a part of the Universe is semiclassical and the rest must be described in a full quantum picture. Vilenkin examined the case in which the quantum variables (labeled by q^ν with $\nu = 1, ..., m$) represent a small quantum subset, with negligible effects on the semiclassical variables (h^a with $a = 1, .., n-m$) dynamics. The full Wheeler–DeWitt equation then becomes

$$\left(-\hbar^2 \nabla^{(c)\,2} + U^{(c)} + \hat{H}^{(q)} \right) \Psi = 0, \tag{15}$$

where using the previous notation $-\hbar^2 \nabla^{(c)\,2} + U^{(c)} = \hat{H}^{(c)}$ is given neglecting the quantum variables and their conjugate momenta, which instead appear in $\hat{H}^{(q)} = -\hbar^2 \nabla^{(q)\,2} + U^{(q)}$ (here the the superscript $^{(q)}$ refers to the quantum components). At the same time, the semiclassical part is assumed to satisfy its own WDW Equation (7), thus obtaining a system of coupled equations for the two sectors dynamics. This separation is backed both by the hypothesis on the smallness of the quantum subsystem, expressed as

$$\frac{\hat{H}^{(q)} \Psi}{\hat{H}^{(c)} \Psi} = \mathcal{O}(\hbar), \tag{16}$$

and by the independence between the two sets, namely

$$G_{ab}(h, q) = G_{ab}(h) + \mathcal{O}(\hbar), \tag{17}$$
$$G_{a\nu} = \mathcal{O}(\hbar). \tag{18}$$

In other words, we are assuming G_{ab} to be dependent on the semiclassical variables only, and the two subspaces to be approximately orthogonal, since any mixed term of the supermetric (being the index a for the semiclassical variables and ν for the quantum variables) is of higher order in the perturbative expansion; it follows that higher order terms will not appear inside $\nabla^{(c)\,2} = G_{ab} \nabla^{(c)\,a} \nabla^{(c)\,b}$ (see for example the applications in Sections 4 and 5). Following these hypotheses, the wave function can be separated in

$$\Psi(h, q) = \psi(h) \chi(q, h) = A(h) e^{\frac{i}{\hbar} S(h)} \chi(h, q). \tag{19}$$

We observe that this ansatz shares similarities with both the WKB and BO approximations. Considering the former, this is due to the presence of a complex exponential with a small parameter that can lead the expansion; instead, the latter is due to a separation of a purely semiclassical sector and the quantum one, as explained in more detail in Section 6. Actually, Vilenkin's proposal can be reformulated as a special case of a BO-like approximation with expansion in $\hbar$, i.e., the semiclassical expansion, as discussed in Section 6.1.

We here use again the Planck constant for the expansion, instead of the parameter proportional to $\hbar$, as mentioned before. Since the previous hypotheses hold, the semiclassical function ψ must satisfy Equation (7), giving Equations (8) and (9) respectively at $\mathcal{O}(\hbar^0)$ and $\mathcal{O}(\hbar)$. Meanwhile, the quantum function χ inherits a different dynamics from Equation (15), that is

$$-\hbar^2 \nabla^{(c)2}\chi - 2\hbar^2 A^{-1}(\nabla^{(c)}A) \cdot \nabla^{(c)}\chi - 2i\hbar\,(\nabla^{(c)}S) \cdot \nabla^{(c)}\chi + \hat{H}^{(q)}\chi = 0, \quad (20)$$

where we can observe that all terms except the last two are of higher order in the expansion parameter ($H^{(q)}$ is of order $\hbar$ due to assumption (16)). Thus, at $\mathcal{O}(\hbar)$, we obtain

$$\hat{H}^{(q)}\chi = 2i\hbar\,(\nabla^{(c)}S) \cdot \nabla^{(c)}\chi. \quad (21)$$

Multiplying (21) by $N(t)$ and using the same time derivative (12) defined for the semiclassical universe, it becomes

$$i\hbar \frac{\partial \chi}{\partial \tau} = N\hat{H}^{(q)}\chi, \quad (22)$$

namely a functional Schrödinger equation for the matter wave function.

It is worth noting that the definition introduced above for ∂_τ is very close to the notion of a composite derivative $\partial_\tau \equiv \frac{dh^a}{d\tau}\partial_{h^a}$ applied to the quantum wave function. This fact can be easily realized by recalling that $\partial_{h^a}S_0$ is just the conjugate momentum p_a and, hence, it is enough to write down the first Hamilton equation (obtained variating the classical action with respect to p_a) to arrive to the desired statement. By other words, the time dependence of the quantum wave function is recovered in the approach proposed in [17], by means of the dependence that the quasi-classical variables h^a acquire, at the leading order, on the label time of the space-time slicing. Clearly, it is also possible and discussed in [17] that one of the h^a themselves is chosen as time coordinate to describe the system evolution., suitably choosing the lapse function $N(t)$. It is also useful to stress that, as we will see later in the considered specific applications, the form of the supermetric G^{ab} as a function of h^a is sensitive to the specific set of adopted configurational variables to describe the studied cosmological model. However, we can observe that any variable among the h^a's, which is related to the Universe volume, acquires a different signature (say a timelike one) different from all the other ones (regarded as space-like coordinates) [1–3,38]. Independently from the specific form of G^{ab} and $H^{(q)}$, the important point to make safe the model self-consistence is that the semiclassical metric G^{ab} and the quantum one, fixed by the form of $H^{(q)}$ itself, live in orthogonal spaces, i.e., cross terms in the supermetric with a classical index and a quantum one must be of higher order in the present formulation, as expressed by (18).

Explicit examples of this classical-from-quantum variable separation are given below, see Sections 4 and 5. We consider here both the situations in which this separation takes place between the gravitational degrees of freedom, e.g., Universe volume taken as a quasi-classical variable and space anisotropies as quantum variables, as well as the case in which the same separation concerns quantum matter living on a quasi-classical space-time. This last situation is of particular physical relevance since, as we shall see below, its analysis to the next order of approximation in the order parameter corresponds to the study of quantum gravity corrections to standard quantum field theory.

Differently from the purely semiclassical case, two probability currents now emerge. The one including the semiclassical sector is

$$j^a = |\chi|^2 |A|^2 \nabla^{(c)a} S \equiv j^{(c)a} \rho_\chi, \tag{23}$$

where $j^{(c)a}$ is the same as (10) and $\rho_\chi = |\chi|^2$ is the probability distribution of the quantum variables computed on the semiclassical trajectories. For the quantum components instead we find

$$j^\nu = -\frac{i}{2}|A|^2 \left(\chi^* \nabla^{(q)\nu} \chi - \chi \nabla^{(q)\nu} \chi^* \right) = \frac{1}{2}|A|^2 j^\nu_\chi, \tag{24}$$

associated to the distribution ρ_χ, where j^ν_χ is a Klein–Gordon-like current. From the conservation of both the total current $\nabla^{(c)}_a j^a + \nabla^{(q)}_\nu j^\nu = 0$ and the semiclassical current $\nabla^{(c)}_a j^{(c)a} = 0$, given by the full WDW Equation (15) and assumption (7) respectively, we can state that at the leading order the following equation holds

$$\frac{\partial \rho_\chi}{\partial \tau} + N \nabla^{(q)}_\nu j^\nu_\chi = 0, \tag{25}$$

which is a continuity equation for the quantum variables. Moreover, both ρ_c and ρ_χ can be normalized on their respective subspaces by requiring $\int d\Sigma^{(c)} \rho^{(c)} = 1$ and $\int d\Omega^{(q)} \rho_\chi = 1$, being $d\Sigma = d\Sigma^{(c)} d\Omega^{(q)}$ the total surface element on the equal-time surfaces identified with the foliation. In this way, the standard probabilistic interpretation is recovered for ψ when such a separation in semiclassical and quantum variables is valid.

However, there is still one case to discuss, that is when in such a framework one (or more) quantum variables become semiclassical at later time. This means that the two subsets change: starting from an initial wave function of the form (19), we have $\phi_k \chi_k \to \sum_l \phi_k(h') \chi_{kl}(h', q')$, the new semiclassical set $\{h'\}$ having increased by one variable and the quantum one $\{q'\}$ decreased by one variable. The sum is explained by the transition during which each semiclassical trajectory branches into many trajectories, each one for a different initial condition of the "new semiclassical" variable. For this reason, one has to impose a unitarity (normalization) condition on the semiclassical current $j^{(c)a}$

$$\int d\Sigma^{(c)}_{ka} j^{(c)a}_k = \sum_l \int d\Sigma^{(c)}_{kla} j^{(c)a}_{kl}, \tag{26}$$

that is satisfied only at an approximate level, i.e., when the cross terms can be neglected. It should be stressed that the division itself between the two subspaces is heavily dependent on the considered case and almost arbitrary in a certain footing, leading to an approximate concept of unitarity for the Universe.

3.1. Boundary Conditions for the Cosmological Wave Function

Vilenkin's work provides a meaningful description at the typical scale of the quantum subsystem of the Universe. One related point concerns how to impose boundary conditions on the wave function (19), which has led to ample discussion in the literature. Vilenkin himself had previously studied this issue [72–74], developing the so-called *tunneling proposal*: he constructed a wave function describing an ensemble of Universes that tunnel from "nothing" to a de Sitter space by implementing a similar expansion of Ψ (19) and choosing the purely expanding solution.

A different implementation is the one by Hartle–Hawking [75], also known as the *no-boundary* proposal. The wave function for a closed Universe is constructed in the Euclidean path integral approach by integrating over all the possible compact 4-geometries corresponding to a certain induced metric h_{ij} on a spacelike boundary (see also discussion in [76]); the resulting wave function can be shown to approximately satisfy the WDW equation, whose corresponding Hamiltonian is required to be a Hermitian operator.

In this respect, the path integral approach [77] represents an alternative formulation of gravity as a quantum field theory and it has been widely discussed in relation to the problems of time and unitarity [46,78–80]. We mention that, actually, Vilenkin's tunneling proposal can be reformulated in the Lorentzian path integral formalism [81]. The WKB implementation also allows to study the probability of tunneling from a false vacuum to a true vacuum state from the Wheeler–deWitt equation (see [82] and references within). We will here focus on the Dirac quantization method only, however an interesting discussion between the two schemes can be found in [83] where, using the Lorentzian path integral, the WDW equation is uniquely recovered in the minisuperspace via a particular gauge fixing on the values of h_{ij} and N, that solves the operator-ordering ambiguity of the Dirac scheme. Some sort of WKB procedure *a la* Vilenkin can also be included in the path integral formalism to study the boundary conditions, see for example [84–86] (Lorentzian), finding in some cases different features with respect to the Hartle–Hawking interpretation.

4. Validation of the Vilenkin Proposal for the Bianchi I Cosmology

One of the most interesting open questions in theoretical cosmology concerns how a primordial quantum universe reaches a classical isotropic limit [28]. The reason to hypothesise a very general morphology of the universe near the singularity (for a big bounce picture of the Bianchi I model see [87–91]) relies on the request to address the quantum cosmological problem within the Bianchi homogeneous framework [6]. These models are characterized by the preservation of the space-line element under a specific group of symmetry, and are collected in the so-called Bianchi classification.

4.1. The Minisuperspace Dynamics of Bianchi Universes

The most general homogeneous model is the Bianchi IX model [28,32,92], also called Mixmaster model [93] (for a recent semiclassical discussion see [94]), that has a relevant role in the study of the cosmological dynamics. Despite its spatial homogeneity, it presents typical features of the generic cosmological solution such as a chaotic time evolution of the cosmic scale factors near the singularity [95]. This corresponds to an infinite sequence of bounces of the point particle, in the Hamiltonian representation, against the time-dependent potential walls which can be shown to induce an ergodic evolution in the Misner–Chitre variables. The standard dynamics in the central region of the potential well are then restored [28] once it escapes the small oscillations configuration. However, in the asymptotic limit to the cosmological singularity, the potential term of Bianchi IX dynamics has the morphology of an equilateral triangle and three open corners appear in the vertices, which correspond to the non-singular Taub cosmology [96], see Figure 1 [97,98].

This kind of cosmology defines the limit of Bianchi IX dynamics when two scale factors are considered equal over the three possible independent ones. The importance of the Hamiltonian formulation of the Mixmaster model (see [93]) using the ADM description, relies on the fact that it is possible to reduce the dynamics to the two-dimensional point particle. We start with the line element of the model in the Misner picture

$$ds^2 = N(t)^2 dt^2 - \eta_{ab}\omega^a\omega^b, \tag{27}$$

where $\omega^a = \omega^a_\alpha dx^\alpha$ is a set of the three invariant differential forms that fixes the geometry of the considered Bianchi model, $N(t)$ is the lapse function and η_{ab} is defined as $\eta_{ab} = e^{2\alpha}(e^{2\beta})_{ab}$. The choice of these variables allows us to separate the isotropic contribution expressing the volume of the universe related to α, i.e., for $\alpha \to -\infty$ the initial singularity is reached, from the gravitational degrees of freedom β_+, β_- contained in the matrix $\beta_{ab} = diag(\beta_+ + \sqrt{3}\beta_-, \beta_+ - \sqrt{3}\beta_-, -2\beta_+)$ acting as the anisotropies of this model. Moreover, the introduction of the Misner variables makes the kinetic term in the Hamiltonian diagonal. We can rewrite the superHamiltonian constraint as

$$H_{IX} = \frac{\kappa}{3(8\pi)^2}e^{-3\alpha}(-p_\alpha^2 + p_+^2 + p_-^2 + \mathcal{V} + \Lambda e^{6\alpha}) = 0, \tag{28}$$

where $\kappa = 8\pi G/c^4$ is the Einstein constant and the potential $\mathcal{V}$ takes the form

$$\mathcal{V} \equiv -\frac{6(4\pi)^4}{\kappa^2}\eta R^{(3)} = \frac{3(4\pi)^4}{\kappa^2}e^{4\alpha}V_{IX}(\beta_\pm), \qquad (29)$$

where the spatial scalar of curvature generates the Bianchi IX potential term depending only on the anisotropies

$$V_{IX}(\beta_\pm) = e^{-8\beta_+} - 4e^{-2\beta_+}\cosh(2\sqrt{3}\beta_-) + 2e^{4\beta_+}[\cosh(4\sqrt{3}\beta_-) - 1]. \qquad (30)$$

This function has the symmetry of an equilateral triangle with steep exponential walls and three open angles. The expressions for the equipotential lines for large values of $|\beta_+|$ and small $|\beta_-|$ are

$$V_{IX}(\beta_\pm) \sim \begin{cases} e^{-8\beta_+} & \beta_+ \to -\infty,\ |\beta_-| \ll 1 \\ 48e^{4\beta_+}\beta_-^2 & \beta_+ \to +\infty,\ |\beta_-| \ll 1 \end{cases} \qquad (31)$$

while close to the origin, for $\beta_\pm \to 0$,

$$V_{IX}(\beta_\pm) \sim \beta_+^2 + \beta_-^2. \qquad (32)$$

The Hamiltonian approach provides the following equations of motion

$$\dot{\alpha} = N\frac{\partial H_{IX}}{\partial \alpha}, \qquad \dot{p}_\alpha = N\frac{\partial H_{IX}}{\partial \alpha}, \qquad (33)$$

$$\dot{\beta}_\pm = N\frac{\partial H_{IX}}{\partial p_\pm}, \qquad \dot{p}_\pm = N\frac{\partial H_{IX}}{\partial \beta_\pm}. \qquad (34)$$

One recognizes that the dynamics of the universe towards the singularity is mapped into the motion of a particle that lives on a plane inside a closed domain and bounces against the potential wall.

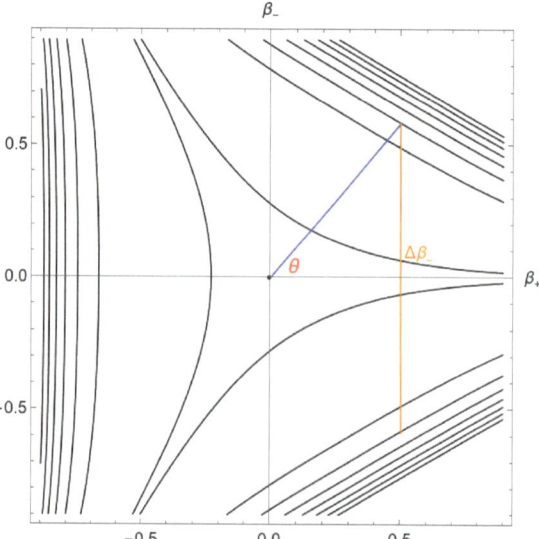

Figure 1. Description of Bianchi IX potential isocurve on which is marked the corner structure. Here, θ describes the width of the β_+ channel.

The canonical quantization of the system consists of the commutation relations

$$[\hat{q}_a, \hat{p}_b] = i\delta_{ab}, \tag{35}$$

which are satisfied for $\hat{p}_a = -i\frac{\partial}{\partial q_a} = -i\partial_a$ where $(a,b = \alpha, \beta_+, \beta_-)$ adopting natural units. By replacing the canonical variables with the corresponding operators, the quantum behaviour of the universe is given by the quantum version of the superHamiltonian constraint (28), i.e., the WDW equation for the Bianchi IX model

$$\hat{H}_{IX}\Psi(\alpha, \beta_\pm) = \left[\partial_\alpha^2 - \partial_+^2 - \partial_-^2 + \frac{3(4\pi)^4}{\kappa^2}e^{4\alpha}V_{IX}(\beta_\pm)\right]\Psi(\alpha, \beta_\pm) = 0, \tag{36}$$

where $\Psi(\alpha, \beta_\pm)$ is the wave function of the universe providing information about its physical state. Therefore, following the step in [17] we can obtain the probability distribution for the wave function of the universe, that reads as

$$\rho(\alpha, \beta_\pm, t) = \rho^{(c)}(\alpha, t)\rho_\chi(\alpha, \beta_\pm(t), t), \tag{37}$$

where in particular $\rho^{(c)}(\alpha, t) = |A(\alpha(t))|^2$ is related to the components of the classical space and $\rho_\chi(\alpha, \beta_\pm(t), t) = |\chi(\alpha, \beta(t), t)|^2$ to those in the quantum subspace, as explained in Section 3.

In the subsection below, we will focus on the Bianchi I model in which the structure constants and so the spatial curvature $R^{(3)}$ vanishes. Hence, the associated superHamiltonian constraint in vacuum read as

$$H_I = \frac{\kappa}{3(8\pi)^2}e^{-3\alpha}(-p_\alpha^2 + p_+^2 + p_-^2) = 0. \tag{38}$$

This cosmology is the natural extension of the FRLW model with $k = 0$ generalizing an homogeneous flat Universe.

4.2. Implementation of the WKB Approach in the Minisuperspace of Bianchi I

Let us consider the case of the Bianchi I model, i.e., with WDW expressed by the constraint (38). As mentioned in Section 2.1, the time definition for the model can be implemented before or after quantization, namely the RPSQ and Vilenkin's proposal, which present striking differences. The Vilenkin proposal (Section 3) is more feasible, namely it avoids the square root non-local Hamiltonian operator emerging from the former but the probabilistic interpretation is achieved only after performing the semiclassical limit and, in this sense, it can not be seen as a fundamental approach. However, the role of the time-like variables itself can make the two schemes comparable. In order to determine under which restrictions the Vilenkin representation of the Universe volume dynamics becomes predictive, it has been shown in [99] a rigorous comparison of the two quantization methods carrying out the probabilistic interpretation of the wave function for the Bianchi I cosmology in which $R^{(3)} = 0$.

We recall that Vilenkin suggested a semiclassical approximation of the wave function to achieve a proper probabilistic interpretation due to the emergence of time. This does not happen for the definition of a scalar product from a conserved current suggested by DeWitt. Hence, considering a Bianchi I model in the presence of a matter contribution and achieving the Schrödinger equation describing the motion of a free particle in the (β_+, β_-) plane, we arrive at the following result

$$\Psi(\alpha, \beta_a) = \frac{e^{-\frac{i}{\hbar}\int_{\alpha_0}^{\alpha} d\alpha' \sqrt{\mu^2(\alpha)}}}{\sqrt[4]{\mu^2(\alpha)}} \int_{\mathbb{R}^2} \frac{d^2p}{2\pi\hbar} e^{-\frac{i}{2\hbar}(p_+^2 + p_-^2)\int_{\alpha_0}^{\alpha} d\alpha' \frac{1}{\sqrt{\mu^2(\alpha)}}} \cdot e^{\frac{i}{\hbar}p_a\beta_a}\tilde{\chi}(\alpha_0, p_a), \tag{39}$$

where the subscript a stands for $(+,-)$ and $\tilde{\chi}(\alpha_0, p_a)$ determines initial conditions. The matter contribution is encoded in the term μ^2 as

$$\mu^2(\alpha) = \sum_w \mu_w^2 \, e^{3(1-w)\alpha}, \tag{40}$$

where the sum contains all the fluids components characterized by different values of w, while μ_w^2 are constants. It is important to stress that with the BO approximation we are assuming α as the slow variable whereas β_+ and β_- are the fast ones. The validity of both Vilenkin's semiclassical expansion and BO approximation, which will be discussed in detail in Sections 6 and 6.2, implies that we admit a decomposition of the wave function as

$$\Psi(\alpha, \beta_a) = \exp\left(\frac{i}{\hbar}\sum_{n=0}(\hbar)^n S_n\right) \tag{41}$$

and the following conditions hold

$$\left|\frac{1}{\mu^3(\alpha)}\frac{d\mu^3(\alpha)}{d\alpha}\right| \ll \frac{4}{\hbar}, \tag{42}$$

$$\hbar|S_2(\alpha)| \ll |S_1(\alpha)| \quad \text{and} \quad \hbar|S_2(\alpha)| \ll 1. \tag{43}$$

Moreover, the integral over the momentum space extends over those values for which

$$(p_+^2 + p_-^2) \neq 0, \tag{44}$$

$$(p_+^2 + p_-^2) \ll \mu^2(\alpha). \tag{45}$$

Now, the BO approximation implies that near a value for which (45) holds, we need a wave packet for the initial conditions sufficiently peaked, for simplicity a Gaussian distribution of the form

$$\tilde{\chi}(\alpha_0, p_a) = \frac{1}{\sqrt{\pi \sigma_+ \sigma_-}} e^{-\frac{(p_+ - \bar{p}_+)^2}{2\sigma_+^2}} e^{-\frac{(p_- - \bar{p}_-)^2}{2\sigma_-^2}}, \tag{46}$$

The aim is to check whether the functional form of the wave functions obtained from the two formalisms, or their associated probabilities, coincide. In order to do this, we need the Klein–Gordon-like time-independent inner product, achieved from the RPSQ approach.

Following the steps described in [6,70] the resulting Schrödinger equation is

$$i\hbar\frac{\partial}{\partial\alpha}\Phi(\beta_a,\alpha) = \sqrt{-\hbar^2\left(\frac{\partial^2}{\partial\beta_+^2} + \frac{\partial^2}{\partial\beta_-^2}\right) + \mu^2(\alpha)}\,\Phi(\beta_a,\alpha), \tag{47}$$

in which $\lim_{\alpha\to-\infty}\mu^2(\alpha) = \mu_1^2$ and we denoted Φ as the wave function of the RPSQ. Hence, via inverse Fourier transform a generic solution can be formally found as

$$\Phi(\beta_a,\alpha) = e^{-\frac{i}{\hbar}\int_{\alpha_0}^{\alpha}d\alpha'\sqrt{-\hbar^2\Delta_\pm + \mu^2(\alpha)}}\Phi(\beta_a,\alpha_0), \tag{48}$$

where $|p_a|^2 = -\hbar^2\Delta_\pm = -\hbar^2\left(\frac{\partial^2}{\partial\beta_+^2} + \frac{\partial^2}{\partial\beta_-^2}\right)$. Now, to compare the two formulations, we need to identify the same time variable. In particular, we need the two lapse functions (one from RPSQ and the other one from Vilenkin's proposal) to be the same

$$\frac{3c\mathcal{K}}{4\pi GT}\frac{e^{3\tau}}{\sqrt{p_+^2 + p_-^2 + \mu^2(\tau)}} = \frac{3c\mathcal{K}}{4\pi GT}\frac{e^{3\tau}}{\sqrt{\mu^2(\tau)}}, \tag{49}$$

where $\mathcal{K} = \int d^3x |\det(e_i^{(a)}(x^k))|$, the vectors $e_i^{(a)}$ constitute the so-called *frame* and $\alpha = t/T = \tau$ in which the constant T can be defined in terms of fundamental constants, e.g., it can be chosen proportional to the Planck length. The above equation is effectively valid if $p_+^2 + p_-^2 \ll \mu^2(\tau)$. An issue arises if we promote β_a to quantum operators since the lapse function N_{RPSQ} (on the left-hand side) becomes an operator acting on the wave function. For this reason, we need to replace it by its expectation value. However, in Bianchi I, p_a are essentially constants of motion and we can treat them as numbers. Now, we are able to choose a range of τ such that the semiclassical approximation is valid, namely τ_S. Hence, by normalizing (48) with respect to the inner product near the singularity, such that $\mu^2 \to \mu_1^2$ becomes time-independent [100], implementing it with the BO approximation, we can write

$$\Phi(\beta_a, \tau) \approx \frac{e^{-\frac{i}{\hbar}\int_{\tau_S}^{\tau} d\tau' \sqrt{\mu^2(\tau')}}}{\sqrt[4]{\mu_1^2}} \int_{\mathbb{R}^2} \frac{d^2p}{(2\pi\hbar)\sqrt{2\pi\sigma_+\sigma_-}} \frac{e^{-\frac{i}{2\hbar}(p_+^2+p_-^2)\int_{\tau_S}^{\tau} d\tau' \frac{1}{\mu^2(\tau')}}}{\sqrt[4]{1+\frac{p_+^2+p_-^2}{\mu_1^2}}} e^{\frac{i}{\hbar}p_a\beta_a} e^{-\frac{(p_+-\bar{p}_+)^2}{2\sigma_+^2} - \frac{(p_--\bar{p}_-)^2}{2\sigma_-^2}}. \quad (50)$$

Two main differences are noticed comparing (50) with (39). In Equation (39), the factor $\left(1 + \frac{p_+^2+p_-^2}{\mu_1^2}\right)^{-1/4}$ is not present and $(\mu_1^2)^{-1/4}$ is replaced by $(\mu^2(\tau))^{-1/4}$. However, we achieve the same probability of finding the Universe in a region of the plane (β_+, β_-) for both approaches, if the spectra of the corresponding momenta span sufficiently small values. In this way, the contribution of the anisotropies to the total energy is negligible with respect to the matter part. In other words, for the Vilenkin approach we need to impose a constraint on the anisotropies variables phase space, namely that they exhibit a "light dynamics".

5. Implementation of the Vilenkin Approach to the Bianchi IX "Corner"

Let us now analyse the well-known Bianchi IX "corner" configuration [28] implementing the WKB idea to separate the quasi-classical component from the "small" variable β_-. Thus, to describe Bianchi IX's dynamics near the singularity using the Misner variables and the Vilenkin approach, we consider as an initial condition for the point-universe the right corner of the potential $V_{IX} \sim 48e^{4\beta_+}\beta_-^2$, where $\beta_+ \to +\infty$ and $|\beta_-| \ll 1$, therefore, α and β_+ have to be semiclassical variables while β_- quantum. In the following analysis (see [101]) we will include a massless scalar field ϕ for which $\dot{\phi} \ll U(\phi)$, and we will assume a synchronous frame $N(t) = 1$.

Substituting the ansatz (19) and using the conditions above in the WDW equation, Equation (9) becomes

$$2(\partial_\alpha A \, \partial_\alpha S - \partial_+ A \, \partial_+ S - \partial_\phi A \, \partial_\phi S) + A\left(\partial_\alpha^2 S - \partial_+^2 S - \partial_\phi^2 S\right) = 0, \quad (51)$$

associated to the probability density, while the dynamics of a harmonic oscillator with time-dependent frequency and unitary mass reads as

$$i\hbar \frac{\partial \chi}{\partial \tau} = (\partial_-^2 + 16e^{4(\alpha+\beta_+)}\beta_-^2)\chi, \quad (52)$$

if we impose $\omega^2(\tau) \equiv 16e^{4(\alpha+\beta_+)}$ and $\tau = c \int e^{-3\alpha} dt$. Note that in what follows time will be rescaled by a factor 2 as in [101]. To solve (52) we make use of the invariant method developed in [102]. The general solution is given by

$$\chi = \sum_n c_n e^{i\alpha_n(\tau)} \phi_n(\beta_-, \tau) = \sum_n c_n \chi_n(\beta_-, \tau), \quad (53)$$

where c_n are numerical coefficients that weight the different χ_n

$$c_n = \int d\beta_- \chi_n(\beta_-, \tau) \chi_0(\beta_-, \tau), \qquad (54)$$

$$\chi_n(\beta_-, \tau) = \frac{e^{i\alpha_n(\tau)}}{\sqrt{\sqrt{\pi} n! 2^n \rho}} h_n\left(\frac{\beta_-}{\rho}\right) e^{\frac{i}{2\hbar}\left(\frac{\dot{\rho}}{\rho} + \frac{i}{\rho^2}\right)\beta_-^2}, \qquad (55)$$

where the index $_0$ states the initial condition, h_n are Hermite polynomials, ρ satisfies the auxiliary equation

$$\ddot{\rho} + \omega^2 \rho - \rho^{-3} = 0, \qquad (56)$$

and

$$\alpha_n(\tau) = -\left(n + \frac{1}{2}\right) \int_0^\tau \frac{1}{\rho^2} d\tau'. \qquad (57)$$

It is usually complicated to analytically solve (56), but in [102] the author developed a method that allows us to have the explicit expression for ρ, linear combination of functions $h(\tau)$ and $r(\tau)$ dependent on the considered model

$$\rho = (\mathcal{W})^{-1}(A^2 r^2 + B^2 h^2 + 2(A^2 B^2 - (\mathcal{W})^2)^{\frac{1}{2}} hr)^{\frac{1}{2}}, \qquad (58)$$

where A^2, B^2 are arbitrary real constants, and $\mathcal{W}$ is the Wronskian.

As a first step, the dynamical evolution of the Mixmaster model could be studied in vacuum, namely the simplest case. For further studies of Bianchi IX considering a vector field, see [103,104].

5.1. Bianchi IX in Vacuum

Starting from (52) and using (36), (33), (34) in particular we find

$$\alpha(\tau) = \frac{1}{3} \log(6|p_\alpha|K) + 2|p_\alpha|\tau, \qquad (59)$$

where $K = \kappa/3(8\pi)^2$. It is worth noting that, in the calculation above, we adopted the absolute value of p_α due to its relation to $\dot{\alpha}$. In fact the expression for $\dot{\alpha}$ (with $\cdot$ referring to the synchronous time t),

$$\dot{\alpha}(t) = -2Kp_\alpha e^{-3\alpha}, \qquad (60)$$

denotes how much the volume of the universe changes with the synchronous time and it has the opposite sign of p_α, so that an *expanding* universe is described by $p_\alpha < 0$. Considering the variable τ, for $0 < t < +\infty$ we have $-\infty < \tau < +\infty$. At the same time, for an expanding universe, the semi-classical variable β_+ increases toward larger values that means $\dot{\beta}_+(t) > 0$ and this, again, translates in $p_+ > 0$. Therefore, the equation for $\beta_+(\tau)$ is

$$\beta_+(\tau) = \beta_0 + 2|p_\alpha|\tau. \qquad (61)$$

Hence, the frequency for the harmonic oscillator becomes $\omega^2(\tau) \sim C e^{m\tau}$, with m and C constants. Now, we can compute the expression for ρ that reads as

$$\rho = \frac{1}{2m} \sqrt{\pi^2 J_0^2\left(\frac{2\sqrt{C}\sqrt{e^{m\tau}}}{m}\right) + 64m^2 N_0^2\left(\frac{2\sqrt{C}\sqrt{e^{m\tau}}}{m}\right) + 8\pi\sqrt{3} m J_0\left(\frac{2\sqrt{C}\sqrt{e^{m\tau}}}{m}\right) N_0\left(\frac{2\sqrt{C}\sqrt{e^{m\tau}}}{m}\right)}, \qquad (62)$$

where J_0 and N_0 represent the Bessel functions of the first and the second kind.

To conclude the study of the probability density, firstly we need to compute (37) using (53). We choose $|\chi_0|^2$ such that it has a Gaussian shape peaked around $\beta_- = 0$. Figure 2 shows the probability density function for different values of the synchronous time variable as a function of the quantum anisotropic variable β_-. We observe that, when the point-universe enters the corner, there is a suppression of the quantum variable β_-, as its standard deviation decays in time. In other words, the Gaussian packet tends to peak around the value $\beta_- = 0$. The corner becomes an attractor for the global system dynamics and the point-universe cannot escape anymore. This is the reason why the universe approaches on a good level the Taub model.

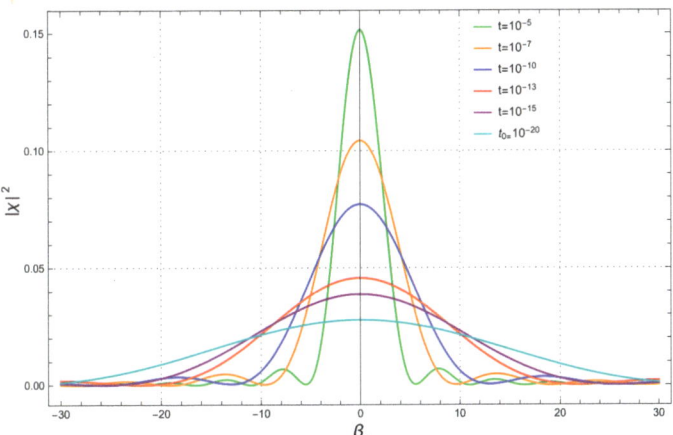

Figure 2. Time evolution of the probability density of the quantum subsystem considering Bianchi IX in the vacuum case with $\beta_+ = 2|p_\alpha|\tau$, for an expanding universe. Figure re-elaborated from [101].

The vacuum case can be analysed also for a *collapsing* behaviour of the universe. The dynamical evolution is represented by a decreasing β_+ for $t \to +\infty$. In this case, the initial assumption is that $\dot{\beta}_+ < 0$, which translates in $p_+ < 0$. Therefore, following the same steps, we achieve

$$\rho(\tau) = \frac{1}{\sqrt{\omega(\tau)}} = \frac{e^{-\beta_0}}{2}, \qquad (63)$$

in which $\omega(\tau) = 4e^{2\beta_0}$ is constant. The eigenfunctions χ_n which depend on time through $\rho(\tau)$ are now constant; hence the probability density distribution $|\chi|^2$ is defined simply by choosing its shape at the initial time. This means that it remains constant as the point-universe moves towards the time singularity, namely the point universe goes deeply inside the corner ($\dot{\beta}_+ < 0$). Here, the backward evolution of the universe would correspond to a Taub universe, which is no longer a singular cosmology in the past, endowed with a small fluctuating anisotropic degree of freedom in addition to the macroscopic classical universe. Hence, the singular behaviour of the Bianchi IX universe can be removed. This result could have a deep implication, under cosmological hypotheses, on the notion of the cosmological singularity as a general property of the Einstein's equations (see Section 5.4 and for the possible removal of the singularity in loop quantum cosmology (LQC) see [105,106]).

5.2. Bianchi IX in the Presence of the Cosmological Constant and a Massless Scalar Field

The aim of this analysis is to mimic the behaviour of the Bianchi IX universe if the de Sitter phase (which is associated to the introduction of the cosmological constant Λ and the scalar field ϕ) takes place when the corner evolution is performed by the point-universe. The quantum part of the superHamiltonian H_q does not change with respect to the previous one but we have extra terms in the classical part, namely

$$H_0 = e^{-3\alpha} K(-p_\alpha^2 + p_+^2 + p_\phi^2 + \Lambda e^{6\alpha}), \qquad (64)$$

where $K = \kappa/3(8\pi)^2$. Now, following the same steps of the previous Section 5.1, expressions for $\tau(t)$, $\alpha(\tau)$ and $\beta_+(\tau)$ are

$$\tau(t) = \frac{1}{6\sqrt{p_+^2 + p_\phi^2}} \log\left[\tanh\left(\frac{1}{2}(6K\sqrt{\Lambda}t + J)\right)\right], \qquad -\infty < \tau < 0 \qquad (65)$$

$$\alpha(\tau) = \frac{1}{3}\log\left[\frac{\sqrt{p_+^2 + p_\phi^2}}{\sqrt{\Lambda}}\sinh\left(2\arctanh\left(e^{6\tau\sqrt{p_+^2+p_\phi^2}}\right)\right)\right], \quad -\infty < \alpha < \infty \quad (66)$$

$$\beta_+(\tau) = \beta_0 + p_+\tau, \quad -\infty < \beta_+ < \beta_0. \quad (67)$$

This time, the evolution of the probability density function is computed numerically for different values of t, as shown in Figure 3.

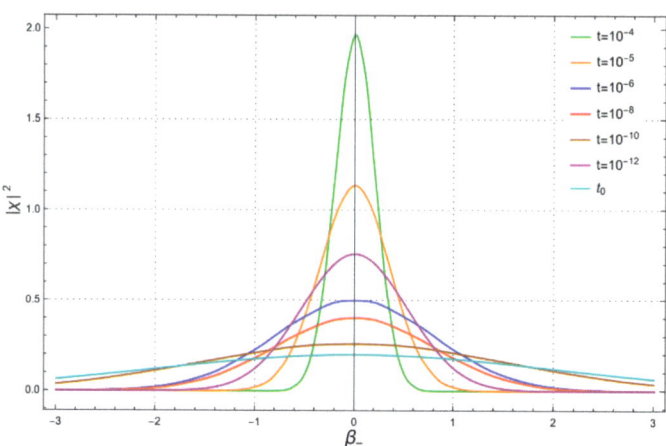

Figure 3. Time evolution of the probability density of the quantum subsystem considering a scalar field and a cosmological constant, in the case of an expanding Universe. Figure re-elaborated from [101].

We can conclude that as the universe evolves in time, the variable β_+ is suppressed while the fully quantum one β_- is characterized by a decaying standard deviation. Hence, in the proposed scheme, the universe naturally isotropizes. In other words, starting with a Gaussian shape, its evolution is then approaching a Dirac δ-function around the zero value of β_-. Thus, this result offers a new paradigm for the Bianchi IX cosmology isotropization based on the idea that the de-Sitter phase is associated with the corner regime of the model.

5.3. Taub Model

Another interesting application [107] could be the case of the Taub model [108] (for a full quantization see [109–112]) that is the natural intermediate step between Friedmann-Lemaître-Robertson-Walker (FLRW), which is invariant under rotations around any axis, and the Bianchi IX universe in which the rotational invariance is absent due to the presence of three different scale factors. Therefore, since the corners of Bianchi IX asymptotically correspond to the equality of two scale factors, e.g., one is fixed by the condition $\beta_- = 0$ and the other two are obtained for the rotational invariance of $2\pi/3$ in the plane (β_+, β_-), this leads to the Taub solution. The line element of the Taub space-time corresponds to (27) but the traceless symmetric matrix which determines the anisotropy via β_+ only is

$$\beta_{ab} = diag(\beta_+, \beta_+, -2\beta_+). \quad (68)$$

Within this study, we consider again a cosmological constant Λ and a free minimally coupled scalar field ϕ to mimic the inflationary scenario. For further studies about inflation in quantum cosmology, see [28,113,114]. The dynamics is summarized by the scalar constraint

$$H_T = Ke^{-3\alpha}(-p_\alpha^2 + p_+^2 + p_\phi^2 + \mathcal{V} + \Lambda e^{6\alpha}) = 0, \quad (69)$$

(we remind that $K = \kappa/3(8\pi)^2$) in which the potential term takes the form

$$V \equiv \frac{3(4\pi)^4}{\kappa^2} e^{4\alpha} V_T(\beta_+), \qquad (70)$$

where $V_T(\beta_+) = e^{-8\beta_+} - 4e^{-2\beta_+}$. The phase space of the system is six-dimensional with coordinates $(\alpha, p_\alpha, \beta_+, p_+, \phi, p_\phi)$ having p_ϕ as a constant of motion because of the absence of a potential term $U(\phi)$. The dynamical picture is completed by taking into account the choice of $N = e^{3\alpha}/K$ which fixes the temporal gauge.

Now, following the same steps of the Vilenkin approach, we can construct the classical and the quantum dynamics. Equations (8) and (9) become

$$-(\partial_\alpha S)^2 + (\partial_\phi S)^2 + \Lambda e^{6\alpha} = 0, \qquad (71)$$

$$\partial_\alpha(A^2 \partial_\alpha S) + \partial_\phi(A^2 \partial_\phi S) = 0. \qquad (72)$$

Equation (22) is instead responsible for the evolution of the quantum subspace, here represented by β_+: introducing the change of variable $e^\alpha = a$, the equation takes the form

$$i\hbar \frac{\partial \chi}{\partial \tau} = \left(-\partial_+^2 + \frac{a^4}{4\kappa^2} V_T(\beta_+)\right)\chi, \qquad (73)$$

where $d\tau = Ke^{-3\alpha}dt$ and the variable α increases with the synchronous time while τ decreases. In this respect we have

$$\frac{d\alpha}{d\tau} = -2Kp_\alpha < 0, \qquad (74)$$

with $p_\alpha \sim \sqrt{\Lambda e^{6\alpha}}$ since in (71) p_ϕ^2 can be neglected for large values of α. We are also taking the positive square root since we consider an expanding universe. The behavior of τ compared to a is then

$$\frac{d\tau}{dt} = -\frac{1}{2K\sqrt{\Lambda}a^4}\frac{da}{dt}. \qquad (75)$$

According to Vilenkin's idea of a small quantum subsystem, the quasi isotropic regime is considered, in which $|\beta_+| \ll 1$, and as a consequence the potential term gets a quadratic form

$$V_T(\beta_+) = -3 + 24\beta_+^2. \qquad (76)$$

It is worth noticing that the zero order of the approximate potential would provide a contribution to the HJ Equation (71) and becomes negligible when the cosmological constant dominates, once substituted into the WDW, i.e., $-3e^{4\alpha} \equiv -3a^4$. Hence, the frequency of the harmonic oscillator reads as $\omega^2(\tau) = 6\tau^{-4/3}/\tilde{k}^2$ where $\tilde{k}^2 = \kappa^2(6\kappa\sqrt{\Lambda})^{4/3}$. Now, with the method used above [102], we can construct an expression for ρ namely

$$\rho(\tau) = \frac{\tilde{k}^3}{324\sqrt{3}} \left\{ \frac{1}{\tilde{k}^2} \left[(9A^2 + 64B^2)(\tilde{k}^2 + 54\,\tau^{2/3}) + \left((-9A^2 + 64B^2)(\tilde{k}^2 - 54\,\tau^{2/3}) - 144\sqrt{24A^2B^2 - \frac{59049}{\tilde{k}^6}}\tilde{k}\tau^{1/3}\right) \right.\right.$$
$$\left.\left. \times \cos\left(\frac{6\sqrt{6}\tau^{1/3}}{\tilde{k}}\right) + 6\sqrt{2}\left(2\sqrt{8A^2B^2 - \frac{19683}{\tilde{k}^6}}\tilde{k}^2 + \sqrt{3}(-9A^2 + 64B^2)\tilde{k}\tau^{1/3} - 108\sqrt{8A^2B^2 - \frac{19683}{\tilde{k}^6}}\tau^{2/3}\right)\right.\right. \qquad (77)$$
$$\left.\left. \times \sin\left(\frac{6\sqrt{6}\tau^{1/3}}{\tilde{k}}\right) \right] \right\}^{1/2},$$

and the probability density for a generic expansion, i.e., $|\chi(\beta_+, \tau)|^2$, is then calculated.

Figure 4 shows that, as the volume of the universe expands, i.e., $\tau \to 0$, the profile of the Gaussian shape becomes more and more peaked. In this case, it cannot reach a real δ-function as stated in [115] but a steady small finite value emerges, namely

$$\rho(\tau \to 0) = \frac{2\sqrt{\frac{2}{3}}}{81}\tilde{k}^3 + \frac{2}{3}\tilde{k}\tau^{2/3} - \frac{3\sqrt{6}\tau^{4/3}}{\tilde{k}} + \mathcal{O}(\tau)^{5/3}. \tag{78}$$

A confirmation of this behaviour is present also considering an asymptotic study of an exact Gaussian solution of the time-dependent Schrödinger equation, which will be discussed below. We can state that the de Sitter exponential expansion of the universe strongly suppresses the quantum anisotropy leaving a small relic at the end of inflation. This surprising result suggests that, although the anisotropy cannot have the same non-suppressed behaviour of a scalar field, a small tensor degree of freedom can be present on a quantum level. In this sense, in the full inhomogeneous scenario, it could originate a smaller tensorial component of the primordial spectrum.

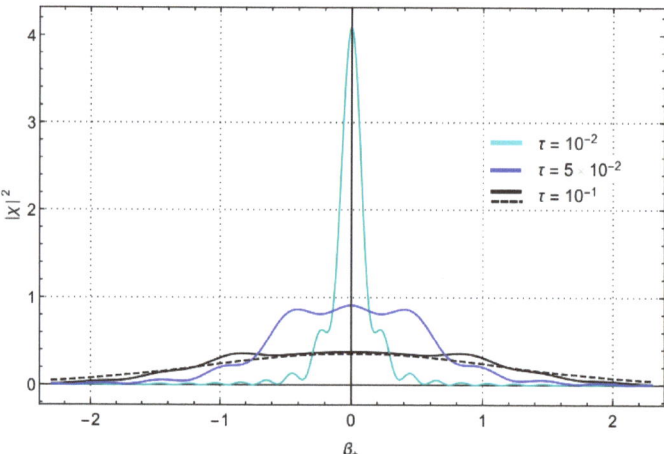

Figure 4. Time evolution of the probability density is highlighted with different colours. The dashed black line represents the initial time τ_i while the continuous line is the solution with Hermite polynomials. The wavy trend is given due to the truncation of the Hermite polynomials. In this plot we used $A = \frac{81\sqrt{3/2}}{2B}$ and $B = 1$. Figure from [107].

To clarify what we anticipated above, we now search for an exact Gaussian solution [116,117] of the time-dependent Schrödinger equation as

$$\chi(\beta_+, \tau) = N(\tau) e^{-\frac{1}{2}\Omega(\tau)\beta_+^2}, \tag{79}$$

since it is evident from the harmonic oscillator eigenfunction that the simplest way to locate the universe is a Gaussian shape. Substituting (79) in (73) and separating all terms of zero and quadratic order in β_+, we get

$$iN'(\tau) = \frac{1}{2}N(\tau)\Omega(\tau), \tag{80}$$

$$i\Omega'(\tau) = \Omega^2(\tau) - \omega^2(\tau). \tag{81}$$

To achieve the modulus of the normalization factor we also request a normalized wave function for any value of time. To obtain the physical information on the anisotropy behaviour we solve (81) since the quantity we need is the inverse Gaussian width. Now, separating Ω into its real and imaginary parts, i.e., $\Omega = f(\tau) + ig(\tau)$, we obtain the following non-linear system

$$2g = \frac{f'}{f}, \tag{82}$$

$$g' = g^2 + \omega^2 - f^2. \tag{83}$$

It is worth noting that (82) and (83) do not admit an analytical solution, but we can easily construct an asymptotic behaviour for which $\tau \to 0$. We achieve

$$g(\tau \to 0) \simeq -\frac{3C^2}{\tau^{1/3}}, \tag{84}$$

$$f(\tau \to 0) \simeq f_0 \, e^{-6C^2 \tau^{2/3}}, \tag{85}$$

where $C^2 = 6/\tilde{k}^2$ and f_0 is an integration constant. The standard deviation of the Gaussian probability distribution is

$$\sigma(\tau \to 0) = \frac{1}{\sqrt{\Re(\Omega)}} \simeq \frac{1}{\sqrt{f_0}} e^{\frac{6}{2} C^2 \tau^{2/3}}. \tag{86}$$

Hence, the standard deviation exponentially decays (Figure 5) when the universe expands, i.e., $\tau \sim 1/a^3$ decreases. However, also here, it approaches a non-zero value. In fact, this feature corresponds to the constant value assumed by ρ in (78). We can state that, if the universe anisotropy is small enough to be in a quantum regime when inflation starts, it is still present at late times.

5.4. Inhomogeneous Extension

We now briefly review the analysis developed in [97], where the ideas presented above have been extended to the generic inhomogeneous cosmological solution, also clarifying the physical conditions under which the WKB scheme becomes applicable.

The analysis of a generic inhomogeneous Universe has been first developed in [118], see also [68,119–121] and it corresponds to the situation in which the functions α, β_+ and β_- acquire a dependence on the spatial coordinates and the 1-forms, describing the geometry of the 3-hypersurfaces, are associated to a generic vector field, whose time dependence is neglected at the higher order.

This scheme allows to implement the so-called "Belinski–Khalatnikov–Lifshitz (BKL) conjecture" (for its validation on a classical level see [28,119]), according to which each region of the order of the averaged cosmological horizon behaves like the homogeneous Bianchi IX and Bianchi VIII models, sufficiently close to the initial singularity. In this picture, the chaotic feature of these two Bianchi models is extended to the dynamics of a generic inhomogeneous Universe as a local concept: each causal region is characterized by the same oscillatory regime and chaotically evolves independently from any other one. Actually, this picture is the result of a more rapid decreasing of the average horizon with respect to the typical inhomogeneous scale, as the initial singularity is approached. Thus, in the limit of the BKL conjecture validity (for the question concerning possible spikes in the spatial gradients see [122]), the Mixmaster scenario described by the triangular potential in Figure 1 can be applied as a point-like model, including the corner dynamics addressed above [28].

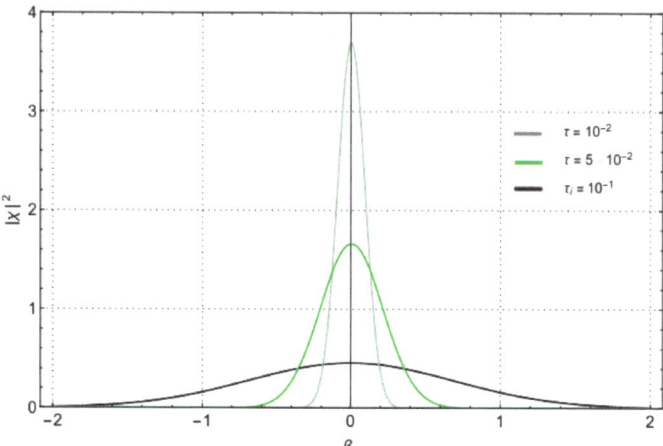

Figure 5. Time evolution of the probability density is highlighted by different colours. The initial time is τ_i. We considered $\frac{1}{f_0} = \frac{2\sqrt{2/3}}{81}$. Figure from [107].

In [97] it has been argued that, inside the corner, the variables α and β_+ are, near the singularity, very large and therefore remain classical degrees of freedom, while the small variable β_- can become a real quantum variable, according to the proposal in [17]. The idea is that, in a long sequence of iterations of the piecewise representation of the evolution in terms of the Kasner-like solution, a deep penetration of the point-universe inside the corner must, soon or later, take place in each spatial point [123].

By other words, it is argued that the uncertainty in the value of the variable β_- in the corner is of the order $\Delta\beta_- \sim 2\beta_+\sqrt{\hbar}$. Hence the uncertainty principle implies that the indetermination on the corresponding momentum is $\Delta p_- \sim 2\sqrt{\hbar}/\beta_+$. Recalling that deeply in the corner β_+ is very large, we deal with a small quantum subsystem associated to the phase space $\{\beta_-, p_-\}$ and the ratio between the quantum Hamiltonian and the classical one is of order $\hbar$. That is, all the assumptions at the ground of the decomposition into two parts of the global system, one classical (here the quantum corrections on α and β_+ are not present at all) and a small quasi-classical subset, considered in [17] are fully satisfied. Hence, the same analysis performed above follows directly in each space point, since the variable β_- dynamics is descried by a time-dependent quantum harmonic oscillator in each locally homogeneous region. However, in the inhomogeneous case, it has to be taken into account the so-called fragmentation of the space [28,120,124,125]. In fact, the chaotic time evolution of the locally homogeneous regions induces a corresponding oscillation of the spatial dependence of the metric functions. As a result, the comoving inhomogeneity scale is not the same during all the evolution toward the singularity, but it also decreases [28,119], although the Mixmaster scenario is preserved.

The important point here is that the corner configuration is then reached in each (even arbitrarily small) space region which contains a rational value of the parameter u, by which the BKL map is described [118,120]. This feature ensures that, as the initial singularity is approached, essentially all the space is (homogeneous patch by homogeneous patch) in the corner configuration (this takes place in different instants of time) and the WKB scenario inferred in Section 3 can be applied. Thus, in the end, since the variable β_- is frozen out to a negligible value (described by a constant standard deviation around the zero mean value), then we deal with a non-singular generic inhomogeneous universe. In this respect, the implementation of the ideas developed in [17] to the inhomogeneous Mixmaster leads to a possible picture to solve the problem of the initial singularity on a very general footing.

6. WKB Expansion for Quantum Gravity Contributions

As seen in Section 3, the work [17] implemented a perturbative expansion in the Planck constant in order to obtain a functional probabilistic interpretation for the wave function of the universe. This procedure can be enclosed as a special case of the WKB method [71], which allows to compute an approximate solution to a differential equation of the WDW type by going to increasing orders of accuracy in a desired parameter, as seen in the study presented in Section 4. To illustrate this method, let us start from the WDW Equation (5); we assume the solution wave function Ψ to be of the form

$$\Psi = e^{iS/\hbar}, \tag{87}$$

with S a complex function, that we expand in some parameter P

$$S = \sum_{n=0}^{\infty} P^n S_n. \tag{88}$$

The application of the superHamiltonian operator to (87) using (88) brings a series of equations, each one at a different order in P and acting as a small perturbation to the previous ones (having chosen P appropriately). Using the Planck constant $\hbar$ as expansion parameter [126] this corresponds to the so-called *semiclassical approximation* in quantum theory; nonetheless, one could implement a different perturbation parameter according to the physical properties of the considered theory. Substituting (87) into (5), one can solve each order in P and, supposing that the universe can be separated as in Section 3, obtain a dynamical description of the quantum subsystem at some level of accuracy, containing corrections from the "semiclassical" sector.

We emphasize the difference between the direct application of this method and the work [17] presented in Section 3, i.e., the ansatz (6) was composed of a semiclassical amplitude $A(\hbar)$, multiplied by an exponential term expanded in $\hbar$. This hypothesis is based on the assumption that the universe can, at some level, be separated between a purely semiclassical sector and the remaining quantum one, as already discussed, but it is not a general feature of the WKB method. However, Vilenkin's work can be recast as a WKB expansion with the ansatz (87) by considering a complex function S and expanding (88) in the parameter $\hbar$, as shown in [22].

The WKB expansion for quantum gravity has been implemented in many works in the literature after [17], mainly focusing on the canonical quantization prescription [127–131] sometimes in different expansion parameters [20,24,116,117,132–139], or taking different paths considering some sort of WKB ansatz [78,140–142]. In several works, the WKB method has been implemented in the context of a BO approximation [143,144] for gravity and matter [22,25,26,145–151]. Reviews regarding the use of the WKB procedure for constructing time in quantum cosmology can be found in [53,55,59,152,153]. In the following section we will explore in more detail some of these works and discuss the emerging problem of non-unitarity for the matter dynamics.

6.1. Time from Gravitational Variables and the Question of Non-Unitarity

The expansion parameter in (88) can also be taken of Planckian size. That is the case of Kiefer and Singh's work [20], who first considered a regime in which the "classical limit" is the absence of matter, i.e., vacuum solutions.

Let us briefly recall this approach. We start by identifying in the system the "subsets" of quantum gravity and quantum matter, such that the WDW equation can be rewritten as

$$\left(-\frac{\hbar^2}{2M}\left(\nabla_g^2 + f \cdot \nabla_g \right) + MV(g) + \hat{H}_m \right) \Psi(g,m) = 0, \tag{89}$$

where M is the Planckian parameter

$$M \equiv \frac{1}{4c^2\kappa} = \frac{cm_P^2}{4\hbar}, \tag{90}$$

being $m_P = \sqrt{\hbar c/8\pi G}$ the reduced Planck mass, the term $f \cdot \nabla_g$ is inserted for generalization to other operator orderings, and H_m is the (scalar) matter superHamiltonian as in (4). An important aspect deriving from the choice of the expansion parameter (90) is that it allows a clear separation between the gravitational and matter subsets, since in the limit $M \to \infty$ ($G \to 0$ as can be seen from (90)) the latter will disappear, leaving only the Einstein's equations in vacuum. Such a choice implies that the WKB expansion will hold for particles with small mass over Compton length ratio, i.e., whose mass is $m \ll m_P$.

Similarly to Section 3, the wave function is taken to be of the WKB form

$$\Psi(g,m) = e^{\frac{i}{\hbar}S(g,m)}, \tag{91}$$

and S is then expanded in powers of M. However, in Vilenkin's work [17], the study was carried out to recover a Schrödinger dynamics for the quantum (here matter) variables, and to formulate a probabilistic interpretation for the complete Ψ, for which the order $\hbar^1$ was enough. In [20] instead, the aim is not only to recover such a dynamics for the matter sector (which will emerge at $\mathcal{O}(M^0)$), but also to investigate its modifications induced by the quantum nature of gravity, i.e., going up to the next order $\mathcal{O}(M^{-1})$. To obtain this, the total function S is first expanded in powers of M and then at each order separated in $a(g) + b(m,g)$, i.e., isolating a purely gravitational function. For the sake of clarity, we here reformulate the approach with that separation from the beginning, writing

$$S(g,m) = MS_0(g) + S_1(g) + \frac{1}{M}S_2(g) + Q_1(m,g) + \frac{1}{M}Q_2(m,g) + \mathcal{O}(M^{-2}), \tag{92}$$

where for consistency the highest function S_0 at $\mathcal{O}(M)$ (Planck scale) depends on gravitational variables only, as can be checked from the perturbative expansion. The matter enters at the next order, such that the gravitational background is naturally recovered without further assumptions. This feature represents a striking difference from the work in Section 3, where the WDW gravitational equation was also imposed. We stress that, in this implementation, the presence of classical matter can only be recovered with some suitable redefinition, for example with a rescaling of the matter fields themselves (see [116,117]).

Expanding in M, the first order M^1 gives

$$\frac{1}{2}(\nabla_g S_0)^2 + V = 0, \tag{93}$$

corresponding to the HJ equation for gravity which provides the classical limit, namely Einstein's equations in vacuum. We note that the coefficient $1/2$ in front of $(\nabla_g S_0)^2$ with respect to Vilenkin's proposal (8) is due to the definition of the expansion parameter M which makes it appear in the starting WDW Equation (89). In this sense, it is not related to any physical properties. The next order M^0 brings

$$\nabla_g S_0 \cdot \nabla_g S_1 + \nabla_g S_0 \cdot \nabla_g Q_1 - \frac{i\hbar}{2}\left(\nabla_g^2 S_0 + f \cdot \nabla_g S_0\right) + \frac{1}{2\sqrt{h}}(\nabla_m Q_1)^2 - \frac{i\hbar}{2\sqrt{h}}\nabla_m^2 Q_1 + U = 0, \tag{94}$$

where we indicate the derivatives with respect to ϕ as ∇_m. Requiring that $S_1(g)$ satisfies

$$\nabla_g S_0 \cdot \nabla_g S_1 - \frac{i\hbar}{2}\left(\nabla_g^2 S_0 + f \cdot \nabla_g S_0\right) = 0, \tag{95}$$

namely a continuity equation for S_1 (being S_0 known from the previous order), the matter wave function $\chi_0 = e^{\frac{i}{\hbar}Q_1}$ satisfies

$$i\hbar \frac{\partial}{\partial \tau} \chi_0 = N \hat{H}_m \chi_0 . \tag{96}$$

Equation (96) is a functional Schrödinger equation where the WKB time is defined by

$$\frac{\partial}{\partial \tau} = N \nabla_g S_0 \cdot \nabla_g, \tag{97}$$

similar to (12), in which the lapse function (that was removed in the original work via a gauge choice) has been reinserted for the general case in order to maintain a parallelism with Vilenkin's definition (12). We emphasize that, in Section 3, the continuity equation was not imposed but obtained from the perturbative procedure since we required the WDW gravitational constraint from the beginning; here instead, there is no such initial assumption. To recover the functional quantum field dynamics we have to impose another condition on S_1, i.e., (95).

Developing the analysis to the next order M^{-1}, one finds

$$\nabla_g S_0 \cdot \nabla_g S_2 + \nabla_g S_0 \cdot \nabla_g Q_2 + \frac{1}{2}\left((\nabla_g S_1)^2 + (\nabla_g Q_1)^2\right) + \nabla_g S_1 \cdot \nabla_g Q_1 - \frac{i\hbar}{2}\left(\nabla_g^2 S_1 + \nabla_g^2 Q_1 + f \cdot \nabla_g S_1 \right. \\ \left. + f \cdot \nabla_g Q_1\right) + \frac{1}{\sqrt{h}} \nabla_m Q_1 \nabla_m Q_2 - \frac{i\hbar}{2\sqrt{h}} \nabla_m^2 Q_2 = 0, \tag{98}$$

which again can be cast in a clearer form once the function S_2 satisfies an analogous continuity equation

$$\nabla_g S_0 \cdot \nabla_g S_2 + \frac{1}{2}(\nabla_g S_1)^2 - \frac{i\hbar}{2}\left(\nabla_g^2 S_1 + f \cdot \nabla_g S_1\right) = 0, \tag{99}$$

thus leaving only

$$\nabla_g S_0 \cdot \nabla_g Q_2 + \frac{1}{2}(\nabla_g Q_1)^2 + \nabla_g S_1 \cdot \nabla_g Q_1 - \frac{i\hbar}{2}(\nabla_g^2 Q_1 + f \cdot \nabla_g Q_1) + \frac{1}{\sqrt{h}} \nabla_m Q_1 \nabla_m Q_2 - \frac{i\hbar}{2\sqrt{h}} \nabla_m^2 Q_2 = 0. \tag{100}$$

We can now decompose the derivatives ∇_g in tangent and normal components to the hypersurfaces $S_0 = const$ and neglect the former by assuming the adiabatic dependence of H_m on the induced metric. Summing (100) with the previous order, the resulting equation for the matter wavefunction $\chi = e^{\frac{i}{\hbar}(Q_1 + \frac{1}{M} Q_2)}$ for $N = 1$ is

$$i\hbar \frac{\partial \chi}{\partial \tau} = \hat{H}_m \chi + \frac{1}{8M\sqrt{h\bar{R}}} \left[\hat{H}_m^2 + i\hbar \left(\frac{\partial H_m}{\partial \tau} - \frac{1}{\sqrt{h\bar{R}}} \frac{\partial(\sqrt{h\bar{R}})}{\partial \tau} \hat{H}_m\right)\right] \chi . \tag{101}$$

Here, the terms after H_m are a modification to the standard quantum matter dynamics and thus they represent quantum gravity corrections. An inspection of these terms reveals that they violate unitarity in the evolution.

It can be noted that, up to the order M^0, the work [20] seems to portray a functional description of the system analogous to the one obtained by Vilenkin (Section 3). Actually, it can be shown that the approaches [17,20] are equivalent to a unique WKB expansion of the WDW equation just by changing the expansion parameter (see reformulation in [22]). As a consequence, Vilenkin's work can also be expanded to the next order in $\hbar$ finding quantum gravity corrections in the functional Schrödinger formalism. However, also in that case, they manifest a non-unitary morphology.

The question of non-unitarity in this kind of approaches has been long discussed in the literature [22,24,25,78,79,150,154,155], with many significant outcomes. As presented in [155], implementing a scalar field clock, the request of unitarity can lead to a quantum recollapse of the model; in [150] an inner product is proposed in relation to the Faddeev–Popov gauge-fixing procedure. We here briefly discuss the proposal [24] to overcome the non-unitarity emerging in Equation (101): the authors construct the set of complex eigenvalues $E(\tau)$ associated to the total non-Hermitian Hamiltonian operator in (101), together with the set of real eigenvalues $\epsilon(\tau)$ of $\hat{H}_m$. In this notation, τ is the only geometrical variable present that is identified as time from the beginning. The functions $E(\tau)$ and $\epsilon(\tau)$ are then expanded in powers of $1/M$. By redefining the quantum wave function with a phase transformation involving the imaginary part of $E(\tau)$, and rescaling the background with the opposite phase, the redefined quantum state gives a contribution in the equation that exactly cancels the non-unitary terms in (101). Thus, the dynamics for the redefined χ at $\mathcal{O}(M^{-1})$ presents only the Hermitian part of the quantum gravity corrections, restoring unitarity; also, a quantum backreaction emerges in the HJ equation due to the rescaling. However, the procedure is built on the assumption that the operators H_{tot} and H_m commute, and thus can be diagonalized simultaneously. This property does not hold in some cases, for instance considering a FLRW model with a cosmological constant and a scalar field. In that setting, H_{tot} at the order $1/M$ contains both H_m and its time derivative $\dot{H}_m$, with H_m including the scale factor a and $\dot{H}_m$ its conjugate momentum, so the two operators cannot commute (for a critical analysis of this restatement, see [22]).

Moreover, the question of non-unitarity has been addressed also in the context of modified theories of gravity, where it can emerge due to renormalizability requirements of the corresponding quantum theory (e.g., [156]). Recent interest has been devoted to the case of massive gravity, where the graviton particle acquires a nonzero mass. Massive gravity was first introduced by the work of Pauli and Fierz [157] and later reformulated with the "gravitational Higgs mechanism" (in which the spontaneously broken symmetry is the one associated to coordinate reparametrization invariance) or via higher-derivative curvature terms [158]. Such theory is however plagued by the emergence of ghost fields, i.e., non-physical states associated to non-dynamical variables, that induce negative probabilities in the theory and so violate unitarity [159,160]. Solutions to this issue have been proposed both in three dimensions, see [161–164] and in four dimensions with the so-called dRGT model [165] (see [166] for some deviations from GR predicted by the model), and also [167–169].

For what concerns the non-unitarity problem in the present General Relativity analysis, the description of quantum gravity corrections to the matter sector dynamics with the WKB procedure leaves some unanswered questions. Another relevant implementation is to regard the gravity and matter system in a Born–Oppenheimer approximation, as mentioned in Section 4, in order to tackle this issue in the canonical quantization framework.

6.2. The Born–Oppenheimer-like Approximation

A further implementation of the DeWitt theory for gravity and matter is the Born–Oppenheimer (BO) extended approach presented in [25], later applied in the context of quantum cosmology in [146,147,149]. In analogy with the BO approximation for molecules, the wave function is separated as $\Psi(g,m) = \psi(g)\chi(m,g)$ since the matter sector is characterized by a lower mass scale with respect to the Planckian one. Hence, the matter can be regarded as the "fast" quantum sector while gravity is the "slow" quantum component. Working in the minisuperspace, the total WDW Equation (5) is averaged over $\chi(m,g)$ and subtracted to the initial equation thus obtaining an equation for the gravitational background ψ and one for the matter sector χ. Both functionals are rescaled making use of the gauge invariance of the system through a phase depending only on the gravitational variables

$$\psi = e^{-\frac{i}{\hbar}\int A\,dg}\widetilde{\psi}, \quad \chi = e^{\frac{i}{\hbar}\int A\,dg}\widetilde{\chi}, \tag{102}$$

where $A = -i\hbar\langle\nabla_g\rangle$. Then, rescaling again χ via $\langle H_m \rangle$ and taking ψ in the WKB form, the HJ Equation (93) is modified by the presence of the matter backreaction $\langle H_m \rangle$. Implementing the time definition (97), the dynamics of the matter sector is given by

$$\left(\hat{H}_m - i\hbar\frac{\partial}{\partial\tau}\right)\chi_s = e^{-\frac{i}{\hbar}\int\langle H_m\rangle\,d\tau - \frac{i}{\hbar}\int A\,dg}\frac{\hbar^2}{2M}\left[\bar{D}^2 - \langle\bar{D}^2\rangle + 2(D\ln\mathcal{N})\bar{D}\right]\chi, \tag{103}$$

where D, $\bar{D}$ are covariant derivatives constructed with A as Berry connection, $1/\mathcal{N}$ is the amplitude associated to the WKB-expanded ψ, and $\chi_s = e^{-\frac{i}{\hbar}\int \langle H_q\rangle d\tau - \frac{i}{\hbar}\int A\, dg}\chi$. As in the previous approaches, in the semiclassical limit the right-hand side vanishes due to the adiabatic approximation and Equation (103) describes the usual Schrödinger dynamics. Furthermore, the authors suggest that the obtaining dynamics is unitary due to the vanishing of

$$i\hbar \frac{\partial}{\partial \tau}\langle \chi_s|\chi_s\rangle = 0. \qquad (104)$$

However, this approach does not completely solve the non-unitarity problem. In fact, while the norm of quantum states preserves unitarity signaling a possible construction of the Hilbert space associated to the matter sector, this might not be true when the quantity (104) is computed between different quantum states. It has also been shown in [22] that, once the gravitational wavefunction ψ is rescaled with $\langle H_m\rangle$ (which is a requirement of the gauge symmetry of the theory), Equation (103) takes a different form, again as a modified Schödinger equation that is unitary only if one considers $\langle \chi_s|\chi_s\rangle$. Moreover, as a consequence of the rescaling, the matter backreaction does not appear at the level of the HJ but goes to the next order where it gets canceled by an opposite term, actually vanishing in the proposed approach.

The presence of the quantum backreaction in these models is also worth discussing [170]. Considering Vilenkin's work, this contribution is absent from the HJ due to the background assumption (7), while in [20] it is forbidden by the choice of expansion parameter, as mentioned above. However, using the same parameter, a matter backreaction term emerges in both [24,25] via some rescaling. In the context of quantum cosmology, when perturbations are present, such backreaction would describe how small scale inhomogeneities influence the large-scale structure of the universe. With this aim, many studies have been carried on considering both semiclassical and quantum backreactions, i.e., with a classical or quantized gravitational sector (see [170] and references within for an overview). In relation to the topics here presented, we mention the implementations based on Space-Adiabatic Perturbation Theory (SAPT) [171], which can be formulated as a generalization of the Born–Oppenheimer procedure aimed at solving the coupled dynamics at a perturbative level [172,173].

7. A Proposal for Unitarity: The Role of the Reference System

The emergence of non-unitarity in the approaches discussed above may signal that the time definitions in (12), (97) are to be reconsidered. Indeed, they bring in the expansion at $\mathcal{O}(M^{-1})$ (or $\mathcal{O}(\hbar)$ in Vilenkin's approach) a squared time derivative coming from ∇_g^2 which leads to non-unitary terms in the modified dynamics [22].

A different implementation of time can follow from exploiting the role of the reference frame, whose presence in the model can be made explicit by adding a suitable term to the action. In the following we will focus on two different types of this implementation, namely the kinematical action and the Gaussian reference frame fixing, discussing their relation and physical meaning.

7.1. The Kinematical Action Proposal

Let us first review Kuchar's discussion presented in [4]. There, the *kinematical action* is defined as the term to be added to the theory, using some Lagrange multipliers, to restore covariance under the ADM foliation and thus under the choice of reference frame. This procedure stems from the observation that, in quantum field theory with an assigned ADM foliation, the relation between points on infinitesimally close hypersurfaces is not evident, i.e., the geometrical meaning of the deformation vector and its components N and N^i is lost, as can be seen in the case of a scalar matter field theory [4,10,22]. In the ADM representation, the kinematical action takes the form

$$S^{kin} = \int dt\, d^3x (p_\mu \partial_t y^\mu - N^\mu p_\mu), \tag{105}$$

where $y^\mu = y^\mu(x^i; x^0)$ define the family of one-parameter hypersurfaces obtained via the foliation, and p_μ are conjugate to y^μ. Adding (105) to the action of the model, further equations of motion (associated to the variations δy^μ, δp_μ and δN^μ) describe the vanishing of the momenta p_μ and restore the geometrical definition of the deformation vector

$$N^\mu = \partial_t y^\mu = N n^\mu + N^i b_i^\mu, \tag{106}$$

being n^μ the timelike direction and b_i^μ the tangent basis to the hypersurfaces identified by the foliation. The superspace constraints are modified by the presence of

$$H^{kin} = n^\mu p_\mu, \tag{107}$$
$$H_i^{kin} = b_i^\mu p_\mu, \tag{108}$$

such that the total superHamiltonian and supermomentum functions must now vanish. We notice that these terms represent a good candidate for the definition of time since Equations (107) and (108) are linear in the momenta p_μ.

Let us now analyze the model following from the definition of time through the kinematical action, as implemented in [22], instead of background variables. Starting from the action

$$S^g + S^m + S^{kin} = \int dx^0 d^3x \left[\Pi_a \dot{h}^a + p_\mu \dot{y}^\mu + \pi \dot{\phi} - N\left(H^g + H^m + H^{kin}\right) - N^i \left(H_i^g + H_i^m + H_i^{kin}\right) \right], \tag{109}$$

and separating the wave function in $\Psi(h, \phi, y^\mu) = \psi(h)\chi(\phi, y^\mu; h)$ as in Section 6.2, the WKB expansion in the Planckian parameter M (90) can be performed

$$\Psi(h, \phi, y^\mu) = e^{\frac{i}{\hbar}(MS_0 + S_1 + \frac{1}{M}S_2)} e^{\frac{i}{\hbar}(Q_1 + \frac{1}{M}Q_2)}, \tag{110}$$

being $S_n = S_n(h)$ and $Q_n = Q_n(\phi, y^\mu; h)$. We stress that, in this separation, the kinematical action (and so the reference frame) is enclosed in the fast quantum sector as are the matter fields, in contrast with the gravitational background; this requirement allows the time parameter to be independent from slow background variables which are related to non-unitarity. In Equation (110), as in (92), the expansion is truncated at order M^{-1} since the aim is to compute quantum gravity corrections to the matter dynamics. The requirements

$$\frac{\langle \hat{H}^m \chi \rangle}{\langle \hat{H}^g \Psi \rangle} = \mathcal{O}(M^{-1}), \tag{111}$$

$$\frac{\delta}{\delta h_{ij}} Q_n(\phi, y^\mu; h) = \mathcal{O}(M^{-1}), \tag{112}$$

are satisfied due to the difference in physical scales and in "velocities" of the two sectors typical of the BO approximation, as discussed in Section 6.2. Following Vilenkin's reasoning, the total WDW equation is imposed together with the analogous equation for the gravitational background, i.e.,

$$\left[-\frac{\hbar^2}{2M}\left(\nabla_g^2 + f \cdot \nabla_g\right) + MV(g) - \hbar^2 \nabla_m^2 + U - i\hbar n^\mu \frac{\delta}{\delta y^\mu} \right] \Psi = 0, \tag{113}$$

$$\left[-\frac{\hbar^2}{2M}\left(\nabla_g^2 + f \cdot \nabla_g\right) + MV(g) \right] \psi = 0, \tag{114}$$

where the term $f \cdot \nabla_g$ has been introduced for generic operator orderings, as in Section 6.1, the matter sector is described by a scalar field ϕ and the gravitational sector potential V possibly includes a cosmological constant term. In the general case, one cannot implement the minisuperspace reduction, thus the theory must take into account also the supermomentum constraints for the total Ψ and for the background respectively

$$\left[2h_i \bar{D}\cdot\nabla_g - \partial_i\phi\cdot\nabla_m - i\hbar b_i^\mu \frac{\delta}{\delta y^\mu}\right]\Psi = 0, \qquad (115)$$

$$[2i\hbar h_i \bar{D}\cdot\nabla_g]\psi = 0, \qquad (116)$$

being $h_i \bar{D}\cdot\nabla_g = h_{ij}\bar{D}_k \frac{\partial}{\partial h_{kj}}$ and $\bar{D}_k$ the (3-dimensional) induced covariant derivative associated to h_{ij}. We stress that, since we are here presenting the more general formalism, i, j, k are explicited spatial indices; we will then implement and discuss the minisuperspace reduction of this model.

Substituting (110), the expansion of the constraints Equations (113)–(116) brings at $\mathcal{O}(M)$

$$\frac{1}{2}\nabla_g S_0 \cdot \nabla_g S_0 + V = 0, \qquad (117a)$$

$$-2h_k \bar{D}\cdot\nabla_g S_0 = 0, \qquad (117b)$$

corresponding to the HJ and the diffeomorphism invariance of S_0. At $\mathcal{O}(M^0)$, from the gravitational constraint, we obtain a relation between S_0 and S_1. Using this link and summing Equations (113) and (115) with coefficients N and N^i respectively, one obtains

$$i\hbar\frac{\partial \chi_0}{\partial \tau} \equiv i\hbar \int d^3x \left(Nn^\mu + N^i b_i^\mu\right)\frac{\delta}{\delta y^\mu}\chi_0 = \hat{\mathcal{H}}^m \chi_0 = \int d^3x \left(N\hat{H}^m + N^i \hat{H}_i^m\right)\chi_0, \qquad (118)$$

where $\chi_0 = e^{\frac{i}{\hbar}Q_1}$ is the matter wavefunction at $\mathcal{O}(M^0)$ and the time derivative, which is defined via the kinematical momenta p_μ, includes the definition of the deformation vector N^μ. At the next order M^{-1}, proceeding in a similar way and making use of the hypothesis (112), the modified matter dynamics is obtained

$$i\hbar\frac{\partial \chi}{\partial \tau} = \hat{\mathcal{H}}^m \chi + \int d^3x \left[N\nabla_g S_0 \cdot (-i\hbar\nabla_g) - 2N^k h_k \bar{D}\cdot(-i\hbar\nabla_g)\right]\chi, \qquad (119)$$

being $\chi = e^{\frac{i}{\hbar}(Q_1 + \frac{1}{M}Q_2)}$. We can observe that the quantum gravity corrections described by the integral terms on the right-hand side are indeed small in the perturbation parameter since they involve the derivative of χ with respect to the gravitational variables, which are of $\mathcal{O}(M^{-1})$ due to the BO approximation (112). Differently from the approaches in Section 6, here the obtained modified dynamics is unitary since the correction terms in Equation (119) involve the conjugate momenta to the gravitational variables and the function S_0 which is constrained to be real from the HJ Equation (117a). A cosmological implementation of this model can be found in [174].

7.2. Fixing a Gaussian Reference Frame

The implementation in Section 7.1 managed to define a time parameter for the matter evolution overcoming the non-unitarity problem, however the connection between the kinematical action (105) and the reference system itself is not straightforward. In this sense Kuchar later proceeded, together with Torre, to study the implementation of a term more clearly related to the reference frame [8]. In this further work, the additional term corresponds to the selection of the Gaussian reference frame $\gamma^{00} = 1$, $\gamma^{0i} = 0$ reparametrized in terms of generic coordinates

$$S_f = \int d^4x \left[\frac{\sqrt{-g}}{2}\mathcal{F}\left(g^{\alpha\beta}\partial_\alpha T(x)\partial_\beta T(x) - 1\right) + \sqrt{-g}\,\mathcal{F}_i\left(g^{\alpha\beta}\partial_\alpha T(x)\partial_\beta X^i(x)\right)\right]. \qquad (120)$$

In Equation (120), $X^i(x^\alpha), T(x^\alpha)$ are the Gaussian coordinates written in terms of the general x^α whose associated metric is $g_{\alpha\beta}$, and $\mathcal{F}, \mathcal{F}_i$ act as Lagrange multipliers. In this notation, $\partial_\alpha X^i = \partial X^i(x^\alpha)/\partial x^\alpha$ and the dependence of the Gaussian coordinates on the x^α will be implied. The choice of the Gaussian coordinates is based on a straightforward implementation of fixing a reference frame (see also [175]), while the case of parametrized unimodular gravity is discussed in [176], see also [177] and the general parametrization process has been addressed in [178]. The so-called Kuchar–Torre model is characterized by the emerging of such Gaussian reference frame as a heat-conducting fluid in the theory. This brings a source term in Einstein's equations

$$T^{\alpha\beta} = \mathcal{F} U^\alpha U^\beta + \frac{1}{2}\left(\mathcal{F}^\alpha U^\beta + \mathcal{F}^\beta U^\alpha\right), \tag{121}$$

being $U^\alpha = g^{\alpha\beta}\partial_\beta T$ the four-velocity of the fluid, $\mathcal{F}$ its energy density, and $\mathcal{F}_\alpha = \mathcal{F}_i\partial_\alpha X^i$ its heat flow. Actually, implementing only the Gaussian time condition in (120), the fluid reduces to an incoherent dust since $\mathcal{F}_i$ is not needed and the stress energy tensor (121) reduces to the typical form $\mathcal{F} U^\alpha U^\beta$. It is clear from Equation (121) that the fluid emerges at the classical level acting as a source term for the gravitational sector; for this reason, the fluid has to satisfy the related energy conditions in order to be physical and not ill-defined. As examined in the original work, this corresponds to the following relation

$$\mathcal{F} \geq 2\sqrt{\gamma^{\alpha\beta}\mathcal{F}_\alpha\mathcal{F}_\beta}. \tag{122}$$

However, this condition is not satisfied in principle and it is also not conserved during the evolution unless the system is closed with an additional constraint that turns the fluid to an incoherent dust and reduces (122) to $\mathcal{F} \geq 0$. Thus, the energy conditions are not satisfied in the general case $\mathcal{F}, \mathcal{F}_i \neq 0$, while it is possible in the incoherent dust case $\mathcal{F}_i = 0$ with some suitable initial conditions.

In the Hamiltonian formalism, the total superspace constraints must vanish, containing the additional functions

$$H^f = W^{-1}P + WW^k P_k, \tag{123}$$
$$H^f_i = P\partial_i T + P_k \partial_i X^k; \tag{124}$$

where the Lagrange multipliers have been written in terms of the momenta P, P_k conjugate to (T, X^k) and the functions W, W^k are defined as

$$W \equiv (1 - h^{jl}\partial_j T\partial_l T)^{-1/2}, \tag{125}$$
$$W^k \equiv h^{jl}\partial_j T\partial_l X^k. \tag{126}$$

As in (107) and (108), the momenta linearly appear in the constraints. Indeed, the authors show that defining the time derivative from the reference fluid variables, the gravity-fluid system is described by a Schrödinger dynamics

$$i\hbar\,\partial_t\Psi = \int_\Sigma d^3x\, \frac{\delta\Psi(T, X^k, h^{jl})}{\delta T(x)}\bigg|_{T=t}\Psi = \hat{\mathcal{H}}\Psi = \int_\Sigma d^3x\, \hat{H}^g\,\Psi. \tag{127}$$

Here, the time derivative is defined in the case of ADM foliation such that the timelike direction coincides with the Gaussian time T one; both the cases in which $x^i \equiv X^i$ and $t \equiv T, x^i \equiv X^i$ are also discussed in the original paper.

Another related approach is the work [9]. There, the added sector is composed of an incoherent dust whose comoving coordinates and proper time identify a "privileged" reference frame which again can be used to overcome the frozen formalism issue. Furthermore, the obtained functional Schrödinger equation is independent from the dust coordinates and a conserved inner product can be defined. However, the square-root form of the dust superHamiltonian, representing the dust scalar energy density, leads to some difficulties in implementing this definition at a WKB perturbative level. For a minisuperspace application of the Kuchar–Brown dust time, using the RPSQ and BO approximation, see [179].

The possibility to implement the same WKB and BO procedure for the model with the Gaussian reference fluid term is investigated in [26]. We start from the WDW equation

$$\left[\left(-\frac{\hbar^2}{2M}\left(\nabla_g^2 + g \cdot \nabla_g\right) + MV\right) + \left(-\hbar^2 \nabla_m^2 + U\right) + \left(W^{-1}P + WW^k P_k\right)\right]\Psi = 0, \quad (128)$$

and the total supermomentum constraint

$$\left[(2i\hbar\, h_i\, \bar{D} \cdot \nabla_g) - (\partial_i \phi)\nabla_m + P\,\partial_i T + P_k\, \partial_i X^k\right]\Psi = 0, \quad (129)$$

for generality. The BO separation is implemented as $\Psi(h_{ij}, \phi, X^\mu) = \psi(h_{ij})\chi(\phi, X^\mu; h_{ij})$, where the inclusion of the Gaussian reference frame into the fast quantum sector is backed by its materialization as a fluid (121). The WKB expansion in M up to $\mathcal{O}(M^{-1})$ corresponds to the same ansatz (110) with functions $Q_n = Q_n(\phi, X^\mu; h_{ij})$. The adiabatic approximation of the BO procedure gives

$$\frac{\delta Q_n}{\delta h_{ij}} = \mathcal{O}(M^{-1}), \quad (130)$$

$$\frac{\langle \hat{H}^m \chi \rangle}{\langle \hat{H}^g \Psi \rangle} = \mathcal{O}(M^{-1}). \quad (131)$$

Considering again the matter backreaction to be negligible at the gravitational scale, as in Section 3, the gravitational constraints (114) and (116) also hold. Expanding the system of Equations (128), (129), (114) and (116) with the ansatz (110), the dynamics at the lowest order $\mathcal{O}(M)$ is described by the same HJ Equation (117a) and diffeomorphism invariance of S_0 (117b). The order M^0 describes a functional Schödinger dynamics with time definition

$$i\hbar \frac{\partial \chi_0}{\partial \tau} = \int d^3x \left[N\left(W^{-1}\frac{\delta}{\delta T} + WW^k\frac{\delta}{\delta X^k}\right) + N^i\left((\partial_i T)\frac{\delta}{\delta T} + (\partial_i X^k)\frac{\delta}{\delta X^k}\right)\right]\chi_0$$
$$= \hat{\mathcal{H}}^m \chi_0, \quad (132)$$

being $\mathcal{H}^m$ the matter Hamiltonian defined as linear combination of superHamiltonian and supermomentum functions as in (118). We stress that the time derivative in (132) is defined for a generic foliation since the general coordinates are left independent from the Gaussian ones, differently from (127). Another key property of the model is that the fluid always emerges at the quantum level, being absent from the HJ Equation (117a), thus it does not suffer from the energy condition problem discussed above. Finally, the order M^{-1} describes quantum gravity corrections in the form

$$i\hbar \frac{\partial \chi}{\partial \tau} = \hat{\mathcal{H}}^m \chi + \int d^3x \left[N \nabla_g S_0 \cdot (-i\hbar \nabla_g) - 2N^i h_i \bar{D} \cdot (-i\hbar \nabla_g)\right]\chi, \quad (133)$$

which, also in this case, are small in the parameter M and unitary due to assumption (130) and the reality of S_0 from the HJ equation.

Actually, we stress that the functional forms of Equations (119) and (133) depict an analogy between the kinematical action and the Gaussian reference frame implementation. Indeed, the time definitions (118) and (132) can be related, if one restricts the kinematical action to the form $\partial_t y^\mu \to \dot{T}$ by selecting the homogeneous setting $N^i = 0$ and the timelike direction $n^\mu = (1, \vec{0})$. Moreover, taking the Gaussian reference frame fixing with only the time condition $\mathcal{F} \neq 0, \mathcal{F}_i = 0$ that is the incoherent dust, the two procedures give the same dynamics both at $\mathcal{O}(M^0)$ and $\mathcal{O}(M^{-1})$. This property signals that the kinematical action is playing the role of the reference frame, acting as a fast quantum matter component and giving a preferred set of variables suitable for the construction of the time parameter. However, the parallelism is not full since the two implementations (118) and (132) differ between each other in the case of a generic foliation. It follows that a direct correlation between the Gaussian reference frame fixing and the kinematical action is not yet understood in the general case.

7.3. Reference Fluid as Time in the Minisuperspace

We now analyze the effects of the modifications in (133) for quantum cosmology in the minisuperspace, focusing on the behaviour of the probability density during the slow-rolling phase of an isotropic universe. Let us consider the minisuperspace reduction of the FLRW model with a free inflaton scalar field ϕ and a positive cosmological constant Λ in the gravitational potential accounting for the slow-roll phase of inflation [26]. This allows us to discard spatial dependencies and restrict the general form (120) to the case of a reference frame having $g^{00} = 1$, i.e., imposing the reparametrized constraint $g^{\mu\nu}\partial_\mu T \partial_\nu T - 1 = 0$ only. In this setting, the WKB expansion in M (due to the related energies being below the Planck scale) and the BO separation (110) are performed, considering a negligible backreaction from the dynamical contributions of the matter scalar field. The line element takes the simple form

$$ds^2 = N^2(t)\, dt^2 - a(t)^2 \left(dx^2 + dy^2 + dz^2\right), \tag{134}$$

in which a is the cosmic scale factor, while the action corresponds to

$$S = \int d^4x \sqrt{-g} \left\{ -\frac{1}{2\kappa}(R + 2\Lambda) + \frac{1}{2} g^{\mu\nu} \partial_\mu \phi\, \partial_\nu \phi + \frac{\mathcal{F}}{2}\left(g^{\mu\nu}\partial_\mu T \partial_\nu T - 1\right) \right\}, \tag{135}$$

where $R = 6\left(\frac{\ddot a}{a} + \frac{\dot a^2}{a^2}\right)$ and the spatial Lagrange multiplier $\mathcal{F}_i$ is discarded. Due to homogeneity $\phi = \phi(t)$, $T = T(t)$, and $\mathcal{F} = \mathcal{F}(t)$. Hence, Equation (135) in the Hamiltonian formulation takes the form

$$S_{RW} = \int dt \left\{ p_a \dot a + p_\phi \dot\phi + p_T \dot T - N\left(-\frac{\kappa}{12}\frac{p_a^2}{a} + \frac{\Lambda}{\kappa}a^3 + \frac{p_\phi^2}{2a^3} + p_T\right) \right\}, \tag{136}$$

where $\dot{} \equiv \partial/\partial t$ and t coincides with the Gaussian time T due to the constraint introduced by $\mathcal{F}$ (also $N = 1$ as a consequence of the Gaussian condition on the metric). The spatial integration has been removed by considering a fiducial volume $V_0 = 1$. In this case, the only contribution from the reference fluid is the momentum p_T present in (136), that is related to $\mathcal{F}$ via $p_T = a^3 \mathcal{F} \dot T / N$. Hence, to recover the lapse function relation $\dot T = N$ it must hold $p_T = \mathcal{F} a^3$. The WDW equation gives

$$\left(\frac{\hbar^2}{48Ma} \partial_a^2 + 4M\Lambda a^3 - \frac{\hbar^2}{2a^3}\partial_\phi^2 - i\hbar\, \partial_T \right) \Psi = 0, \tag{137}$$

where we have considered the natural ordering with $f \cdot \nabla_g \equiv 0$.

Following the procedure of Section 7, Equation (137) and the gravitational constraint (114) are expanded at each order in M. We emphasize that the supermomentum constraints are automatically satisfied due to homogeneity of the model, such that the respective equations are not included in the minisuperspace reduction. Numerical solutions for the gravitational functions S_n are computed, selecting the ones corresponding to an expanding universe

$$S_0(a) = -\frac{8\sqrt{3}}{3}\sqrt{\Lambda}\left(a^3 - a_0^3\right), \tag{138a}$$

$$S_1(a) = i\hbar \log\left(\frac{a}{a_0}\right), \tag{138b}$$

$$S_2(a) = -\frac{\hbar^2}{24\sqrt{3}\sqrt{\Lambda}}\left(a^{-3} - a_0^{-3}\right), \tag{138c}$$

where a_0 is an integration constant corresponding to the reference value of the scale factor at the beginning of the slow-rolling phase. The matter sector at order M^0 follows the dynamics

$$-\frac{\hbar^2}{2a^3}\partial_\phi^2 \chi_0 = \hat{H}^m \chi_0 = i\hbar \frac{\partial \chi_0}{\partial T}, \tag{139}$$

which is the minisuperspace reduction of (132). Solutions to (139) are, in Fourier space, the plane waves

$$\tilde{\chi}_0 = e^{-i\hbar \frac{p_\phi^2}{2a^3}T}, \tag{140}$$

corresponding to standard field theory evolution on curved background. At $\mathcal{O}(M^{-1})$ such dynamics is modified by quantum gravity corrections, such that summing with the previous order and taking into account the expansion parameter the equation becomes

$$\left(-\frac{\hbar^2}{2a^3}\partial_\phi^2 + i\hbar \frac{1}{24a}(\partial_a S_0)\partial_a\right)\chi = i\hbar \frac{\partial \chi}{\partial T}, \tag{141}$$

for which explicit solutions can be computed in Fourier space by changing the time variable to a re-scaled time $d\tau = \frac{dT}{a^3}$ evolving with the universe volume. Then, the solution to (141) reads as

$$\tilde{\chi} = \exp\left(-i\hbar \frac{p_\phi^2}{2}\tau + i\frac{p_a(-\tau)^{7/3}}{7(3\Lambda)^{1/6}}\right), \tag{142}$$

where the smallness of the corrections is ensured by the hypothesis $|p_a| < M^{-1}$ deriving from the adiabatic approximation (130). As discussed before, the kinematical action implementation would describe the same modified dynamics in the minisuperspace setting. The solution (142) corresponds to a time-dependent shift in the matter energy spectrum

$$E = E_0 + \frac{\hbar p_a(-\tau)^{7/3}}{3(3\Lambda)^{1/6}}. \tag{143}$$

To better investigate its effects, it is useful to construct an initial Gaussian wavepacket

$$\chi(a,\phi,T) = \int dp_\phi \int dp_a\, \tilde{\chi}(p_\phi, p_a, T)\frac{1}{\sqrt{(2\pi)^{1/2}\sigma_a}}\exp\left(-\frac{(p_a - \bar{p}_a)^2}{4\sigma_a^2}\right)\frac{1}{\sqrt{(2\pi)^{1/2}\sigma_\phi}}\exp\left(-\frac{(p_\phi - \bar{p}_\phi)^2}{4\sigma_\phi^2}\right), \tag{144}$$

where $\sigma_a, \bar{p}_a$ and $\sigma_\phi, \bar{p}_\phi$ describe the standard deviation and mean value of the wavepacket associated to the gravitational and matter variable respectively. The wavefunction has a small dependence on the scale factor a, due to the condition (130) on p_a, as shown in Figure 6.

The probability density associated to (144) using the solution $\tilde{\chi}$ in (142) can be investigated at different values of the rescaled time τ. Figure 6 illustrates the effects of the time-dependent modifications (142), which are computed for the maximum τ in the allowed domain. We stress that near the Planck scale, that is outside of this domain, the previous approximations break down and one should consider an alternative algorithm to infer the evolution of the matter dynamics.

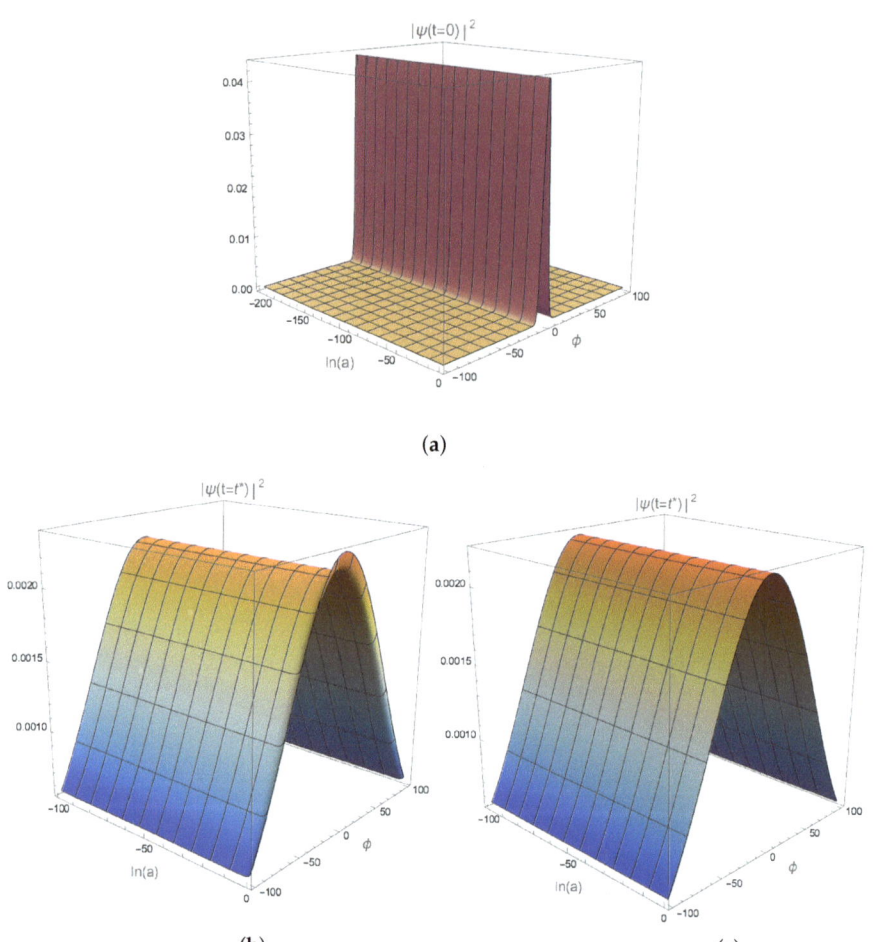

Figure 6. Evolution of the probability density with and without quantum gravity effects. Plot (**a**) represents the initial probability density at $\tau = 0$ associated to the wavepacket (144) Gaussian in the variables a, ϕ and satisfying the condition $|p_a| \ll M^{-1}$ (we used $M = 100$, $\Lambda = 10^{-2}$, $ln(a)_0 = 10$, $\bar{p}_\phi = 0, \sigma_\phi = 3, \bar{p}_a = 0, \sigma_a = 2 \cdot 10^{-2}$). Plots (**b**) and (**c**) show the spreading of the wave packet at later times without and with the quantum gravity effects computed in (141) respectively. We note that the quantum gravity corrections cause a deformation along the a axis when a approaches a reference value. Each wavefunction has been normalized on a suitable interval of values for the logarithm of the cosmic scale factor $\ln(a)$. Figures re-elaborated from [26].

8. Discussion and Conclusions

We analysed different aspects concerning the separation of a system phase space into a quasi-classical part and a small quantum subsystem. We first discussed the original idea in [17] about the possibility to re-construct a Schrödinger equation for the quantum variables and then we considered also the possibility to include quantum effects of the quasi-classical system into the quantum evolution of the small subsystem.

The first analysis had the main task to show how, when applied to the mini-superspace of the Bianchi models, this approach is able to provide interesting implications on the nature of the so-called corner configuration [101,107]. In particular, the possibility of a non-singular picture of the Bianchi VIII and IX dynamics, as well as for the generic cosmological solution, emerged. The crucial point was here the non-singular behavior of the Bianchi I dynamics when the variable β_- is vanishing. Since, according to the method in [17] (see also [97]), this variable dynamics is described via the Schrödinger equation of a time-dependent harmonic oscillator, we arrive to describe the corner dynamics via a steady classical universe over which a very small quantum anisotropy still lives. Actually, β_- has a probability distribution peaked around its zero value and characterized by a constant small anisotropic standard deviation. As extended to a generic inhomogeneous cosmological solution, this picture offers an intriguing paradigm to solve the problem of the initial singularity.

It is also an interesting achievement to have demonstrated that, comparing the Bianchi I model described in the ADM quantization procedure (also known as reduced phase space quantization [6,28]) with Vilenkin's formulation [17], the coincidence of the two approaches emerged when the quantum phase space of the anisotropic variables is sufficiently small. This has confirmed the consistency of the original proposal, where such an hypothesis on the quantum phase space was considered a basic statement.

In the second part of this review, we studied the various approaches proposed in the literature to determine the possible quantum gravity corrections to quantum field theory. By other words, we consider the small quantum subsystem coinciding with matter fields, while the quasi-classical component was the background gravitational field.

With respect to the original analysis in [17], the WKB procedure has been developed to the next order of approximation when quantum gravity corrections to the standard matter quantum dynamics have to arise, as in [20]. In particular, we re-analyzed the emerging problem that, at such further order of approximation, the Schrödinger equation for the matter fields acquires non-unitary (non-physical) contributions. The analyses in [17] and in [20] have been compared, showing, on one hand, that they are essentially equivalent and, on the other hand, that some proposed solutions to the non-unitary problem [24,25] are not consistently viable. The delicate point emerged to be the construction of a time evolution in terms of the classical dependence of the gravitational field on the label time. On the base of this argument, we eventually revised two different approaches in which the time coordinate belongs to the fast (matter) component of a Born–Oppenheimer scheme. In particular, we re-analyzed two related proposals, one based on the introduction of the so-called kinematical action [4] and one on the "materialization" of a fixed reference frame as a fluid, first investigated in [8]. These formulations led to the unitary Schrödinger equation amended for quantum gravity effects on the quantum matter dynamics. A cosmological implementation of the analysis in [26] (de facto valid also for the proposal in [22]) shows the consistency of the procedure and outlined some delicate questions concerning the dependence of the matter wave function on an intrinsic quantum gravity effect (there corresponding to the presence of the cosmic scale factor of the isotropic Universe), actually absent in the cosmological applications [116,117] of the study [20]. The present review had the scope to collect together some different efforts to amend quantum field theory for quantum gravity corrections. Our presentation elucidated that, as far as we limit our attention to the first two orders of approximation in the WKB expansion of the theory, the procedure remains consistent and it gives interesting insight on the primordial universe evolution. On the contrary, when the next order of approximation has been included, the one really introducing quantum gravity effects, then we have to move on a rather pioneering topic in which basic inconsistencies and intriguing proposals co-exist calling attention for further investigation in the future.

We conclude by observing that the analysis in Section 7 provides an interesting framework to search for phenomenological fingerprints of the quantum gravity corrections to quantum field theory. In particular, the determination of the primordial spectrum of the inflaton field is a natural arena to test the predictivity of such kind of reformulations toward observations of the microwave background radiation (see for instance [180,181]). Furthermore, the analysis in Section 5.3 gives a significant insight on the possibility that pre-inflationary tensor perturbations survive in the later universe and can leave a trace in the B-modes of the microwave background spectrum [182].

Author Contributions: All authors contributed to all parts of the conceptual definition of the review and to the writing of the manuscript. All authors have read and agreed to the published version of the manuscript.

Funding: This research received no external funding.

Data Availability Statement: Not applicable

Acknowledgments: G. Maniccia thanks the TAsP INFN initiative for support.

Conflicts of Interest: The authors declare no conflict of interest.

References

1. DeWitt, B.S. Quantum Theory of Gravity. I. The Canonical Theory. *Phys. Rev.* **1967**, *160*, 1113–1148. [CrossRef]
2. DeWitt, B.S. Quantum Theory of Gravity. II. The Manifestly Covariant Theory. *Phys. Rev.* **1967**, *162*, 1195–1239. [CrossRef]
3. DeWitt, B.S. Quantum Theory of Gravity. III. Applications of the Covariant Theory. *Phys. Rev.* **1967**, *162*, 1239–1256. [CrossRef]
4. Kuchař, K.V. *Canonical Methods of Quantization*; Oxford Conference on Quantum Gravity; Clarendon Press: Oxford, UK, 1980; pp. 329–376.
5. Thiemann, T. Solving the Problem of Time in General Relativity and Cosmology with Phantoms and k–Essence. *arXiv* **2006**, arXiv:astro-ph/0607380. [CrossRef]
6. Cianfrani, F.; Lecian, O.M.; Lulli, M.; Montani, G. *Canonical Quantum Gravity*; World Scientific: Singapore, 2014. [CrossRef]
7. Rovelli, C. Time in quantum gravity: An hypothesis. *Phys. Rev. D* **1991**, *43*, 442–456. [CrossRef]
8. Kuchař, K.V.; Torre, C.G. Gaussian reference fluid and interpretation of quantum geometrodynamics. *Phys. Rev. D* **1991**, *43*, 419–441. [CrossRef]
9. Brown, J.D.; Kuchař, K.V. Dust as a standard of space and time in canonical quantum gravity. *Phys. Rev. D* **1995**, *51*, 5600–5629. [CrossRef]
10. Montani, G. Canonical quantization of gravity without "frozen formalism". *Nuclear Phys. B* **2002**, *634*, 370–392. [CrossRef]
11. Mercuri, S.; Montani, G. Dualism between Physical Frames and Time in Quantum Gravity. *Mod. Phys. Lett. A* **2004**, *19*, 1519–1527. [CrossRef]
12. Zonetti, S.; Montani, G. Parametrizing Fluids in Canonical Quantum Gravity. *Int. J. Mod. Phys. A* **2008**, *23*, 1240–1243. [CrossRef]
13. Castellana, M.; Montani, G. Physical state condition in quantum general relativity as a consequence of BRST symmetry. *Class. Quantum Gravity* **2008**, *25*, 149802. [CrossRef]
14. Cianfrani, F.; Montani, G.; Zonetti, S. Definition of a time variable with entropy of a perfect fluid in canonical quantum gravity. *Class. Quantum Gravity* **2009**, *26*, 125002. [CrossRef]
15. Montani, G.; Cianfrani, F. General relativity as classical limit of evolutionary quantum gravity. *Class. Quantum Gravity* **2008**, *25*, 065007. [CrossRef]
16. Thiemann, T. *Modern Canonical Quantum General Relativity*; Cambridge Monographs on Mathematical Physics; Cambridge University Press: Cambridge, UK, 2007. [CrossRef]
17. Vilenkin, A. Interpretation of the wave function of the Universe. *Phys. Rev. D* **1989**, *39*, 1116–1122. [CrossRef]
18. Lapchinsky, V.G.; Rubakov, V.A. Canonical Quantization of Gravity and Quantum Field Theory in Curved Space-time. *Acta Phys. Polon. B* **1979**, *10*, 1041–1048.
19. Banks, T. TCP, quantum gravity, the cosmological constant and all that... *Nucl. Phys. B* **1985**, *249*, 332–360. [CrossRef]
20. Kiefer, C.; Singh, T.P. Quantum gravitational corrections to the functional Schrödinger equation. *Phys. Rev. D* **1991**, *44*, 1067–1076. [CrossRef]
21. Singh, T.P. Gravity induced corrections to quantum mechanical wavefunctions. *Class. Quantum Gravity* **1990**, *7*, L149–L154. [CrossRef]
22. Di Gioia, F.; Maniccia, G.; Montani, G.; Niedda, J. Nonunitarity problem in quantum gravity corrections to quantum field theory with Born-Oppenheimer approximation. *Phys. Rev. D* **2021**, *103*, 103511. [CrossRef]
23. Arnowitt, R.L.; Deser, S.; Misner, C.W. Republication of: The dynamics of general relativity. *Gen. Relativ. Gravit.* **2008**, *40*, 1997–2027. [CrossRef]
24. Kiefer, C.; Wichmann, D. Semiclassical approximation of the Wheeler-DeWitt equation: Arbitrary orders and the question of unitarity. *Gen. Relativ. Gravit.* **2018**, *50*, 66. [CrossRef]
25. Bertoni, C.; Finelli, F.; Venturi, G. The Born-Oppenheimer approach to the matter-gravity system and unitarity. *Class. Quantum Gravity* **1996**, *13*, 2375–2383. [CrossRef]
26. Maniccia, G.; Montani, G. Quantum gravity corrections to the matter dynamics in the presence of a reference fluid. *Phys. Rev. D* **2022**, *105*, 086014. [CrossRef]
27. Misner, C.; Thorne, K.; Wheeler, J.; Kaiser, D. *Gravitation*; Princeton University Press: Princeton, NJ, USA, 2017.
28. Montani, G.; Battisti, M.V.; Benini, R.; Imponente, G. *Primordial Cosmology*; World Scientific: Singapore, 2011. [CrossRef]

29. Capozziello, S.; Lambiase, G. Selection rules in minisuperspace quantum cosmology. *Gen. Relativ. Gravit.* **2000**, *32*, 673–696. [CrossRef]
30. Arnowitt, R.; Deser, S.; Misner, C.W. Canonical Variables for General Relativity. *Phys. Rev.* **1960**, *117*, 1595–1602. [CrossRef]
31. Montani, G.; Battisti, M.V.; Benini, R.; Imponente, G. Classical and quantum features of the Mixmaster singularity. *Int. J. Mod. Phys. A* **2008**, *23*, 2353–2503. [CrossRef]
32. Landau, L.; Lifshitz, E. *The Classical Theory of Fields*, 4th ed.; Course of Theoretical Physics; Pergamon Pr: Oxford, UK, 1975; Volume 2.
33. Faraoni, V.; Capozziello, S. *Beyond Einstein Gravity: A Survey of Gravitational Theories for Cosmology and Astrophysics*; Springer: Dordrecht, The Netherlands, 2011. [CrossRef]
34. Capozziello, S.; Bajardi, F. Minisuperspace Quantum Cosmology in Metric and Affine Theories of Gravity. *Universe* **2022**, *8*, 177. [CrossRef]
35. Cianfrani, F.; Lulli, M.; Montani, G. Solution of the noncanonicity puzzle in General Relativity: A new Hamiltonian formulation. *Phys. Lett. B* **2012**, *710*, 703–709. [CrossRef]
36. Bjorken, J.; Drell, S. *Relativistic Quantum Mechanics*; International series in pure and applied physics; McGraw-Hill: New York, NY, USA, 1964.
37. Wald, R.M. Proposal for solving the "problem of time" in canonical quantum gravity. *Phys. Rev. D* **1993**, *48*, R2377–R2381. [CrossRef]
38. Giovannetti, E.; Montani, G. Is Bianchi I a bouncing cosmology in the Wheeler-DeWitt picture? *Phys. Rev. D* **2022**, *106*, 044053. [CrossRef]
39. Caderni, N.; Martellini, M. Third quantization formalism for Hamiltonian cosmologies. *Int. J. Theor. Phys.* **1984**, *23*, 233–249. [CrossRef]
40. McGuigan, M. Third quantization and the Wheeler-DeWitt equation. *Phys. Rev. D* **1988**, *38*, 3031–3051. [CrossRef]
41. Hawking, S. Quantum coherence down the wormhole. *Phys. Lett. B* **1987**, *195*, 337–343. [CrossRef]
42. Giddings, S.B.; Strominger, A. Axion-induced topology change in quantum gravity and string theory. *Nucl. Phys. B* **1988**, *306*, 890–907. [CrossRef]
43. Coleman, S. Why there is nothing rather than something: A theory of the cosmological constant. *Nucl. Phys. B* **1988**, *310*, 643–668. [CrossRef]
44. Rubakov, V. On third quantization and the cosmological constant. *Phys. Lett. B* **1988**, *214*, 503–507. [CrossRef]
45. Giddings, S.B.; Strominger, A. Baby universe, third quantization and the cosmological constant. *Nucl. Phys. B* **1989**, *321*, 481–508. [CrossRef]
46. Vilenkin, A. Approaches to quantum cosmology. *Phys. Rev. D* **1994**, *50*, 2581–2594. [CrossRef]
47. Isham, C.J. Canonical Quantum Gravity and the Problem of Time. In *Integrable Systems, Quantum Groups, and Quantum Field Theories*; Springer: Dordrecht, The Netherlands, 1993; pp. 157–287. [CrossRef]
48. Kiefer, C.; Sandhöfer, B. Quantum cosmology. *Z. für Naturforschung A* **2022**, *77*, 543–559. [CrossRef]
49. Halliwell, J.J. Introductory lectures on Quantum Cosmology. In *7th Jerusalem Winter School for Theoretical Physics: Quantum Cosmology and Baby Universes*; World Scientific: Singapore, 1991. [CrossRef]
50. Wiltshire, D.L. An Introduction to quantum cosmology. In *8th Physics Summer School on Cosmology: The Physics of the Universe*; World Scientific: Singapore, 1997; pp. 473–531. [CrossRef]
51. Dirac, P.A.M. *Lectures on Quantum Mechanics*; Belfer Graduate School of Science, Dover Publications: Mineola, NY, USA, 2001; first published 1964.
52. Mercuri, S.; Montani, G. Revised Canonical Quantum Gravity via the Frame Fixing. *Int. J. Mod. Phys. D* **2004**, *13*, 165–186. [CrossRef]
53. Kiefer, C. Conceptual Problems in Quantum Gravity and Quantum Cosmology. *ISRN Math. Phys.* **2013**, *2013*, 509316. [CrossRef]
54. Feinberg, J.; Peleg, Y. Self-adjoint Wheeler-DeWitt operators, the problem of time, and the wave function of the Universe. *Phys. Rev. D* **1995**, *52*, 1988–2000. [CrossRef] [PubMed]
55. Kuchař, K.V. Time and interpretations of Quantum Gravity. *Int. J. Mod. Phys. D* **2011**, *20*, 3–86. [CrossRef]
56. Bojowald, M.; Halnon, T. Time in quantum cosmology. *Phys. Rev. D* **2018**, *98*, 066001. [CrossRef]
57. Gielen, S. Frozen formalism and canonical quantization in group field theory. *Phys. Rev. D* **2021**, *104*, 106011. [CrossRef]
58. Gorobey, N.; Lukyanenko, A.; Goltsev, A.V. Wave Functional of the Universe and Time. *Universe* **2021**, *7*, 452. [CrossRef]
59. Kiefer, C.; Peter, P. Time in Quantum Cosmology. *Universe* **2022**, *8*, 36. [CrossRef]
60. Altaie, M.B.; Hodgson, D.; Beige, A. Time and Quantum Clocks: A Review of Recent Developments. *Front. Phys.* **2022**, *10*, 460. [CrossRef]
61. Kehagias, A.; Partouche, H.; Toumbas, N. Probability distribution for the quantum universe. *J. High Energy Phys.* **2021**, *2021*, 165. [CrossRef]
62. He, D.; Gao, D.; Cai, Q.-y. Dynamical interpretation of the wavefunction of the universe. *Phys. Lett. B* **2015**, *748*, 361–365. [CrossRef]
63. Rovelli, C. Quantum mechanics without time: A model. *Phys. Rev. D* **1990**, *42*, 2638–2646. [CrossRef]
64. Rovelli, C. Quantum reference systems. *Class. Quantum Gravity* **1991**, *8*, 317–331. [CrossRef]

65. Briggs, J.S. Equivalent emergence of time dependence in classical and quantum mechanics. *Phys. Rev. A* **2015**, *91*, 052119. [CrossRef]
66. Guven, J.; Ryan, M.P. Functional integrals and canonical quantum gravity. *Phys. Rev. D* **1992**, *45*, 3559–3576. [CrossRef]
67. Misner, C.W. Quantum Cosmology. I. *Phys. Rev.* **1969**, *186*, 1319–1327. [CrossRef]
68. Benini, R.; Montani, G. Inhomogeneous quantum Mixmaster: From classical towards quantum mechanics. *Class. Quantum Gravity* **2006**, *24*, 387–404. [CrossRef]
69. Henneaux, M.; Teitelboim, C. *Quantization of Gauge Systems*; Princeton paperbacks; Princeton University Press: Princeton, NJ, USA, 1992.
70. Thiemann, T. Reduced phase space quantization and Dirac observables. *Class. Quantum Gravity* **2006**, *23*, 1163–1180. [CrossRef]
71. Dunham, J.L. The Wentzel-Brillouin-Kramers Method of Solving the Wave Equation. *Phys. Rev.* **1932**, *41*, 713–720. [CrossRef]
72. Vilenkin, A. Creation of universes from nothing. *Phys. Lett. B* **1982**, *117*, 25–28. [CrossRef]
73. Vilenkin, A. Birth of inflationary universes. *Phys. Rev. D* **1983**, *27*, 2848–2855. [CrossRef]
74. Vilenkin, A. Boundary conditions in quantum cosmology. *Phys. Rev. D* **1986**, *33*, 3560–3569. [CrossRef] [PubMed]
75. Hartle, J.B.; Hawking, S.W. Wave function of the Universe. *Phys. Rev. D* **1983**, *28*, 2960–2975. [CrossRef]
76. Page, D.N. Susskind's challenge to the Hartle–Hawking no-boundary proposal and possible resolutions. *J. Cosmol. Astropart. Phys.* **2007**, *2007*, 004. [CrossRef]
77. Feynman, R.; Hibbs, A.; Styer, D. *Quantum Mechanics and Path Integrals*; Dover Books on Physics; Dover Publications: Mineola, NY, USA, 2010.
78. Barvinsky, A. The general semiclassical solution of the wheeler-dewitt equations and the issue of unitarity in quantum cosmology. *Phys. Lett. B* **1990**, *241*, 201–206. [CrossRef]
79. Barvinsky, A. Unitarity approach to quantum cosmology. *Phys. Rep.* **1993**, *230*, 237–367. [CrossRef]
80. Amaral, M.; Bojowald, M. A path-integral approach to the problem of time. *Ann. Phys.* **2018**, *388*, 241–266. [CrossRef]
81. Vilenkin, A. Quantum creation of universes. *Phys. Rev. D* **1984**, *30*, 509–511. [CrossRef]
82. Kristiano, J.; Lambaga, R.; Ramadhan, H. Coleman-de Luccia tunneling wave function. *Phys. Lett. B* **2019**, *796*, 225–229. [CrossRef]
83. Halliwell, J.J. Derivation of the Wheeler-DeWitt equation from a path integral for minisuperspace models. *Phys. Rev. D* **1988**, *38*, 2468–2481. [CrossRef]
84. Lehners, J.L. Classical inflationary and ekpyrotic universes in the no-boundary wavefunction. *Phys. Rev. D* **2015**, *91*, 083525. [CrossRef]
85. Bramberger, S.F.; Farnsworth, S.; Lehners, J.L. Wavefunction of anisotropic inflationary universes with no-boundary conditions. *Phys. Rev. D* **2017**, *95*, 083513. [CrossRef]
86. Jonas, C.; Lehners, J.L.; Quintin, J. Cosmological consequences of a principle of finite amplitudes. *Phys. Rev. D* **2021**, *103*, 103525. [CrossRef]
87. Ashtekar, A.; Wilson-Ewing, E. Loop quantum cosmology of Bianchi type I models. *Phys. Rev. D* **2009**, *79*, 083535. [CrossRef]
88. Cianfrani, F.; Marchini, A.; Montani, G. The picture of the Bianchi I model via gauge fixing in Loop Quantum Gravity. *EPL (Europhys. Lett.)* **2012**, *99*, 10003. [CrossRef]
89. Moriconi, R.; Montani, G. Behavior of the Universe anisotropy in a big-bounce cosmology. *Phys. Rev. D* **2017**, *95*, 123533. [CrossRef]
90. Montani, G.; Marchi, A.; Moriconi, R. Bianchi I model as a prototype for a cyclical Universe. *Phys. Lett. B* **2018**, *777*, 191–200. [CrossRef]
91. Giovannetti, E.; Montani, G.; Schiattarella, S. On the semiclassical and quantum picture of the Bianchi I polymer dynamics. *arXiv* **2021**, arXiv:2110.13141. [CrossRef]
92. Belinsky, V.A.; Khalatnikov, I.M.; Lifshitz, E.M. Oscillatory approach to a singular point in the relativistic cosmology. *Adv. Phys.* **1970**, *19*, 525–573. [CrossRef]
93. Misner, C.W. Mixmaster Universe. *Phys. Rev. Lett.* **1969**, *22*, 1071–1074. [CrossRef]
94. Brizuela, D.; Uria, S.F. Semiclassical study of the mixmaster model: The quantum Kasner map. *Phys. Rev. D* **2022**, *106*, 064051. [CrossRef]
95. Imponente, G.; Montani, G. Covariance of the mixmaster chaoticity. *Phys. Rev. D* **2001**, *63*, 103501. [CrossRef]
96. Misner, C.W.; Taub, A.H. A singularity-free empty universe. *Sov. Phys. JETP* **1969**, *28*, 122.
97. Montani, G.; Chiovoloni, R. Scenario for a singularity-free generic cosmological solution. *Phys. Rev. D* **2021**, *103*, 123516. [CrossRef]
98. De Angelis, M.; Montani, G. On the emergence of a classical Isotropic Universe from a Quantum $f(R)$ Bianchi Cosmology in the Jordan Frame. *arXiv* **2022**, arXiv:2207.14683. [CrossRef]
99. Agostini, L.; Cianfrani, F.; Montani, G. Probabilistic interpretation of the wave function for the Bianchi I model. *Phys. Rev. D* **2017**, *95*, 126010. [CrossRef]
100. Mostafazadeh, A. Hilbert space structures on the solution space of Klein Gordon-type evolution equations. *Class. Quantum Gravity* **2002**, *20*, 155–171. [CrossRef]
101. Chiovoloni, R.; Montani, G.; Cascioli, V. Quantum dynamics of the corner of the Bianchi IX model in the WKB approximation. *Phys. Rev. D* **2020**, *102*, 083519. [CrossRef]

102. Lewis, H.R. Classical and Quantum Systems with Time-Dependent Harmonic-Oscillator-Type Hamiltonians. *Phys. Rev. Lett.* **1967**, *18*, 510–512. [CrossRef]
103. Berkowitz, D. Bianchi IX and VIII Quantum Cosmology with a Cosmological Constant, Aligned Electromagnetic Field, and Scalar Field. *arXiv* **2021**, arXiv:2102.02343. [CrossRef]
104. Benini, R.; Kirillov, A.; Montani, G. Vector Field Induced Chaos in Multi-dimensional Homogeneous Cosmologies. In *The Eleventh Marcel Grossmann Meeting*; World Scientific Publishing Company: Singapore, 2008. [CrossRef]
105. Ashtekar, A.; Pawlowski, T.; Singh, P. Quantum Nature of the Big Bang. *Phys. Rev. Lett.* **2006**, *96*, 141301. [CrossRef]
106. de Haro, J. Does loop quantum cosmology replace the big rip singularity by a non-singular bounce? *J. Cosmol. Astropart. Phys.* **2012**, *2012*, 037–037. [CrossRef]
107. De Angelis, M.; Montani, G. Dynamics of quantum anisotropies in a Taub universe in the WKB approximation. *Phys. Rev. D* **2020**, *101*, 103532. [CrossRef]
108. Taub, A.H. Empty Space-Times Admitting a Three Parameter Group of Motions. *Ann. Math.* **1951**, *53*, 472–490. [CrossRef]
109. Battisti, M.V.; Lecian, O.M.; Montani, G. GUP vs polymer quantum cosmology: The Taub model. *arXiv* **2009**, arXiv:0903.3836. [CrossRef]
110. Catren, G.; Ferraro, R. Quantization of the Taub model with extrinsic time. *Phys. Rev. D* **2000**, *63*, 023502. [CrossRef]
111. Berkowitz, D. Applying the Euclidean-signature semi-classical method to the quantum Taub models with a cosmological constant and aligned electromagnetic field. *J. Math. Phys.* **2021**, *62*, 083510. [CrossRef]
112. Berkowitz, D. Towards Uncovering Generic Effects Of Matter Sources In Anisotropic Quantum Cosmologies Via Taub Models. *arXiv* **2020**, arXiv:2011.04229. [CrossRef]
113. Vilenkin, A. Quantum cosmology and eternal inflation. In *Workshop on Conference on the Future of Theoretical Physics and Cosmology in Honor of Steven Hawking's 60th Birthday*; Cambridge University Press: Cambridge, UK, 2002; pp. 649–666. [CrossRef]
114. Weinberg, S. *Cosmology*; OUP Oxford: Oxford, UK, 2008.
115. Battisti, M.V.; Belvedere, R.; Montani, G. Semiclassical suppression of weak anisotropies of a generic Universe. *EPL (Europhys. Lett.)* **2009**, *86*, 69001. [CrossRef]
116. Brizuela, D.; Kiefer, C.; Krämer, M. Quantum-gravitational effects on gauge-invariant scalar and tensor perturbations during inflation: The de Sitter case. *Phys. Rev. D* **2016**, *93*, 104035. [CrossRef]
117. Brizuela, D.; Kiefer, C.; Krämer, M. Quantum-gravitational effects on gauge-invariant scalar and tensor perturbations during inflation: The slow-roll approximation. *Phys. Rev. D* **2016**, *94*, 123527. [CrossRef]
118. Belinskii, V.; Khalatnikov, I.; Lifshitz, E. A general solution of the Einstein equations with a time singularity. *Adv. Phys.* **1982**, *31*, 639–667. [CrossRef]
119. Kirillov, A. On the nature of the spatial distribution of metric inhomogeneities in the general solution of the Einstein equations near a cosmological singularity. *J. Exp. Theor. Phys.* **1993**, *76*, 355–358.
120. Montani, G. On the general behaviour of the universe near the cosmological singularity. *Class. Quantum Gravity* **1995**, *12*, 2505–2517. [CrossRef]
121. Benini, R.; Montani, G. Frame independence of the inhomogeneous mixmaster chaos via Misner-Chitré-like variables. *Phys. Rev. D* **2004**, *70*, 103527. [CrossRef]
122. Heinzle, J.M.; Uggla, C.; Lim, W.C. Spike oscillations. *Phys. Rev. D* **2012**, *86*, 104049. [CrossRef]
123. Khalatnikov, I.; Lifshitz, E.; Khanin, K.; Shchur, L.; Sinai, Y. On the stochasticity in relativistic cosmology. *J. Stat. Phys.* **1985**, *38*, 97–114. [CrossRef]
124. Belinskii, V.A. Turbulence of the gravitational field near a cosmological singularity. *Pisma v Zhurnal Eksperimentalnoi i Teoreticheskoi Fiziki* **1992**, *56*, 437–440.
125. Barrow, J.D. Multifractality in the general cosmological solution of Einstein's equations. *Phys. Rev. D* **2020**, *102*, 041501. [CrossRef]
126. Landau, L.D.; Lifshitz, E.M. *Quantum Mechanics: Non-Relativistic Theory*, 3rd ed.; Course on Theoretical Physics; Pergamon Pr: Oxford, UK, 1981; Volume 3.
127. Lifschytz, G.; Mathur, S.D.; Ortiz, M. Note on the semiclassical approximation in quantum gravity. *Phys. Rev. D* **1996**, *53*, 766–778. [CrossRef]
128. Castagnino, M.A.; Lombardo, F. Origin and measurement of time in quantum cosmology. *Phys. Rev. D* **1993**, *48*, 1722–1735. [CrossRef]
129. Barbour, J.B. The timelessness of quantum gravity: II. The appearance of dynamics in static configurations. *Class. Quantum Gravity* **1994**, *11*, 2875–2897. [CrossRef]
130. Ohkuwa, Y. Time in the semi-classical approximation to quantum cosmology. *Nuovo C. B Ser.* **1995**, *110B*, 53–60. [CrossRef]
131. Damour, T.; Vilenkin, A. Quantum instability of an oscillating universe. *Phys. Rev. D* **2019**, *100*, 083525. [CrossRef]
132. Castagnino, M.A.; Mazzitelli, F.D. Notion of time and the semiclassical regime of quantum gravity. *Phys. Rev. D* **1990**, *42*, 482–487. [CrossRef] [PubMed]
133. Moffat, J.W. Quantum gravity, the origin of time and time's arrow. *Found. Phys.* **1993**, *23*, 411–437. [CrossRef]
134. Barvinsky, A.O.; Kiefer, C. Wheeler-DeWitt equation and Feynman diagrams. *Nucl. Phys. B* **1998**, *526*, 509–539. [CrossRef]
135. Bolotin, A. Concerning Infeasibility of the Wave Functions of the Universe. *Int. J. Theor. Phys.* **2015**, *54*, 3215–3221. [CrossRef]
136. Giulini, D.; Großardt, A. The Schrödinger–Newton equation as a non-relativistic limit of self-gravitating Klein–Gordon and Dirac fields. *Class. Quantum Gravity* **2012**, *29*, 215010. [CrossRef]

137. Kiefer, C.; Kwidzinski, N.; Piontek, D. Singularity avoidance in Bianchi I quantum cosmology. *Eur. Phys. J. C* **2019**, *79*, 686. [CrossRef]
138. Rotondo, M. The Functional Schrödinger Equation in the Semiclassical Limit of Quantum Gravity with a Gaussian Clock Field. *Universe* **2020**, *6*, 176. [CrossRef]
139. Rotondo, M. A Wheeler-DeWitt Equation with Time. *arXiv* **2022**, arXiv:2201.00809. [CrossRef]
140. Halliwell, J.J. Decoherence in quantum cosmology. *Phys. Rev. D* **1989**, *39*, 2912–2923. [CrossRef]
141. Barbour, J.B. Time and complex numbers in canonical quantum gravity. *Phys. Rev. D* **1993**, *47*, 5422–5429. [CrossRef]
142. Robles-Pérez, S.J. Quantum Cosmology with Third Quantisation. *Universe* **2021**, *7*, 404. [CrossRef]
143. Born, M.; Oppenheimer, R. Zur Quantentheorie der Molekeln. *Ann. der Phys.* **1927**, *389*, 457–484. [CrossRef]
144. Bransden, B.; Joachain, C. *Physics of Atoms and Molecules*; Prentice Hall: Hoboken, NJ, USA, 2003.
145. Massar, S.; Parentani, R. Particle creation and non-adiabatic transitions in quantum cosmology. *Nucl. Phys. B* **1998**, *513*, 375–401. [CrossRef]
146. Kamenshchik, A.Y.; Tronconi, A.; Venturi, G. The Born–Oppenheimer method, quantum gravity and matter. *Class. Quantum Gravity* **2017**, *35*, 015012. [CrossRef]
147. Kamenshchik, A.Y.; Tronconi, A.; Venturi, G. Quantum cosmology and the inflationary spectra from a nonminimally coupled inflaton. *Phys. Rev. D* **2020**, *101*, 023534. [CrossRef]
148. Chataignier, L. Construction of quantum Dirac observables and the emergence of WKB time. *Phys. Rev. D* **2020**, *101*, 086001. [CrossRef]
149. Kamenshchik, A.Y.; Tronconi, A.; Venturi, G. The Born–Oppenheimer approach to quantum cosmology. *Class. Quantum Gravity* **2021**, *38*, 155011. [CrossRef]
150. Chataignier, L.; Krämer, M. Unitarity of quantum-gravitational corrections to primordial fluctuations in the Born-Oppenheimer approach. *Phys. Rev. D* **2021**, *103*, 066005. [CrossRef]
151. Chataignier, L. Beyond semiclassical time. *Z. für Naturforschung A* **2022**, *77*, 805–812. [CrossRef]
152. Unruh, W.G.; Wald, R.M. Time and the interpretation of canonical quantum gravity. *Phys. Rev. D* **1989**, *40*, 2598–2614. [PubMed]
153. Kiefer, C. The semiclassical approximation to quantum gravity. In *Canonical Gravity: From Classical to Quantum*; Ehlers, J., Friedrich, H., Eds.; Lecture Notes in Physics; Springer: Berlin/Heidelberg, Germany, 1994; Volume 434, pp. 170–212. [CrossRef]
154. Mostafazadeh, A. Quantum mechanics of Klein–Gordon-type fields and quantum cosmology. *Ann. Phys.* **2004**, *309*, 1–48. [CrossRef]
155. Gielen, S.; Menéndez-Pidal, L. Unitarity, clock dependence and quantum recollapse in quantum cosmology. *Class. Quantum Gravity* **2022**, *39*, 075011. [CrossRef]
156. Fradkin, E.; Tseytlin, A. Renormalizable asymtotically free quantum theory of gravity. *Phys. Lett. B* **1981**, *104*, 377–381. [CrossRef]
157. Fierz, M.; Pauli, W. On relativistic wave equations for particles of arbitrary spin in an electromagnetic field. *Proc. Roy. Soc. Lond. A* **1939**, *173*, 211–232. [CrossRef]
158. Hinterbichler, K. Theoretical aspects of massive gravity. *Rev. Mod. Phys.* **2012**, *84*, 671–710. [CrossRef]
159. Boulware, D.G.; Deser, S. Can Gravitation Have a Finite Range? *Phys. Rev. D* **1972**, *6*, 3368–3382. [CrossRef]
160. Creminelli, P.; Nicolis, A.; Papucci, M.; Trincherini, E. Ghosts in massive gravity. *J. High Energy Phys.* **2005**, *2005*, 003. [CrossRef]
161. Bergshoeff, E.A.; Hohm, O.; Townsend, P.K. Massive Gravity in Three Dimensions. *Phys. Rev. Lett.* **2009**, *102*, 201301. [CrossRef]
162. Nakasone, M.; Oda, I. On Unitarity of Massive Gravity in Three Dimensions. *Prog. Theor. Phys.* **2009**, *121*, 1389–1397. [CrossRef]
163. Arvanitakis, A.S.; Townsend, P.K. Minimal massive 3D gravity unitarity redux. *Class. Quantum Gravity* **2015**, *32*, 085003. [CrossRef]
164. Setare, M. On the generalized minimal massive gravity. *Nucl. Phys. B* **2015**, *898*, 259–275. [CrossRef]
165. de Rham, C.; Gabadadze, G.; Tolley, A.J. Resummation of Massive Gravity. *Phys. Rev. Lett.* **2011**, *106*. [CrossRef]
166. Arraut, I. On the apparent loss of predictability inside the de Rham-Gabadadze-Tolley non-linear formulation of massive gravity: The Hawking radiation effect. *EPL (Europhys. Lett.)* **2015**, *109*, 10002. [CrossRef]
167. Park, M. Quantum aspects of massive gravity II: Non-Pauli-Fierz theory. *J. High Energy Phys.* **2011**, *2011*, 130. [CrossRef]
168. Paulos, M.F.; Tolley, A.J. Massive Gravity theories and limits of ghost-free bigravity models. *J. High Energy Phys.* **2012**, *2012*, 2. [CrossRef]
169. Einhorn, M.B.; Jones, D.R.T. Renormalizable, asymptotically free gravity without ghosts or tachyons. *Phys. Rev. D* **2017**, *96*, 124025. [CrossRef]
170. Schander, S.; Thiemann, T. Backreaction in Cosmology. *Front. Astron. Space Sci.* **2021**, *8*. [CrossRef]
171. Panati, G.; Spohn, H.; Teufel, S. Space-Adiabatic Perturbation Theory. *arXiv* **2002**, arXiv:math-ph/0201055. [CrossRef]
172. Stottmeister, A.; Thiemann, T. Coherent states, quantum gravity, and the Born-Oppenheimer approximation. I. General considerations. *J. Math. Phys.* **2016**, *57*, 063509. [CrossRef]
173. Schander, S.; Thiemann, T. Quantum cosmological backreactions. I. Cosmological space adiabatic perturbation theory. *Phys. Rev. D* **2022**, *105*, 106009. [CrossRef]
174. Maniccia, G.; Montani, G. WKB approach to the gravity-matter dynamics: A cosmological implementation, 2021. In Proceedings of the 16th Marcel Grossmann Meeting, Rome, Italy, 5–10 July 2021. To be published by World Scientific. [CrossRef]
175. Cianfrani, F.; Montani, G. Synchronous Quantum Gravity. *Int. J. Mod. Phys. A* **2008**, *23*, 1149–1156. [CrossRef]

176. Kuchař, K.V. Does an unspecified cosmological constant solve the problem of time in quantum gravity? *Phys. Rev. D* **1991**, *43*, 3332–3344. [CrossRef]
177. Magueijo, J. Connection between cosmological time and the constants of Nature. *arXiv* **2021**, arXiv:2110.05920. [CrossRef]
178. Isham, C.; Kuchar, K. Representations of spacetime diffeomorphisms. II. Canonical geometrodynamics. *Ann. Phys.* **1985**, *164*, 316–333. [CrossRef]
179. Giesel, K.; Tambornino, J.; Thiemann, T. Born–Oppenheimer decomposition for quantum fields on quantum spacetimes. *arXiv* **2009**, arXiv:0911.5331. [CrossRef]
180. Ade, P.A.R. et al. [Planck collaboration] Planck 2015 results XX. Constraints on inflation. *Astron. Astrophys.* **2016**, *594*, A20.
181. Cabass, G.; Di Valentino, E.; Melchiorri, A.; Pajer, E.; Silk, J. Constraints on the running of the running of the scalar tilt from CMB anisotropies and spectral distortions. *Phys. Rev. D* **2016**, *94*, 023523. [CrossRef]
182. Ade, P.A.R. et al. [Planck collaboration] Planck intermediate results XLI. A map of lensing-induced B-modes. *Astron. Astrophys.* **2016**, *596*, A102.

Article

Constraints on the Duration of Inflation from Entanglement Entropy Bounds

Suddhasattwa Brahma [1,2]

[1] Higgs Centre for Theoretical Physics, School of Physics & Astronomy, University of Edinburgh, Edinburgh EH9 3FD, UK; suddhasattwa.brahma@gmail.com
[2] Department of Physics, McGill University, Montreal, QC H3A 2T8, Canada

Abstract: Using the fact that we only observe those modes that exit the Hubble horizon during inflation, one can calculate the entanglement entropy of such long-wavelength perturbations by tracing out the unobservable sub-Hubble fluctuations they are coupled with. On requiring that this perturbative entanglement entropy, which increases with time, obey the covariant entropy bound for an accelerating background, we find an upper bound on the duration of inflation. This presents a new perspective on the (meta-)stability of de Sitter spacetime and an associated lifetime for it.

Keywords: inflation; entanglement entropy; scrambling time; Trans-Planckian Censorship Conjecture

Although cosmic inflation is widely regarded as the standard paradigm for the early universe, its embedding into a fundamental theory of quantum gravity (QG) remains an open question. Recently, there have been different arguments against long-lived accelerating spacetimes, especially in the context of string theory (ST) [1–3]. One such conjecture states that trans-Planckian modes should never cross the Hubble horizon during inflation, leading to an upper bound on the number of e-foldings [4]:

$$N < \ln\left(\frac{M_{\text{Pl}}}{H}\right), \qquad (1)$$

where H denotes the Hubble parameter during inflation. Although the physical motivation behind this conjecture—a trans-Planckian mode should never become part of late-time macroscopic inhomogeneities—has been heavily debated [5,6], it does find some connections to other aspects of the ST 'swampland' [7–13]. As a whole, there seem to be various obstructions to finding a quantum gravity completion for long-lived accelerating backgrounds. Although the specific technical difficulties have been realized in the context of ST, many of the arguments apply much more generally to any quantum gravity model. In particular, a corollary of this is that only extremely short-lived de Sitter (dS) spaces can arise in a UV-complete theory [2,4].

Indeed, it has been long argued that dS space is metastable from different points of view[1]. There are three time-scales often associated with the lifetime of dS — the scrambling time $\sim H^{-1}\ln(S_{\text{dS}})$ [21] corresponding to (1), the quantum breaking time $\sim H^{-1}S_{\text{dS}}$, and the Poincaré recurrence time $\propto e^{S_{\text{dS}}}$, where the Gibbons-Hawking entropy for dS is given by $S_{\text{dS}} \sim (M_{\text{Pl}}/H)^2$ [22]. Clearly, (1) puts an upper bound on the number of e-foldings N that is much smaller than the other two time-scales, with drastic implications for inflation [23].

In this essay, we present a different argument for finding the maximum amount of e-foldings allowed for inflation and, therefore, set an upper bound on the lifetime of dS. Instead of invoking any QG reasoning, we employ a bottom-up argument by requiring that the entanglement entropy (EE) of scalar perturbations during inflation be bounded by the Gibbons–Hawking entropy. We note that the first arguments in favor of the so-called dS conjecture also followed from an application of the covariant entropy bound (CEB) and the distance conjecture [2]. However, that derivation was (i) intimately tied to details of

ST and (ii) valid only in asymptotic regions of moduli space. Here, we circumvent both these obstructions.

Discussions of EE have become ubiquitous in the context of gravity. However, in most cases, one considers the EE between different geometric regions of space — in the context of black holes [24], Minkowski [25], or dS space [26]. Moreover, the EE of cosmological backgrounds have sometimes been carried out using holographic methods [27,28]. Nevertheless, it is not necessary to define a subsystem, which is separated out in the position space domain, e.g., demarcation by a black hole horizon. In cosmology, it is more instructive to consider EE between different bands in momentum space, since it is the correlation functions of the momentum modes of cosmological perturbations, which are generally probed. For momentum space, the vacuum of the free field theory is factorized, and any EE come from the interactions that lead to mode coupling.

One can calculate the perturbative EE in momentum space for a scalar in flat spacetime as outlined, for example, in [29]. The full Hilbert space can be partitioned into two parts separated by some fiducial momentum scale μ such that $\mathcal{H} = \mathcal{H}_S \otimes \mathcal{H}_\mathcal{E}$. The Hamiltonian of the system is decomposed as

$$H = H_S \otimes 1 + 1 \otimes H_\mathcal{E} + \lambda H_{\text{int}}, \qquad (2)$$

where $H_{\mathcal{E},S}$ are the free Hamiltonians of the respective subsystems, and the interacting Hamiltonian H_{int} has a coupling parameter λ. The ground state is the product of the individual harmonic vacua of H_S and $H_\mathcal{E}$, i.e., $|0,0\rangle = |0\rangle_S \otimes |0\rangle_\mathcal{E}$. The energy eigenbasis of S and $\mathcal{E}$ are denoted by $|n\rangle$ and $|N\rangle$, respectively, while the corresponding energy eigenvalues are E_n and $\tilde{E}_N$. The (perturbative) interacting vacuum can be written as (up to normalization)

$$|\Omega\rangle = |0,0\rangle + \sum_{n\neq 0} A_n |n,0\rangle + \sum_{n\neq 0} B_N |0,N\rangle + \sum_{n,N\neq 0} C_{n,N} |n,N\rangle, \qquad (3)$$

where the matrix elements $A_n, B_N,$ and $C_{n,N}$ are calculated using standard perturbation theory. The reduced density matrix corresponding to subsystem S is obtained by tracing out the $\mathcal{E}$ modes. From that, one can extract the leading order contribution to the (von Neumann) EE:

$$S_{\text{ent}} = -\lambda^2 \log(\lambda^2) \sum_{n,N\neq 0} \frac{|\langle n,N|H_{\text{int}}|0,0\rangle|^2}{(E_0 + \tilde{E}_0 - E_n - \tilde{E}_N)^2} + \mathcal{O}(\lambda^2), \qquad (4)$$

where it is understood that at least one momentum in the matrix element is below μ, and at least one momentum is above.

The main calculation of [30] was to extend this result to an inflating background, which we now present. Considering the density perturbations in the comoving gauge

$$ds^2 = -a^2(\tau)\left[d\tau^2 - (1+2\zeta)dx^2\right], \qquad (5)$$

where a and ϵ are the usual symbols for the scale factor and the (first) slow-roll parameter for a quasi-dS expansion written as functions of the conformal time τ. The quadratic action for ζ, in terms of momentum modes, describes a collection of harmonic oscillators with a time-dependent mass term. This implies a difference between the quasi-dS geometry and Minkowski background [30] — the sub-Hubble modes ($k \gg aH$) are in their quantum (Bunch-Davies) vacuum, while the super-Hubble ones ($k \ll aH$) are in a (two-mode) squeezed state $|SQ(k,\tau)\rangle$ [31]. Thus, the ground state factorizes as

$$|0,0\rangle = |0\rangle_{\mathcal{E}:k>aH} \otimes |SQ\rangle_{S:k<aH}. \qquad (6)$$

Since the modes which exit the horizon during inflation can only be later observed, we consider the super-Hubble modes as our system (S), while the sub-Hubble ones are the environment ($\mathcal{E}$) (Figure 1). (The dynamics of this system were studied in [32].) Finally, the nonlinearity of GR also provides us with an interaction term which couples the sub- and super-Hubble modes; thus, this interaction is universal and can never be turned off.

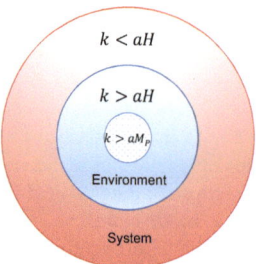

Figure 1. Schematic illustration of the system and environment modes for this setup. Gravity plays the role of providing a natural scale — the comoving Hubble horizon — which demarcates "long" and "short" degrees of freedom (dofs). We impose the Planck mass as the natural cutoff for the UV-modes and assume that these can be properly accounted for within some QG theory.

Mathematically, this translates into having a Hilbert space: $\mathcal{H} = \mathcal{H}_\mathcal{S} \otimes \mathcal{H}_\mathcal{E}$ where $\mathcal{H}_\mathcal{S} = \prod_k \mathcal{H}_k$, $|k| < aH$ and similarly for the sub-Hubble modes. The full Hamiltonian is given by $H = H_\mathcal{S}^{(2)} + H_\mathcal{E}^{(2)} + H_{\text{int}}$, where

$$H_{\text{int}} = \frac{M_{\text{Pl}}^2}{2} \int d^3x \, \epsilon^2 \, a \, \zeta (\partial \zeta)^2 , \tag{7}$$

is the leading order cubic non-Gaussian term (since ζ "freezes" outside the horizon) out of all the available interactions [33]. We now need to apply time-dependent perturbation theory to calculate the matrix elements since $\lambda = \sqrt{\epsilon}/(2\sqrt{2} a M_{\text{Pl}})$ is time dependent. Moreover, there is no well-defined notion of energy for the squeezed state, but we only need energy differences in (4). This is a rather important point which deserves further explanation. Definitions of entanglement entropy in flat space can heavily depend on notions of energy which, in turn, is dependent on the Hamiltonian (and the corresponding vacuum) for the system. It is well known that there are ambiguities in defining the vacuum (or initial) state for inflation. In this particular case, we assume that all the scalar quantum fluctuations started out in their Bunch–Davies vacuum, while the modes that crossed the Hubble horizon correspond to the squeezed state. Although there is no good notion of energy for the squeezed vacuum, we can nevertheless define excited states (as N-particle states by acting with the relevant number of creation operators over the squeezed vacuum). Furthermore, we only need the difference in the energy between these excited states and the squeezed vacuum, which can be expressed in terms of the physical momenta of the ladder operators. This is why we are able to generalize the standard definition of EE in flat space to that for inflation [30].

Using these inputs, one can carefully evaluate (4) for inflationary perturbations [30], resulting in the EE (per unit physical volume):

$$s_{\text{ent}} \sim \epsilon \, H^2 \, M_{\text{Pl}} \, (a_f/a_i)^2 , \tag{8}$$

where (a_f) a_i is the scale factor at the (end) beginning of inflation. In this calculation, it was assumed that there were no super-Hubble modes at the beginning of inflation, and both (H, ϵ) remain constant. It was shown that the dominant contribution comes from the squeezed configuration, and in the large squeezing limit, we present the leading order estimate omitting some $\mathcal{O}(1)$ factors as well as small logarithmic corrections [30].

Firstly, we note that the resulting EE between the sub- and super-Hubble modes is sensitive to the UV-cutoff M_{Pl}, as expected. Note that this is not an artificial cutoff introduced in the system, but rather one should take the view that the Planck mass is an energy scale beyond which we should not expect standard cosmological perturbation theory on a classical inflationary background to be valid any longer. However, what is remarkable about this result is that the EE increases secularly with time, as signified by the $(a_f/a_i)^2$ term. The intuitive reason for this is that the dimension of $\mathcal{H}_S$ increases with time as modes get stretched outside the horizon. It is not surprising that the EE is increasing with time, as it does for many systems with dynamical backgrounds. This is indeed what we would expect to happen for any cosmological (expanding) spacetimes. However, what is remarkable for inflation is that the rate of increase is very high due to the exponential expansion of the background. Thus, although we have calculated the EE as a perturbative quantity here, it will necessarily become very "large" over time.

Let us quantify the last statement made above. We used standard perturbation theory to calculate the first-order matrix element, which has given us the leading order EE for the density fluctuations during inflation. Given that this is a perturbative calculation, it is automatically a small quantity suppressed by factors of the interaction parameter of the cubic nonlinear term. However, we find that given enough time, this perturbative EE will soon become quite large and can overcome any entropy bound. To give us an idea of this, we can consider different entropy bounds. It was shown in [30] that if we demand that the EE remains subdominant to the thermal entropy, then we end up with a bound for the duration of inflation that is very close to the one derived from the TCC.

In this work, we want to use the background Gibbons–Hawking entropy as the upper bound for it. The main idea is that the growth of entropy often leads to deep puzzles in theoretical physics, and we wish to make our most crucial observation in this context. We require that the *total EE obeys the CEB* [34], i.e., the EE in a Hubble patch can, at most, saturate the entropy corresponding to the apparent horizon (S_{dS}). The reader might be a bit confused here as to why we are comparing our momentum space EE with the Gibbons–Hawking entropy, which is something calculated in position space. However, we are not actually claiming that the momentum space EE calculated here *must* satisfy the CEB. In fact, the CEB is a bound for real space entropy for a causal patch, whereas we are calculating a momentum space EE. Therefore, in order to be more precise, one would have to take our momentum space result and try to "Fourier transform" it to real space in order to obtain a strict bound from the CEB.

However, our goal is somewhat different here. We are merely interested in generalizing the results of [30] by using the CEB as a measure to point out how rapidly the EE is increasing, and even a perturbative quantity like itself can overcome the CEB within a few e-foldings. In fact, our main argument is that any bound on the EE — whether it is the thermal entropy during reheating or the CEB — will soon be saturated. Even more interestingly, the quantitative measure of how soon the EE saturates any such entropy bound is seemingly always related to the scrambling time of dS, as shown below.

More explicitly, requiring that the EE saturates the CEB (where the total $S_{GH} \sim M_{Pl}^2/H^2$ for inflation), we find the relation:

$$\epsilon\, e^{2N} < \left(\frac{M_{Pl}}{H}\right) \quad \Rightarrow \quad N < \frac{1}{2}\ln\left(\frac{M_{Pl}}{H}\right) - \frac{1}{2}\ln \epsilon. \tag{9}$$

With the observed power spectrum $P_\zeta \sim 10^{-9}$, one finds $\epsilon \sim 10^9 (H/M_{Pl})^2$, and therefore,

$$N < \frac{3}{2}\ln\left(\frac{M_{Pl}}{H}\right) - \frac{9}{2}\ln 10. \tag{10}$$

This bound on the number of *e*-foldings is very similar to the one in (1), without requiring *any* QG input, and it predicts an upper limit on the lifetime of dS given by

$$T < \frac{1}{H}\left[\frac{3}{2}\ln\left(\frac{M_{Pl}}{H}\right) - \frac{9}{2}\ln 10\right], \tag{11}$$

closely related to the scrambling time up to small factors. Our result has far-reaching implications both for the UV-completions of dS space as well as the phenomenological predictions of inflation[2]. Interestingly, a very similar bound on the duration of inflation was derived from measures of complexity and chaos for inflationary perturbations [35], adding more evidence that the EE grows to such values on scrambling times that the standard EFT of inflation fails on these scales.

An immediate question is the implication for our bound on specific models of inflation. What is clear for from (11) is that the lifetime is larger for low-scale models of inflation. In other words, models with lower energy scales of inflation (given by the Hubble scale H) will be preferred according to this. Apart from the fact that such models will have a larger lifetime according to (10), they are also the ones which require a lower number of e-folds in order to solve the horizon and flatness problems. Therefore, GUT-scale models (which generically produce a large *tensor-to-scalar* ratio) are disfavored by our bounds. This, by itself, is not a large problem for inflation since there are plenty of small-field models (which would be the ones obeying the above condition if we allow for single-field models only). However, it is known that these models lose the preferred "attractor" feature of inflation [23,36], since it is clear that only small-field models are preferred by (10), which rules out polynomial potentials (such as the quadratic one) and only allows for hilltoplike potentials (which have a very small tensor-to-scalar ratio). As an example, for typical GUT-scale models $H \sim 10^{-5} M_{Pl}$, the number of e-foldings allowed would be $\mathcal{O}(10)$. This is, of course, completely incompatible with the required number of e-foldings to explain, for example, the horizon problem. This gives us an insight into why low-scale models of inflation are the only ones allowed by this bound.

To summarize, for a dS geometry, an observer has access to only part of the entire spacetime. In particular for inflation, tracing out the unobservable sub-Hubble modes leads to a non-zero EE for the curvature perturbations that increases with time. However, since the EE can, at best, saturate the CEB, this puts an upper limit on the duration of inflation. Our calculation provides an universal limit since we take the simplest case of a minimally-coupled scalar field — any additional fields or extra-couplings (which give rise to stronger non-Gaussian) would only enhance the EE and strengthen our result. We emphasize that our bound does not arise from demanding a finite-dimensional Hilbert space for dS [37] or that we live in an asymptotically dS universe [38]. Finally, note that EE for other early-universe scenarios do not produce such bounds on the lifetime [39].

Funding: S.B. is supported in part by the Higgs Fellowship.

Institutional Review Board Statement: Not applicable.

Informed Consent Statement: Not applicable.

Data Availability Statement: Not applicable.

Acknowledgments: I am grateful to Robert Brandenberger for multiple discussions. I am also thankful to an anonymous referee for suggestions that led to improving the draft.

Conflicts of Interest: The author declares no conflict of interest.

Notes

[1] These arguments, to name a few, are based on the finiteness of dS entropy [14,15], the typical lifetime of 4-dimensional dS vacua in ST [16], nonperturbative effects in the context of eternal inflation [17], or treating dS as a coherent state [18–20] and so on.

[2] A direct evaluation of EE in momentum space for a scalar field on pure dS, as well as a more sophisticated calculation of the inflationary system allowing for a slowly varying H and ϵ, shall be carried out in the future.

References

1. Palti, E. The swampland: Introduction and review. *Fortschritte Physik* **2019**, *67*, 1900037. [CrossRef]
2. Ooguri, H.; Palti, E.; Shiu, G.; Vafa, C. Distance and de Sitter Conjectures on the Swampland. *Phys. Lett. B* **2019**, *788*, 180–184. [CrossRef]
3. Garg, S.K.; Krishnan, C. Bounds on slow roll and the de Sitter swampland. *J. High Energy Phys.* **2019**, *2019*, 75. [CrossRef]

4. Bedroya, A.; Vafa, C. Trans-Planckian censorship and the swampland. *J. High Energy Phys.* **2020**, *2020*, 123. [CrossRef]
5. Burgess, C.P.; de Alwis, S.P.; Quevedo, F. Cosmological trans-Planckian conjectures are not effective. *J. Cosmol. Astropart. Phys.* **2021**, *2021*, 37. [CrossRef]
6. Dvali, G.; Kehagias, A.; Riotto, A. Inflation and decoupling. *arXiv* **2020**, arXiv:2005.05146.
7. Bedroya, A. de Sitter Complementarity, TCC, and the Swampland. *arXiv* **2020**, arXiv:2010.09760.
8. Rudelius, T. Dimensional reduction and (Anti) de Sitter bounds. *J. High Energy Phys.* **2021**, *2021*, 41. [CrossRef]
9. Andriot, D. Tachyonic de Sitter solutions of 10d type II supergravities. *Fortschritte Physik* **2021**, *69*, 2100063. [CrossRef]
10. Andriot, D.; Cribiori, N.; Erkinger, D. The web of swampland conjectures and the TCC bound. *J. High Energy Phys.* **2020**, *2020*, 162. [CrossRef]
11. Berera, A.; Brahma, S.; Calderón, J.R. Role of trans-Planckian modes in cosmology. *J. High Energy Phys.* **2020**, *2020*, 71. [CrossRef]
12. Brahma, S. Trans-Planckian censorship conjecture from the swampland distance conjecture. *Phys. Rev. D* **2020**, *101*, 046013. [CrossRef]
13. Aalsma, L.; Shiu, G. Chaos and complementarity in de Sitter space. *J. High Energy Phys.* **2020**, *2020*, 152. [CrossRef]
14. Goheer, N.; Kleban, M.; Susskind, L. The trouble with de Sitter space. *J. High Energy Phys.* **2003**, *2003*, 056. [CrossRef]
15. Arkani-Hamed, N.; Dubovsky, S.; Nicolis, A.; Trincherini, E.; Villadoro, G. A measure of de Sitter entropy and eternal inflation. *J. High Energy Phys.* **2007**, *2007*, 55. [CrossRef]
16. Kachru, S.; Kallosh, R.; Linde, A.; Trivedi, S.P. De Sitter vacua in string theory. *Phys. Rev. D* **2003**, *68*, 046005. [CrossRef]
17. Dubovsky, S.; Senatore, L.; Villadoro, G. The volume of the universe after inflation and de Sitter entropy. *J. High Energy Phys.* **2009**, *2009*, 118. [CrossRef]
18. Dvali, G.; Gómez, C.; Zell, S. Quantum break-time of de Sitter. *J. Cosmol. Astropart. Phys.* **2017**, *2017*, 028. [CrossRef]
19. Brahma, S.; Dasgupta, K.; Tatar, R. de Sitter space as a Glauber-Sudarshan state. *J. High Energy Phys.* **2021**, *2021*, 104. [CrossRef]
20. Brahma, S.; Dasgupta, K.; Tatar, R. Four-dimensional de Sitter space is a Glauber-Sudarshan state in string theory. *J. High Energy Phys.* **2021**, *2021*, 114. [CrossRef]
21. Susskind, L. Addendum to fast scramblers. *arXiv* **2011**, arXiv:1101.6048.
22. Gibbons, G.W.; Hawking, S.W. Cosmological event horizons, thermodynamics, and particle creation. In *Euclidean Quantum Gravity*; World Scientific: Singapore, 1993; pp. 281–294.
23. Bedroya, A.; Brandenberger, R.; Loverde, M.; Vafa, C. Trans-Planckian censorship and inflationary cosmology. *Phys. Rev. D* **2020**, *101*, 103502. [CrossRef]
24. Solodukhin, S.N. Entanglement entropy of black holes. *Living Rev. Relativ.* **2011**, *14*, 8. [CrossRef] [PubMed]
25. Casini, H.; Huerta, M. Entanglement entropy in free quantum field theory. *J. Phys. A Math. Theor.* **2009**, *42*, 504007. [CrossRef]
26. Maldacena, J.; Pimentel, G.L. Entanglement entropy in de Sitter space. *J. High Energy Phys.* **2013**, *2013*, 38. [CrossRef]
27. Giataganas, D.; Tetradis, N. Entanglement entropy in FRW backgrounds. *Phys. Lett. B* **2021**, *820*, 136493. [CrossRef]
28. Giantsos, V.; Tetradis, N. Entanglement entropy in a four-dimensional cosmological background. *arXiv* **2022**, arXiv:2203.06699.
29. Balasubramanian, V.; McDermott, M.B.; Van Raamsdonk, M. Momentum-space entanglement and renormalization in quantum field theory. *Phys. Rev. D* **2012**, *86*, 045014. [CrossRef]
30. Brahma, S.; Alaryani, O.; Brandenberger, R. Entanglement entropy of cosmological perturbations. *Phys. Rev. D* **2020**, *102*, 043529. [CrossRef]
31. Albrecht, A.; Ferreira, P.; Joyce, M.; Prokopec, T. Inflation and squeezed quantum states. *Phys. Rev. D* **1994**, *50*, 4807. [CrossRef]
32. Brahma, S.; Berera, A.; Calderón-Figueroa, J. Universal signature of quantum entanglement across cosmological distances. *arXiv* **2021**, arXiv:2107.06910.
33. Maldacena, J. Non-Gaussian features of primordial fluctuations in single field inflationary models. *J. High Energy Phys.* **2003**, *2003*, 13. [CrossRef]
34. Bousso, R. A covariant entropy conjecture. *J. High Energy Phys.* **1999**, *1999*, 4. [CrossRef]
35. Bhattacharyya, A.; Das, S.; Haque, S.S.; Underwood, B. Rise of cosmological complexity: Saturation of growth and chaos. *Phys. Rev. Res.* **2020**, *2*, 033273. [CrossRef]
36. Brandenberger, R. Initial conditions for inflation—A short review. *Int. J. Mod. Phys. D* **2017**, *26*, 1740002. [CrossRef]
37. Banks, T. Cosmological breaking of supersymmetry? *Int. J. Mod. Phys. A* **2001**, *16*, 910–921. [CrossRef]
38. Banks, T.; Fischler, W. An upper bound on the number of e-foldings. *arXiv* **2003**, arXiv:astro-ph/0307459.
39. Brahma, S.; Brandenberger, R.; Wang, Z. Entanglement entropy of cosmological perturbations for S-brane Ekpyrosis. *J. Cosmol. Astropart. Phys.* **2021**, *2021*, 94. [CrossRef]

Article

Phenomenological Inflationary Model in Supersymmetric Quantum Cosmology

Nephtalí E. Martínez-Pérez [1,†], Cupatitzio Ramírez-Romero [1,*,†] and Víctor M. Vázquez-Báez [2,†]

[1] Facultad de Ciencias Físico Matemáticas, Benemérita Universidad Autónoma de Puebla, Puebla 72000, Mexico
[2] Facultad de Ingeniería, Benemérita Universidad Autónoma de Puebla, Puebla 72000, Mexico
* Correspondence: cramirez@fcfm.buap.mx
† These authors contributed equally to this work.

Abstract: We consider the effective evolution of a phenomenological model from FLRW supersymmetric quantum cosmology with a scalar field. The scalar field acts as a clock and inflaton. We examine a family of simple superpotentials that produce an inflation whose virtual effect on inhomogeneous fluctuations shows very good agreement with PLANCK observational evidence for the tensor-to-scalar ratio and the scalar spectral index.

Keywords: quantum cosmology; local supersymmetry; cosmic inflation

1. Introduction

It is widely accepted that the observable universe originates from an early homogeneous phase beginning presumably around the Planck scale, just after a less understood phase of quantum spacetime. This homogeneous phase has a classical description by an effective FLRW model with scalar matter. The subsequent inhomogeneity is attributed to quantum fluctuations of space that induce matter inhomogeneities that grow, and loss coherence due to inflation. Thus, a quantum treatment seems to be natural for the homogeneous phase.

Standard quantum cosmology is based on the canonical quantization of general relativity [1]. Following Dirac, the Hamiltonian constraint, generator of time reparametrizations, is implemented as a time-independent Schrödinger equation, the Wheeler–DeWitt (WDW) equation. Additionally, the solution to the WDW equation, i.e., the wave function of the universe, must be supplemented with a prescription to compute probabilities considering the singular character of the universe. Further, as the Hamiltonian operator annihilates the wave function, this is a timeless theory. However, the universe exhibits a marked one-way evolution (arrow of time), and internal observers have clocks [2–5].

In recent years, there have been several proposals in standard quantum cosmology that mainly relate time to the scale factor, essentially by a gauge fixing. These proposals involve mostly approximations of the semiclassical type, as well as Born–Openheimer with a weakly coupled matter sector and a large-scale factor. In [6], a general gauge fixing was analyzed, which was applied in [7] to several minisuperspace models with a classical 'extrinsic' time. In [8], a Born–Oppenheimer approximation was used to separate the scale factor from a weakly coupled matter sector. This allowed for defining a time as in the WKB approach, proportional to the square of the scale factor, and parametrize matter evolution with an effective Hamiltonian; see also [9,10]. Further, in [8] and a subsequent series of works, increasingly complex settings were analyzed, an inflaton with a mass term in [8], additional generic matter in [11], Mukhanov–Sasaki scalar perturbations for a de Sitter evolution in [12], and tensor perturbations in a general slow-roll inflationary setting in [13]. The parameter values of the last work were restricted in [14] by comparison with observational data, and in [15], they were considered effects on the spectra of primordial perturbations. In [16] the consequences of the interference of the wave functions before and

after a bounce were analyzed. For supersymmetric quantum cosmology, in [9,10,17], in the semiclassical approach, the effect of supersymmetry was explored for the solution.

The purely bosonic Wheeler–DeWitt equation is a second-order partial differential equation (PDE) in (mini)superspace. To single out a solution that corresponds to the wave function of the universe, one must impose suitable boundary conditions. Defining the appropriate boundary conditions that give rise to the universe with which we are familiar constitutes a well-known fundamental problem in quantum cosmology. There is also the possibility that a more complete theory introduces additional restrictions that uniquely determine a quantum state. An example of this is quantum supersymmetric cosmology; see, e.g., [9,10]. In this case, the most general state has multiple components, and the Hamiltonian constraint amounts to a system of coupled second-order PDEs that is equivalent to a system of first-order PDEs, the supersymmetric constraints. In [18], we worked out a supersymmetric model leading to an analytic solution to the WDW equation. In the present work, with a slightly simpler WDW equation by a different operator ordering, we consider the time choice of [18] by analyzing the wave function. The general expression of this wave function allows for the identification of a curve of most probable values in configuration space (superspace), parametrized by the scalar field, and suggests to choose it as time. Such a type of choice has been known for a long time [19]. Here, we analyze in detail the wave function regarding the choice in [18], of an effective time-dependent wave function with a probability density that corresponds to the conditional probability of measuring a value of the scale factor for a given value of the scalar field. Thus, the mean values computed with this wave function were time-dependent, in particular for the scale factor, which gives a trajectory that follows closely the previously mentioned curve of most probable values. Further, we consider the quantum cosmology of this approach, and explore a family of inflationary solutions.

We consider supersymmetric models because, even if at LHC energies, supersymmetry has not been found, it might be broken at higher energies, and it continues to be part of important candidates of ultraviolet completions for quantum gravity, supergravity, and string theory. Supersymmetry nontrivially relates fermions and bosons; in a supersymmetric quantum field theory, fermionic and bosonic divergences are cancelled [20]. Thus, supersymmetric quantum cosmology [21–24] is a relevant option for the study of quantum cosmology. Supersymmetry can be formulated by the extension of spacetime translations to translations in a Grassmann-extended spacetime, which includes fermionic coordinates, called superspace[1]. The fields on this supersymmetry superspace are called superfields, and supergravity can be formulated as a general relativity theory on a supermanifold [20,25]. There are several formulations for supersymmetric extensions of homogeneous cosmological models [22,23]. One class of such formulations comes from a dimensional reduction of four- or higher-dimensional supergravity theories by considering homogeneous fields depending only on time and integrating the space coordinates [26]. The other class is obtained by supersymmetric extensions of homogeneous models, invariant under general reparametrizations of time [27–29], or invariant under general reparametrizations on a superspace, with anticommutative coordinates besides time [30,31]. In [18,31], we followed [25], where one of us gave a geometric Lorentz covariant superfield model building tool for supergravity that can be straightforwardly applied to any number of spacetime dimensions. In particular, a homogeneous formulation could be straightforwardly traced back to four-dimensional supergravity.

As usual in theories with fermionic degrees of freedom, their conjugate momenta are eliminated by solving a number of second-class constraints, leaving an algebra of Dirac brackets. The corresponding quantum fermionic operators can be represented in several ways, e.g., by a matrix representation [21,32], representing half the fermionic degrees of freedom by either differential operators [33], or by creation operators acting on a vacuum state [18,29]. The wave function is spinorial in the first case, or a finite expansion of possible states in the second case. In any case, the Wheeler–DeWitt equation is equivalent to a system of first-order partial differential equations. In many cases, they have exact solutions [30–32], and the integration constants can be assimilated in the normalization. Therefore no initial conditions are required, although the state depends on the model. For supersymmetric extensions of higher-order theories, such as $f(R)$ theories, the differential equations might not be first-order [34].

In Section 2, we review the superfield formulation of supersymmetric cosmology. In Section 3, we review the quantization of the models from [18]; supersymmetric Wheeler–DeWitt equations have an analytic solution that depends on the scale factor and the superpotential. In Section 4, we discuss the problem of time. The identification of high-probability paths in configuration space to which mean trajectories correspond leads to the identification of the scalar field as time. Thus, following [18], a time-dependent effective wave function can be given. This effective wave function allows for computing mean values of the scale factor that give a time evolution. This scale factor is inversely proportional to the cubic root of the superpotential, and we obtain inflationary behavior for a family of superpotentials, as shown in Section 5. These superpotentials depend on three parameters, namely, μ, λ, and p; the first determines the time scale, the second the number of e-folds, and the last is a power that modifies the initial conditions of the scale factor. After fixing parameters μ and λ, we compute the tensor/scalar ratio and the scalar spectral index with remarkably good agreement with the observational bounds of PLANCK observatory without having to adjust parameters for this purpose. Lastly, in Section 6, we present a short discussion with some remarks and future work perspectives. In Appendix A, we give the effective wave function and the scale factor for $k=1$.

2. Supersymmetric FLRW Model with a Scalar Field

The large-scale observable universe has been modelled in general relativity by the FLRW metric with scalar fields. This is quite a general setting that could follow from a fundamental theory, and can account for inflation, primordial matter generation and structure formation, and dark energy. We consider the most studied model, the simplest, with a single minimally coupled scalar field $\frac{1}{2\kappa^2} \int \sqrt{-g} R d^4x + \int \sqrt{-g}[\frac{1}{2}\partial^\mu \phi \partial_\mu - V(\phi)]d^4x$, where $\kappa^2 = \frac{8\pi G}{c^4}$. For the FLRW metric, it reduces to well-known form

$$I = \frac{1}{\kappa^2} \int \left\{ -\frac{3}{c^2} N^{-1} a \dot{a}^2 + 3Nka - Na^3\Lambda + \kappa^2 a^3 \left[\frac{1}{2c^2} N^{-1} \dot{\phi}^2 - NV(\phi)\right] \right\} dt. \quad (1)$$

This Lagrangian is invariant under general time reparametrizations. From this action, follow the Friedmann equations and the conservation equation for a perfect fluid described by scalar field $\phi(t)$, i.e., in natural units and comoving gauge, $\frac{\dot{a}^2}{a^2} - \frac{\Lambda}{3} + \frac{k}{a^2} = \frac{\kappa^2}{3}\rho$, $\frac{2\ddot{a}}{a} + \frac{\dot{a}^2}{a^2} - \Lambda + \frac{k}{a^2} = -\kappa^2 p$ and $\dot{\rho} + \frac{3\dot{a}}{a}(p+\rho) = 0$, with $\rho = \frac{1}{2}\dot{\phi}^2 + V(\phi)$ the energy density, and $p = \frac{1}{2}\dot{\phi}^2 - V(\phi)$ the pressure for the perfect fluid $\phi(t)$. The momenta are $\pi_a = -\frac{6}{c^2\kappa^2} N^{-1} a \dot{a}$ and $\pi_\phi = -\frac{1}{c^2} N^{-1} a^3 \dot{\phi}$. The Hamiltonian is $H = NH_0$, where H_0 is the Hamiltonian constraint, which generates time reparametrizations.

2.1. Supersymmetric Cosmology

Supergravity, the supersymmetric version of gravity, can be formulated as general relativity in superspace, an extension of spacetime by anticommutative spinorial variables, in four dimensions, $x^m \rightarrow (x^m, \theta_\alpha, \bar{\theta}^{\dot{\alpha}})$, where θ are Weyl spinors and $\bar{\theta}$ their conjugates [20], $\alpha, \dot{\alpha} = 1, 2$. Supersymmetric field theory on superspace realizes supersymmetry algebra $\{Q_\alpha, \bar{Q}_{\dot{\alpha}}\} = 2i\sigma^m_{\alpha\dot{\alpha}} \partial_m$, which extends the Poincaré algebra. Thus, for homogeneous fields, the charges decompose into two copies

$$\{Q_\alpha, \bar{Q}_{\dot{\alpha}}\} = 2i\sigma^0_{\alpha\dot{\alpha}} \partial_0 = 2i\delta_{\alpha\dot{\alpha}} \partial_0. \quad (2)$$

Therefore, a minimal version of homogeneous supersymmetric field theory can be given by extending time by one complex anticommuting coordinate, which amounts to supersymmetric quantum mechanics. This theory can be obtained, in general, by dimensional reduction from higher-dimensional models to one (time) dimension [35]. Thus, supersymmetric cosmology can be obtained from one-dimensional supergravity [30].

For the sake of clarity, here we shortly review the derivation of the supersymmetric Wheeler–DeWitt equation following [18,31]. In these works, we formulated it as general relativity on supersymmetry–superspace, $t \to z^M = (t, \Theta, \bar{\Theta})$, where Θ and $\bar{\Theta}$ are the anticommuting coordinates, the so-called "new" Θ-variables [20]. Hence, under $z^M \to z'^M = z^M + \zeta^M(z)$, the superfields, see e.g., [20,25], transform as $\delta_\zeta \Phi(z) = -\zeta^M(z) \partial_M \Phi(z)$, and their covariant derivatives are $\nabla_A \Phi = \nabla_A^M(z) \partial_M \Phi$. $\nabla_A^M(z)$ is the superspace vielbein, whose superdeterminant gives the invariant superdensity $\mathcal{E} = \operatorname{Sdet} \nabla_M{}^A$, $\delta_\zeta \mathcal{E} = (-1)^m \partial_M(\zeta^M \mathcal{E})$. For the supersymmetric extension of the FLRW metric, we have $\mathcal{E} = -N - \frac{i}{2}(\Theta \bar{\psi} + \bar{\Theta} \psi)$ [31]. In this formulation, to the scale factor and the scalar field correspond real scalar superfields [20,30].

$$\mathcal{A}(t, \Theta, \bar{\Theta}) = a(t) + \Theta \lambda(t) - \bar{\Theta} \bar{\lambda}(t) + \Theta \bar{\Theta} B(t), \tag{3}$$

$$\Phi(t, \Theta, \bar{\Theta}) = \phi(t) + \Theta \eta(t) - \bar{\Theta} \bar{\eta}(t) + \Theta \bar{\Theta} G(t). \tag{4}$$

The supersymmetric extension of Action (1) for $k = 0, 1$ is $I = I_G + I_M$, where I_G is the supergravity action, and I_M is the matter term [18,30,36]

$$I_G = \frac{3}{\kappa^2} \int \mathcal{E} \left(\mathcal{A} \nabla_{\bar{\Theta}} \mathcal{A} \nabla_\Theta \mathcal{A} - \sqrt{k} \mathcal{A}^2 \right) d\Theta d\bar{\Theta} dt, \tag{5}$$

$$I_M = \int \mathcal{E} \mathcal{A}^3 \left[-\frac{1}{2} \nabla_{\bar{\Theta}} \Phi \nabla_\Theta \Phi + W(\Phi) \right] d\Theta d\bar{\Theta} dt, \tag{6}$$

where W is the superpotential.

2.2. Component Formulation

The component action follows from (5) and (6). After performing the Grassmann integrals, integrating out auxiliary fields B and G, and producing the redefinitions $\lambda \to a^{1/2} \lambda$, $\bar{\lambda} \to a^{1/2} \bar{\lambda}$, $\eta \to a^{3/2} \eta$, and $\bar{\eta} \to a^{3/2} \bar{\eta}$, the Lagrangian reads, see [18]

$$L = -\frac{3a\dot{a}^2}{c^2 N \kappa^2} + \frac{a^3 \dot{\phi}^2}{2c^2 N} + \frac{3kNa}{\kappa^2} - 3\sqrt{k} Na^2 W + \frac{3N\kappa^2}{4} a^3 W^2 - \frac{1}{2} a^3 NW'^2 + \frac{3i}{c\kappa^2}\left(\lambda \dot{\bar{\lambda}} + \bar{\lambda}\dot{\lambda}\right) - \frac{i}{2c}(\eta \dot{\bar{\eta}} + \bar{\eta}\dot{\eta})$$

$$+ \frac{3\sqrt{a}\dot{a}}{c\kappa^2 N}(\psi \lambda - \bar{\psi}\bar{\lambda}) - \frac{a^{3/2}\dot{\phi}}{2cN}(\psi \eta - \bar{\psi}\bar{\eta}) + \frac{3i\dot{\phi}}{2c}(\lambda \bar{\eta} + \bar{\lambda} \eta) + 3N\left(\frac{\sqrt{k}}{\kappa^2 a} - \frac{3}{2}W\right) \lambda \bar{\lambda} + N\left(-\frac{3\sqrt{k}}{2a} + \frac{3\kappa^2}{4}W - W''\right) \eta \bar{\eta}$$

$$+ 3ia^{3/2}\left(\frac{\sqrt{k}}{\kappa^2 a} - \frac{1}{2}W\right)(\psi \lambda + \bar{\psi}\bar{\lambda}) - \frac{ia^{3/2}}{2}W'(\psi \eta + \bar{\psi}\bar{\eta}) - \frac{3NW'}{2}(\lambda \bar{\eta} - \bar{\lambda}\eta) - \frac{3}{2N\kappa^2}\psi\bar{\psi}\lambda\bar{\lambda} + \frac{1}{4N}\psi\bar{\psi}\eta\bar{\eta},$$

where $W \equiv W(\phi)$, $W' \equiv \partial_\phi W(\phi)$, and $W'' \equiv \partial_\phi^2 W(\phi)$.

The Hamiltonian is of the form $H = NH_0 + \frac{1}{2}\psi S - \frac{1}{2}\bar{\psi}\bar{S}$, with the components of the one-dimensional supergravity multiplet $(N, \psi, \bar{\psi})$ as Lagrangean multipliers enforcing the Hamiltonian ($H_0 \approx 0$) and supersymmetric ($S \approx 0$, $\bar{S} \approx 0$) constraints [18].

The basic nonvanishing Dirac brackets are $\{a, \pi_a\} = \{\phi, \pi_\phi\} = 1$, $\{\lambda, \bar{\lambda}\}_+ = \frac{c\kappa^2}{6}$, $\{\eta, \bar{\eta}\}_+ = -c$. With these brackets, one can verify the following algebra of constraints

$$\{S, \bar{S}\}_+ = -2H_0, \tag{7}$$

$$\{H_0, S\} = \{H_0, \bar{S}\} = 0. \tag{8}$$

The scalar potential in the Hamiltonian H_0 is [18]

$$V_S = \frac{3\sqrt{k}}{a}W - \frac{3\kappa^2}{4}W^2 + \frac{1}{2}W'^2. \tag{9}$$

For $k = 0$, the sign of the superpotential does not matter for the scalar potential.

3. Quantization

Homogeneous cosmology is a mechanical system; hence, it can be quantized with the formalism of ordinary quantum mechanics. There are, however, several well-known problems. On the one hand, the probabilistic character of measurements in quantum mechanics clashes with the uniqueness of the system as there is no ensemble of universes to perform a series of tests, in identical, observer shaped conditions [37]. Nevertheless, observables such as the Hubble parameter can be determined by a set of observations. On the other hand, since the Hamiltonian vanishes, a time parameter cannot be introduced by means of the Schrödinger equation, or for the wave function or for the observables. Nonetheless, the Wheeler–DeWitt equation gives a time-independent Schrödinger equation with zero eigenvalue whose solution depends on the superspace variables, the minisuperspace in the homogeneous case, and gives the probability amplitude for the universe to be found in certain superspace configurations. The fact that the theory does not give a time evolution is the well-known consequence of invariance under time reparametrizations. Time is argued to be an internal property that can be determined by the choice of a clock [3]. On the other side, the observed universe is classical [37]; hence, its description is given by mean values of the quantum operators. We further discuss the time problem in Section 4.

3.1. Supersymmetric Wheeler–DeWitt Equations

For the derivation of supersymmetric Wheeler–DeWitt equations, we follow [18], but with a different ordering for fermions which yields somewhat simpler solutions. For consistency, the Hamiltonian operator must be Hermitian; hence, the supercharges must satisfy $\bar{S} = S^\dagger$ and $S = \bar{S}^\dagger$. The only nonzero (anti)commutators are

$$[a, \pi_a] = [\phi, \pi_\phi] = i\hbar, \quad \{\lambda, \bar{\lambda}\} = \frac{4\pi}{3} l_P^2, \quad \{\eta, \bar{\eta}\} = -\hbar c, \quad (10)$$

where $l_P^2 = \frac{\hbar G}{c^3}$ is the Planck length. For the quantization, we redefine the fermionic degrees of freedom as $\lambda = \sqrt{\frac{\hbar c \kappa^2}{6}} \alpha$, $\bar{\lambda} = \sqrt{\frac{\hbar c \kappa^2}{6}} \bar{\alpha}$, $\eta = \sqrt{\hbar c} \beta$ and $\bar{\eta} = \sqrt{\hbar c} \bar{\beta}$. Hence, the anticommutators are

$$\{\alpha, \bar{\alpha}\} = 1, \quad \{\beta, \bar{\beta}\} = -1. \quad (11)$$

as well as $\alpha^2 = \beta^2 = \bar{\alpha}^2 = \bar{\beta}^2 = 0$. The bosonic momenta are represented by derivatives, α and β are annihilation operators, and $\bar{\alpha}$ and $\bar{\beta}$ are creation operators. We fixed the ordering ambiguities by Weyl ordering, which is antisymmetric for fermions. Hence, the supersymmetric constraint operators read

$$\frac{1}{\sqrt{\hbar c}} S = \frac{c\kappa}{2\sqrt{6}} \left(a^{-\frac{1}{2}} \pi_a + \pi_a a^{-\frac{1}{2}} \right) \alpha + c a^{-\frac{3}{2}} \pi_\phi \beta + \frac{3i\kappa}{\sqrt{6}} a^{\frac{3}{2}} W \alpha + i a^{\frac{3}{2}} W' \beta - i \frac{\sqrt{6k}}{\kappa} a^{\frac{1}{2}} \bar{\alpha} - \frac{i\sqrt{3}}{4\sqrt{2}} \hbar c \kappa a^{-\frac{3}{2}} \alpha [\bar{\beta}, \beta], \quad (12)$$

$$\frac{1}{\sqrt{\hbar c}} \bar{S} = \frac{c\kappa}{2\sqrt{6}} \left(a^{-\frac{1}{2}} \pi_a + \pi_a a^{-\frac{1}{2}} \right) \bar{\alpha} + c a^{-\frac{3}{2}} \pi_\phi \bar{\beta} - \frac{3i\kappa}{\sqrt{6}} a^{\frac{3}{2}} W \bar{\alpha} - i a^{\frac{3}{2}} W' \bar{\beta} + i \frac{\sqrt{6k}}{\kappa} a^{\frac{1}{2}} \bar{\alpha} + \frac{i\sqrt{3}}{4\sqrt{2}} \hbar c \kappa a^{-\frac{3}{2}} \bar{\alpha} [\bar{\beta}, \beta]. \quad (13)$$

Anticommutator $\{S, \bar{S}\} = -2\hbar c H_0$ gives the quantum Hamiltonian

$$H_0 = -\frac{c^2 \kappa^2}{24} \left(a^{-1} \pi_a^2 + \pi_a^2 a^{-1} \right) + \frac{c^2}{2} a^{-3} \pi_\phi^2 - \frac{\sqrt{3}i}{2\sqrt{2}} \hbar c^2 \kappa a^{-3} \pi_\phi (\alpha \bar{\beta} + \bar{\alpha} \beta) - \frac{3k}{\kappa^2} a - \frac{\sqrt{k}}{4} \hbar c a^{-1} [\alpha, \bar{\alpha}] + \frac{3\sqrt{k}}{4} \hbar c a^{-1} [\beta, \bar{\beta}]$$
$$- \frac{3\kappa^2}{4} a^3 W^2 + 3\sqrt{k} a^2 W + \frac{1}{2} a^3 W'^2 + \frac{3}{8} \hbar c \kappa^2 W [\alpha, \bar{\alpha}] - \frac{3}{8} \hbar c \kappa^2 W [\beta, \bar{\beta}] + \frac{\sqrt{3}}{2\sqrt{2}} \hbar c \kappa W' (\alpha \bar{\beta} - \bar{\alpha} \beta) + \frac{1}{2} \hbar c W'' [\beta, \bar{\beta}]$$
$$+ \frac{3}{16} (\hbar c \kappa)^2 a^{-3} (\bar{\alpha} \alpha \beta \bar{\beta} + \alpha \bar{\alpha} \bar{\beta} \beta). \quad (14)$$

The Hilbert space is generated from the vacuum state $|1\rangle$, which satisfies $\alpha|1\rangle = \beta|1\rangle = 0$. Hence, there are four orthogonal states

$$|1\rangle, \quad |2\rangle = \bar{\alpha}|1\rangle, \quad |3\rangle = \bar{\beta}|1\rangle \quad \text{and} \quad |4\rangle = \bar{\alpha}\bar{\beta}|1\rangle, \qquad (15)$$

which have norms $\langle 2|2\rangle = \langle 1|1\rangle$, $\langle 3|3\rangle = -\langle 1|1\rangle$ and $\langle 4|4\rangle = -\langle 1|1\rangle$. Hence, a general state has the form

$$|\psi\rangle = \psi_1(a,\phi)|1\rangle + \psi_2(a,\phi)|2\rangle + \psi_3(a,\phi)|3\rangle + \psi_4(a,\phi)|4\rangle. \qquad (16)$$

Therefore, from constraint equation $S|\psi\rangle = 0$, we obtain

$$a\left(\partial_a - \frac{3}{\hbar c}a^2 W + \frac{6\sqrt{k}}{\hbar c \kappa^2}a + \frac{1}{2}a^{-1}\right)\psi_2 - \frac{\sqrt{6}}{\kappa}\left(\partial_\phi - a^3 W'\right)\psi_3 = 0, \qquad (17)$$

$$\left(\partial_a - \frac{3}{\hbar c}a^2 W + \frac{6\sqrt{k}}{\hbar c \kappa^2}a - a^{-1}\right)\psi_4 = 0 \quad \text{and} \quad \left(\partial_\phi - \frac{1}{\hbar c}a^3 W'\right)\psi_4 = 0, \qquad (18)$$

while from $\bar{S}\psi = 0$

$$a\left(\partial_a + \frac{3}{\hbar c}a^2 W - \frac{6\sqrt{k}}{\hbar c \kappa^2}a + \frac{1}{2}a^{-1}\right)\psi_3 - \frac{\sqrt{6}}{\kappa}\left(\partial_\phi + a^3 W'\right)\psi_2 = 0, \qquad (19)$$

$$\left(\partial_a + \frac{3}{\hbar c}a^2 W - \frac{6\sqrt{k}}{\hbar c \kappa^2}a - a^{-1}\right)\psi_1 = 0, \quad \text{and} \quad \left(\partial_\phi + \frac{1}{\hbar c}a^3 W'\right)\psi_1 = 0, \qquad (20)$$

The terms with a^{-1} in (17)–(20) differ from those in [18] due to a different operator ordering in the Hamiltonian. In fact, the classical Hamiltonian in [18] has a term $a^{-1}\pi_a^2$ that can be ordered in many ways to give a Hermitian operator as $a^{-1}\pi_a^2 \to \frac{1}{2k+l}(ka^{-1}\pi_a^2 + l\pi_a^{-1}\pi_a + k\pi_a a^{-1})$.

3.2. Solutions

As the Wheeler–DeWitt equation is second-order, its solutions require boundary conditions. However, in supersymmetric theory, Equations (17)–(20) are first-order and have unique solutions that can be fixed by consistency and normalization. The equations for ψ_1 and ψ_4 can be straightforwardly solved yielding the unique solutions up to constant factors [30]

$$\psi_1(a,\phi) = Aa \exp\left[-\frac{1}{\hbar c}\left(a^3 W(\phi) - \frac{3\sqrt{k}a^2}{\kappa^2}\right)\right], \qquad (21)$$

$$\psi_4(a,\phi) = Aa \exp\left[\frac{1}{\hbar c}\left(a^3 W(\phi) - \frac{3\sqrt{k}a^2}{\kappa^2}\right)\right], \qquad (22)$$

The power of factor a of these solutions differs from the solutions in [18] due to the different operator ordering mentioned end of the preceding section. As shown in [18], the solutions of Equations (17) and (19) are not defined at $a = 0$ unless they are trivial. Thus, we chose the solutions $|\psi\rangle = C_1 \psi_1(a,\phi)|1\rangle + C_4 \psi_4(a,\phi)|4\rangle$, where C_1 and C_4 are arbitrary constants [18]. The norm of this state is

$$\langle\psi|\psi\rangle = \left[|C_1|^2 \int |\psi_1(a,\phi)|^2 da d\phi - |C_4|^2 \int |\psi_4(a,\phi)|^2 da d\phi\right]\langle 1|1\rangle. \qquad (23)$$

Classically, $a \geq 0$ and could be a problem for quantization, e.g., it could require an infinite wall [4]. However, solutions (21) and (22) already vanish at $a = 0$. For a positive superpotential, ψ_1 has a bell form and tends to zero as a increases; see Figure 1. In this case, solution ψ_2 must be set to be the trivial one. Oppositely, for a negative superpotential, ψ_2 tends to zero as a increases, and ψ_1 must be discarded. For $\phi \to \pm\infty$, the behavior of (21) and (22) depends on the form of the superpotential. Therefore, we considered only positive or negative superpotentials, and we chose $\langle 1|1 \rangle = 1$ for positive superpotentials, and $\langle 1|1 \rangle = -1$ for negative superpotentials. Hence,

$$|\psi\rangle = C\psi_1(a,\phi)|1\rangle, \quad \text{if } W(\phi) > 0, \tag{24}$$
$$|\psi\rangle = C\psi_4(a,\phi)|4\rangle, \quad \text{if } W(\phi) < 0. \tag{25}$$

From Expansions (16), and (15), we see that these states correspond to scalars. By construction, these states are invariant under supersymmetry transformations.

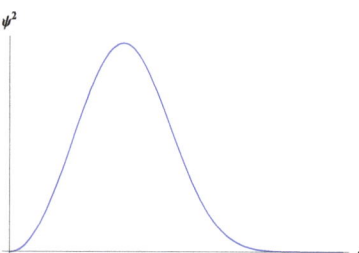

Figure 1. Profile of $\psi^2(a,\phi)$, for ϕ constant.

Further, for $\phi \to \pm\infty$, the behavior of (21) and (22) depends on the form of the superpotential. For a localized particle, the wave function has well-defined position probabilities and probability conservation. These conditions also guarantee hermicity of operators. On the other side, the wave functions of free particles do not vanish at infinity, but can be given a meaning by considering relative probabilities. If we restrict the superpotential to be an even function of ϕ, then operators π_ϕ and H_0 are self-adjoint even if the wave function does not vanish at $\phi \to \pm\infty$. Otherwise, if the wave function vanishes at $\phi \to 0$, the domain of ϕ can be taken to be $[0,\infty)$. A self-adjoint Hamiltonian constraint is consistent with the lack of evolution in the Heisenberg picture

$$\langle \psi| \frac{da}{dt} |\psi\rangle = \frac{i}{\hbar} \langle \psi| [H_0, a] |\psi\rangle = 0. \tag{26}$$

In the following, unless otherwise stated, we consider $k = 0$. In this case, we can write either (24) or (25) as

$$\psi(a,\phi) = Ca \exp\left[-\frac{1}{\hbar c} a^3 |W(\phi)|\right]. \tag{27}$$

This wave function differs from the one in [18], by the power of the a in front of the exponential, due to a different operator ordering, as mentioned in the preceding section.

In Appendix A we give the expressions for $k = 1$.

4. Time

Ordinary quantum mechanics assigns operators with real spectra to observables and probability amplitudes for their occurrence. A time dependence of the wave function is generated by the Hamiltonian operator. However, in an actually occurring state, a definite time entails certain energy indeterminacy. Hence, the energy spectrum cannot be restricted to one value, and time-dependent mean values arise by interference among different energy states. In general relativity, the Hamiltonian is constrained to vanish, and time is undetermined. This is consistent, since different energy states can manifest only if there is an environment that makes possible transitions among them. Nonetheless, the universe certainly has different components interacting at least gravitationally, and one such component, suitable chosen, can play the role of a clock [19]; see also [3,5].

In [18], we chose the scalar field as time parameter, and defined an effective time-dependent wave function that allows for computing time-dependent mean values. Here, we analyze Wave Function (27) and its probability density regarding the motivation of this choice of time.

We adopted the standard interpretation of quantum mechanics for the solutions of the Wheeler–DeWitt equation. Hence, the square modulus of the wave function gives the probability density for the possible three-geometries. Further, invariance under time reparametrizations ensures that the superspaces of each of the space slices are equivalent. Thus, this wave function describes, in a quantum mechanical sense, the space geometries of the whole spacetime. Furthermore, if it is possible to identify mean trajectories in superspace along regions around the maxima of the wave function, then it should be possible to parametrize these trajectories following the idea of Misner's supertime [38]. These trajectories should be around classical trajectories [37] corresponding to effective theories. Strictly speaking, measurements should give random values around these mean values.

In the model of this paper, the configuration space is given by the scale factor and the scalar field. Further, from (27), we see that, if we keep ϕ constant, the probability density of Wave Function (27) has the generic bell form of Figure 1, with the maximum at

$$a_{\max}(\phi) = \left[\frac{\hbar c}{3W(\phi)}\right]^{1/3}. \tag{28}$$

Allowing for ϕ to vary, these maxima trace out a curve of most probable values of the scale factor. The probability density along this curve is

$$\psi^2(a_{\max}, \phi) = e^{-2/3} a_{\max}^2(\phi), \tag{29}$$

Therefore, higher values of the scale factor along this curve have a higher relative probability, and we could think of an expanding universe for this wave function. On the other side, in quantum mechanics mean values are generally time-dependent, and we can speculate that later times correspond to higher probabilities here. Hence, time would increase monotonically with the scale factor. Nevertheless, $a_{\max}$ is driven by Scalar Field (28); hence, a natural choice for time is the scalar field, and the superpotential should ensure that $a_{\max}$ increases properly. Such a choice corresponds to a gauge where the scalar field is constant on the spacial slices. In this case, the universe is not localized in the ϕ direction, and the wave function is not normalizable in this direction, but relative probabilities can be considered, i.e., quotients of probabilities. As the value of ϕ is highly uncertain, a measurement gives random values, and we can ask then for the probability that the scale factor takes a value, which is given by the conditional probability of obtaining this value of a, given a value of ϕ [18]

$$|\Psi(a,\phi)|^2 = \frac{|\psi(a,\phi)|^2}{\int_0^\infty da\, |\psi(a,\phi)|^2}. \tag{30}$$

This probability must be used if there is a correlation between both fields, as required for a clock in [19], where it is called relative probability. Probability (30) can be obtained [18] from a wave function normalized for a given value of ϕ

$$\Psi(a,\phi) = \frac{1}{\sqrt{\int_0^\infty da\, |\psi(a,\phi)|^2}} \psi(a,\phi). \tag{31}$$

Thus, making $\kappa\phi \to t/\mu$ where μ accounts for the time scale, and taking Solution (27), Wave Function (31) becomes

$$\Psi(a,t) = \sqrt{\frac{6|W(t)|}{\hbar c}}\, a \exp\left[-\frac{a^3|W(t)|}{\hbar c}\right], \tag{32}$$

and satisfies a conservation equation [18]. Further, as the observed universe is classical, what we can give a meaning to, following the Ehrenfest theorem, is mean values. Thus, under the preceding ansatz, for the scale factor, we obtain [18]

$$a(t) = \int_0^\infty a|\Psi(a,t)|^2 da = \Gamma(4/3)\left[\frac{\hbar c}{2|W(t)|}\right]^{1/3}, \tag{33}$$

which is close to (28), as their quotient is $\Gamma(4/3)(3/2)^{1/3} \sim 1.02$.

The quantum fluctuations produce standard deviations that satisfy the Heisenberg relation [18].

In this work, we are not producing any approximation, semiclassical or of another type. As the fermionic degrees of freedom do not have classical counterparts, the mean values do not necessarily correspond to trajectories that approximate classical solutions, see e.g., [17]. Actually, (33) can be inserted into the Friedmann equations, from which a potential can be read out. If we consider the effective FLRW model with a scalar, obtained by inserting (33) into the Friedmann equations, the corresponding potential can be read out as a time function

$$\frac{1}{3c^4\kappa^2}\left(\frac{2W'^2}{W^2} - \frac{W''}{W}\right), \tag{34}$$

which can be compared with the scalar potential (9), which follows when the fermionic terms in the Hamiltonian are eliminated.

The results for this section can also be given analytically for $k = 1$; see Appendix A. They involve hypergeometric, AiryBi, and AiryBi' functions that have exponential behavior, and their numerical evaluation is troublesome. As we were interested in qualitative features here, we restricted ourselves to $k = 0$. In this case, considering that the sign of the superpotential did not have consequences for the wave function or the scalar potential, we chose the superpotential as positive definite in the following.

5. Inflationary Model

In this section, we consider a class of phenomenological superpotentials that give inflationary dynamics that satisfactorily agrees with the observational bounds. Potentials of this form appear, e.g., in string-inspired tachyon models. With these superpotentials, the wave function tends to zero as $\phi \to 0$, and we can restrict $\phi \geq 0$, as mentioned in Section 3.2. Thus, we consider universes with origin at $t = 0$, described by homogeneous quantum cosmology from its very beginning, although above Planck scale, full gravitational interactions become relevant, and some ultraviolet completion would be required.

In [39,40], we performed a qualitative discussion of several superpotentials that have diverse drawbacks, as phantom matter. Here, we consider a family of superpotentials

$$W(\phi) = \frac{c^4 M_p^3}{\hbar^2}\left[\frac{\sqrt{8\pi}\,\Gamma(4/3)}{2^{1/3}n}\right]^{1/3}\left\{\frac{(\kappa\phi)^{-p}}{\left[1+e^{(\kappa\phi)^{1/2}}\right]} + \lambda\right\}, \tag{35}$$

for $p > 0$. The wave function fulfils $\lim_{a\to 0}\psi(a,\phi) = 0$ and $\lim_{\phi\to 0}\psi(a,\phi) = 0$.

Thus, from (33), we have

$$a(t) = n\ell_p \left[\frac{t^p \left(1 + e^{\sqrt{t/\mu}}\right)}{\mu^p + \lambda t^p \left(1 + e^{\sqrt{t/\mu}}\right)} \right]^{1/3}, \qquad (36)$$

where ℓ_p is the Planck length. In the following, we set the normalization constant equal to one, $n = 1$, and we redefined the scale factor to be dimensionless, $a/\ell_p \to a$. Constant μ accounts for the time scale. With respect to λ, (33) shows that, if the value of W decreases, the scale factor grows, and, in order to finish this growth, W must stop decreasing; this is the role of λ, which must be positive. Thus, in order to have a sufficiently large increase in scale factor, λ must be small enough. This parameter could be seen as the remainder of a term responsible for dark energy in the superpotential

$$\frac{\lambda}{1 + e^{\beta(\phi - \phi_d)}}, \qquad (37)$$

we do not pursue this idea further in this work.

For the verification of the predictions of the evolution of Scale Factor (36), we consider $N = 60$ e folds and observational data of PLANCK observatory [41]

- Tensor-to-scalar ratio bound $r < 0.064$,
- Scalar spectral index $n_s = 0.9649 \pm 0.0042$.

The acceleration, as usual, can be written by identity

$$\ddot{a}(t) = aH^2(t)(1 - \varepsilon(t)), \qquad (38)$$

where $\varepsilon(t) = -\frac{\dot{H}(t)}{H^2(t)}$ is the slow roll parameter.

The enhancing effect of inflation on inhomogeneous quantum fluctuations is well-known [42]. As a consequence, the fluctuations last after inflation and produce structure seeds, and gravitational waves are generated in the process. These primordial effects last in the CMB, and their study led to stringent bounds on virtual predictions of inflationary models, such as the tensor-to-scalar ratio r and scalar spectral index n_s, that can be evaluated directly from the characteristics of the inflationary evolution.

The tensor-to-scalar ratio is the quotient of the tensor-to-scalar power spectra and is given by

$$r = 16\varepsilon(t). \qquad (39)$$

Scalar spectral index n_s follows from scalar power spectrum

$$\Delta_s^2(t) = \frac{1}{8\pi^2 M_P^2} \left. \frac{H^2(t)}{\varepsilon(t)} \right|_{a(t)H(t)=k}, \qquad (40)$$

as

$$n_s(t) = 1 + \left. \frac{d \ln \Delta_s(t)^2}{d \ln k} \right|_{a(t)H(t)=k}, \qquad (41)$$

where k are the wavenumbers of the scalar perturbations that, during inflation, exit the horizon at time t, i.e., $a(t)H(t) = \dot{a}(t) = k$. Thus, (41) can be written as

$$n_s(t) = 1 + k \frac{dt}{dk} \left. \frac{d \ln \Delta_s(t)^2}{dt} \right|_{a(t)H(t)=k} = 1 + \frac{\dot{a}(t)}{\ddot{a}(t)} \left. \frac{d \ln \Delta_s(t)^2}{dt} \right|_{a(t)H(t)=k}. \qquad (42)$$

which can also be written in terms of the Hubble parameter

$$n_s = \frac{\dot{H}(H^2 + 5\dot{H}) - H\ddot{H}}{\dot{H}(H^2 + \dot{H})}. \qquad (43)$$

Therefore, in order to estimate the values of r and n_s predicted by our model during inflation, we can evaluate them in the dependence of time.

A first approximation for λ can be obtained from the limit of very large times of the scale factor, $\lim_{t\to\infty} a(t) = 1/\lambda \sim e^N$. N are the e folds generated by inflation for the scale factor

$$e^N = \frac{a(t_{\text{exit}})}{a(t_{\text{in}})}, \tag{44}$$

where t_{in} and t_{exit} are the beginning and end times of inflation. Then, if we compute (44) with the scale factor (36), we can solve for λ

$$\lambda = \frac{\mu^p}{e^{3N}-1}\left[\frac{1}{t_{\text{in}}{}^p\left(1+e^{\sqrt{t_{\text{in}}/\mu}}\right)} - \frac{e^{3N}}{t_{\text{exit}}{}^p\left(1+e^{\sqrt{t_{\text{exit}}/\mu}}\right)}\right] \tag{45}$$

Further, as $\lambda > 0$, from (45) we obtain a condition for μ. For instance, as $t_{\text{in}} \ll t_{\text{exit}}$, if we set $t_{\text{in}} \ll \mu \ll t_{\text{exit}}$, we can approximate $e^{t_{\text{in}}/\mu} \ll 1$ and $e^{t_{\text{exit}}/\mu} \gg 1$. In this case,

$$\mu < \frac{t_{\text{exit}}}{[3N + p(\ln t_{\text{in}} - \ln t_{\text{exit}})]^2}. \tag{46}$$

Strictly speaking, we must still define how to define the entry and exit times of inflation depending on the evolution of the scale factor. Relations (45) and (46) help in checking the consistency for choices of μ and λ. With this in mind, we can fix the values of these parameters and evaluate the behavior in each case.

In order to get criteria for the entry and exit times, we first analyse the limits of the scale factor, its velocity, and acceleration as $t \to 0$. It turns out as follows. The scale factor tends zero for $p > 0$. The velocity tends to ∞ for $0 < p < 3$, for $p = 3$ it tends to $\frac{2^{1/3}}{\mu}$, and for $p > 3$ it tends to zero. The acceleration tends to $-\infty$ for $0 < p < 3$, for $3 \leq p < 6$ it tends to ∞, for $p = 6$ it tends to $\frac{2^{4/3}}{\mu^2}$, and for $p > 6$ it is zero. Additionally, for $t \to \infty$, the scale factor tends to $1/\lambda$, the velocity tends to 0^+, and the acceleration tends to 0^-.

Considering now the acceleration, if it is negative at $t = 0$, for $0 < p < 3$, it increases until it becomes zero and then positive; we took this zero as the entry time for inflation, see Figure 2. Further, the acceleration continues increasing until a maximum later, and after that becomes zero again, and then negative. We took this second zero as the exit time. Afterwards, the acceleration has a minimum, and then tends to zero from the negative numbers; see Figure 3. For cases $3 \leq p < 6$ where the acceleration tends to infinity at $t = 0$, it decreases to a very small minimum that can be taken as the entry time, see Figure 2. Further, as in the previous case, it has a maximum and then a zero, and we took it as exit time. For $p \geq 6$, the acceleration at $t = 0$ is equal to or larger than zero and increases from the beginning; hence, the entry time is zero, and, as $a(0) = 0$, there is no precise way to give an initial time for inflation. For this reason, we considered only $0 \leq p < 5$.

Further, as a first input, we took the initial time for inflation at a scale lower than the GUT scale 10^{15} GeV. Hence, $t_{\text{in}} \gtrsim 5 \times 10^6$ $T_P \approx 10^{-37}$ s. For the exit of inflation, we took the generally accepted interval $t \sim 10^{-33}$–10^{-32} s, and we set $t_R = 5 \times 10^{-33}$ s $= 3.7 \times 10^{10}$ T_P, as reference time for exit $t_{\text{exit}} \approx t_R$. With these times, for a given value of p, we estimated from (46) and set $\mu \sim \times 10^6$ T_P. From it, we computed λ from (45) for $N = 60$, and then evaluated the entry and exit times from the acceleration $\ddot{a}(t)$. Following this procedure, we adjusted μ and λ, so that $t_{\text{exit}} \approx t_R$ and $N = \ln a(t_{\text{exit}})/a(t_{\text{in}}) \approx 60$. With these data, we computed the tensor to scalar ratio (39), and the scalar spectral index (41). We performed these steps for $p = 1, 2, 3, 4, 5$.

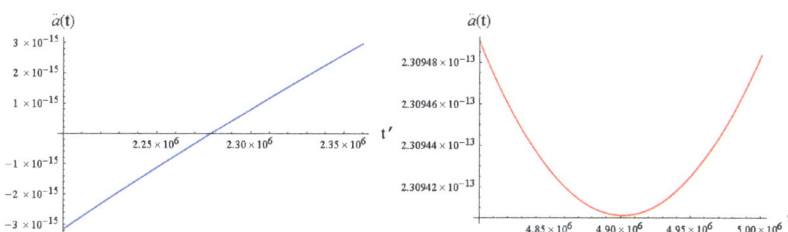

Figure 2. Inflation entry times for $p = 2$ and $p = 3$.

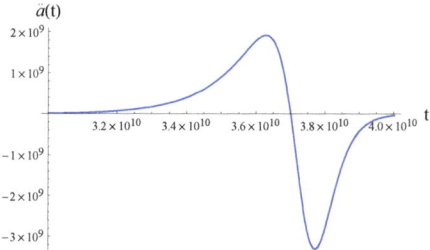

Figure 3. Inflation exit time for $p = 1$.

It is remarkable that, with the parameters fixed in the previous way, for given $0 < p \leq 5$, the entry time is of the order of 10^6 T$_P$, and the tensor/scalar ratio and the scalar spectral index agree quite well with the observational constrains.

Tensor/scalar Ratio (39) (see Figure 4) satisfies bound $r < 0.032$ [40] very well and depends slightly on p.

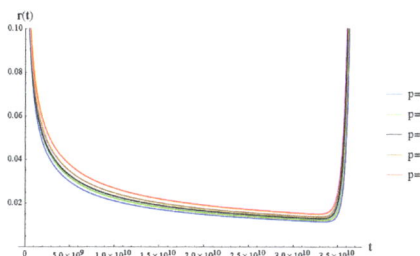

Figure 4. Tensor/scalar ratio.

Further, from (41), we computed the scalar spectral index whose behavior is shown in Figure 5.

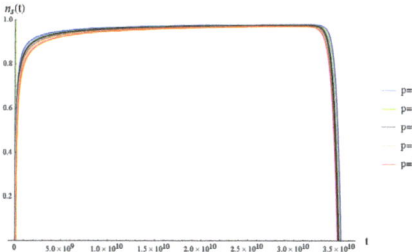

Figure 5. Scalar spectral index.

Here, the agreement with observational bound 0.9649 ± 0.0042 (with 68% CL) [41], was very good for most of the duration of inflation, as shown in Figure 6. It is also slightly dependent of p. Note that the running of the scalar spectral index $\frac{dn_s}{d \ln k}$ is positive, with a value around 0.0004, with a a negative running of running.

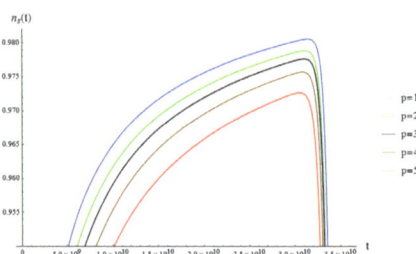

Figure 6. Detail of scalar spectral index.

6. Discussion

Quantum cosmology gives a canonical quantization of general relativity in the Schrödinger picture, with the Wheeler–DeWitt equation as a time-independent Schrödinger equation. The wave function gives the probability amplitudes for the occurrence of all possible spatial three-geometries and field distributions, at any time. On the other hand, the observed universe is classical and has time, and classical physics follows from quantum physics, hence an effective time evolution should follow from the quantum description. Further, the Wheeler–DeWitt equation is a second-order scalar equation. In the supersymmetric case, there is a system of first-order equations whose solutions are spinorial wave functions, see, e.g., [22,23]. A particular meaning of the components of these wave functions has not been given; it would quickly lead to a Machian discussion. In many cases, there are only two nonvanishing components, see, e.g., [32], given by real exponentials with opposite exponent sign. Here, we considered an FLRW theory with a minimally coupled scalar field. Single-field models are extremely effective to account for the inflationary era.

From the four components of the solution of the supersymmetric Wheeler–DeWitt equation, only one tends to zero as $a \to \infty$, and the other can be taken to be trivial. The form of Solution (27) suggests the ansatz that, in a certain gauge, time can be given by the scalar field, and trajectories for the observables are mean values on an effective wave function that corresponds to conditional probabilities Thus, for the scale factor, we obtain an evolution $a(t)$, and we can perform time reparametrizations considering that it is scalar $a'(t') = a(t)$. The resulting evolution is nonperturbative. These trajectories are classical if the quantum fluctuations are negligible. The time proposed here does not follow from a semiclassical approach.

We considered in this formulation one family of inflationary models that depends on the parameter $p > 0$; as for $p > 5$ there is eternal inflation, we restrict it to $p < 5$ and for simplicity consider only integer values. There are other two parameters, μ and λ; the first fixes the time scale. Parameter λ corresponds to the extent of the inflation, the e folds. We convened to set the entry of inflation, depending on the value of p, at zeros or minima of the acceleration, and exit times as zeros. In this way, for all p, we adjusted parameters μ and λ, so that the exit time coincided with a reference time $t_R = 5 \times 10^{-33} s$, and $N \approx 60$. With this setting, the corresponding tensor-to-scalar ratio and scalar spectral index can be computed with values that agree with the observational bounds remarkably well.

For indepth analysis, inhomogeneous perturbations should be introduced to evaluate their evolution and the the production of primordial gravitational waves. These computations should be performed at best in the superfield formalism. An interesting perspective is also the study of this formalism for dark energy.

Author Contributions: All authors have equally contributed to the research and manuscript preparation. All authors have read and agreed to the published version of the manuscript.

Funding: This research was funded by BUAP grant number 11171277-VIEP2021. N.E. Martínez-Pérez thanks CONACyT for studies grant.

Acknowledgments: C.R. thanks O. Obregón and H. García-Compeán, and V. V. thanks G. García-Arroyo and J.A. Vázquez for the interesting discussions.

Conflicts of Interest: The authors declare no conflict of interest.

Appendix A

In this appendix, we give, for $k = 1$, the expressions for the normalization factor for the wave function (21) for $k = 1$, and the time-dependent scale factor (33). The normalization factor of

$$\psi(a,\phi) = |C|a \exp\left[\frac{1}{\hbar c}\left(-a^3|W(\phi)| + \frac{3\sqrt{k}a^2}{\kappa^2}\right)\right], \tag{A1}$$

is given by

$$|C|^{-2} = \frac{c\hbar}{6}\int_{-\infty}^{\infty}\frac{1}{|W(\phi)|}\left[{}_2F_2\left(\frac{1}{2},1;\frac{1}{3},\frac{2}{3};\frac{8}{c\hbar\kappa^6 W^2(\phi)}\right) + 4\pi\left[\frac{2}{9c\hbar\kappa^6 W(\phi)^2}\right]^{1/3}e^{\frac{4}{c\hbar\kappa^6 W^2(\phi)}}\right.$$

$$\left.\times\left[\text{Bi}'\left(\left[\frac{6}{c\hbar\kappa^6 W(\phi)^2}\right]^{2/3}\right) + \left[\frac{6}{c\hbar\kappa^6 W(\phi)^2}\right]^{1/3}\text{Bi}\left(\left[\frac{6}{c\hbar\kappa^6 W(\phi)^2}\right]^{2/3}\right)\right]\right]d\phi, \tag{A2}$$

where Bi is the Airy function of second kind.

For the denominator of (31)

$$\int_0^\infty da\,|\psi(a,\phi)|^2 = \frac{c\hbar}{18W(\phi)}\times\left\{3\;{}_2F_2\left(\frac{1}{2},1;\frac{1}{3},\frac{2}{3};-\frac{8}{c\hbar\kappa^6 W(\phi)^2}\right) + 4\times 6^{1/3}\pi\left[\frac{1}{c\hbar\kappa^6 W(\phi)^2}\right]^{2/3}e^{-\frac{4}{c\hbar\kappa^6 W(\phi)^2}}\right.$$

$$\left.\times\left[6^{1/3}\text{Bi}\left(\left[\frac{6}{c\hbar\kappa^6 W(\phi)^2}\right]^{2/3}\right) - \left[c\hbar\kappa^6 W(\phi)^2\right]^{1/3}\text{Bi}'\left(\left[\frac{6}{c\hbar\kappa^6 W(\phi)^2}\right]^{2/3}\right)\right]\right\},$$

from which follows

$$a(t) = \left\{9\sqrt[3]{3}\kappa^4(c\hbar)^{2/3}W(t)^{4/3}\,{}_2F_2\left(1,\frac{3}{2};\frac{2}{3},\frac{4}{3};\frac{8}{c\kappa^6\hbar W(t)^2}\right)e^{-\frac{4}{c\hbar\kappa^6 W(t)^2}} + 2^{2/3}\pi\left[24 + c\kappa^6\hbar W(t)^2\right]\text{Bi}\left(\left[\frac{6}{c\hbar\kappa^6 W(t)^2}\right]^{2/3}\right)\right.$$

$$\left.+ 8\sqrt[3]{2}2^{2/3}\pi\kappa^2\sqrt[3]{c\hbar}W(t)^{2/3}\,\text{Bi}'\left(\left[\frac{6}{c\hbar\kappa^6 W(t)^2}\right]^{2/3}\right)\right\} / \left\{3\sqrt[3]{3}\kappa^6(c\hbar)^{2/3}W(t)^{7/3}\,{}_2F_2\left(\frac{1}{2},1;\frac{1}{3},\frac{2}{3};\frac{8}{c\kappa^6\hbar W(t)^2}\right)e^{-\frac{4}{c\kappa^6\hbar W(t)^2}}\right.$$

$$\left.+ 4\sqrt[3]{2}2^{2/3}\pi\kappa^4\sqrt[3]{c\hbar}W(t)^{5/3}\,\text{Bi}'\left(\left[\frac{6}{c\hbar\kappa^6 W(t)^2}\right]^{2/3}\right) + 12\cdot 2^{2/3}\pi\kappa^2 W(t)\,\text{Bi}\left(\left[\frac{6}{c\hbar\kappa^6 W(t)^2}\right]^{2/3}\right)\right\}.$$

Due to the exponential behavior of the hypergeometric and Airy functions that appear in the numerator and denominator in these expressions, it is difficult to handle them numerically.

Note

1 Should not be confused with the superspace of geometrodynamics.

References

1. Hartle, J.B.; Hawking, S.W. Wave function of the Universe. *Phys. Rev. D* **1983**, *28*, 2960. [CrossRef]
2. DeWitt, B.S. Dynamical Theory in Curved Spaces. I. A Review of the Classical and Quantum Action Principles. *Rev. Mod. Phys.* **1957**, *29*, 377. [CrossRef]
3. Kuchař, K.V. Time and interpretations of quantum gravity. *Int. J. Mod. Phys. D* **2011**, *20*, 3. 11019347. [CrossRef]
4. Isham, C.J. Canonical quantum gravity and the problem of time. In Proceeding of the NATO Advanced Study Institute "Recent Problems in Mathematical Physics", Salamanca, Spain, 15–27 June 1992. [CrossRef]
5. Anderson, E. Problem of time in quantum gravity. *Ann. Der Phys.* **2012**, *524*, 757. [CrossRef]
6. Barvinsky, A.O. Unitarity approach to quantum cosmology. *Phys. Rep.* **1993**, *230*, 237. [CrossRef]
7. Barvinsky, A.O.; Kamenshchik, A.Y. Selection rules for the Wheeler-DeWitt equation in quantum cosmology. *Phys. Rev. D* **2014**, *89*, 043526. [CrossRef]
8. Finelli, F.; Vacca, G.P.; Venturi, G. Chaotic inflation from a scalar field in non-classical states. *Phys. Rev. D* **1998**, *58*, 103514. [CrossRef]
9. Kiefer, C.; Lück, T.; Moniz, P. The semiclassical approximation to supersymmetric quantum gravity. *Phys. Rev. D* **2005**, *72*, 045006. [CrossRef]
10. Fathi, M.; Jalalzadeh, S.; Moniz, P.V. Classical Universe emerging from quantum cosmology without horizon and flatness problems. *Eur. Phys. J. C* **2016**, *76*, 527. [CrossRef]
11. Tronconi, A.; Vacca, G.P.; Venturi, G. The Inflaton and Time in the Matter-Gravity System. *Phys. Rev. D* **2003**, *67*, 063517. [CrossRef]
12. Kamenshchik, A.Y.; Tronconi, A.; Venturi, G. Inflation and Quantum Gravity in a Born-Oppenheimer Context. *Phys. Lett. B* **2013**, *726*, 518. [CrossRef]
13. Kamenshchik, A.Y.; Tronconi, A.; Venturi, G. Signatures of Quantum Gravity in a Born-Oppenheimer Context. *Phys. Lett. B* **2014**, *734*, 72. [CrossRef]
14. Kamenshchik, A.Y.; Tronconi, A.; Venturi, G. Quantum Gravity and the Large Scale Anomaly. *J. Cosmol. Astropart. Phys.* **2015**, *4*, 46. [CrossRef]
15. Kamenshchik, A.Y.; Tronconi, A.; Venturi, G. Quantum Cosmology and the Evolution of Inflationary Spectra. *Phys. Rev. D* **2016**, *94*, 123524. [CrossRef]
16. Kamenshchik, A.Y.; Tronconi, A.; Vardanyan, T.; Venturi, G. Quantum Gravity, Time, Bounces and Matter. *Phys. Rev. D* **2018**, *97*, 123517. [CrossRef]
17. Escamilla-Rivera, C.; Obregon, O.; Urena-Lopez, L.A. Supersymmetric classical cosmology. *J. Cosmol. Astropart. Phys.* **2010**, *12*, 11. [CrossRef]
18. Ramírez, C.; Vázquez-Bxaxez, V. Quantum supersymmetric FRW cosmology with a scalar field. *Phys. Rev. D* **2016**, *93*, 043505. [CrossRef]
19. Banks, T. TCP, Quantum Gravity, the Cosmological Constant, and all That. *Nucl. Phys. B* **1985**, *249*, 332. 0550-3213(85)90020-3. [CrossRef]
20. Wess, J.; Bagger, J. *Supersymmetry and Supergravity*; Princeton University Press: Princeton, PJ, USA, 1992.
21. Macías, A.; Obregón, O.; Ryan, M.P. Quantum cosmology: The supersymmetric square root. *Class. Quantum Gravity* **1987**, *4*, 1477. [CrossRef]
22. D'Eath, P.D. *Supersymmetric Quantum Cosmology*; Cambridge University Press: Cambridge, UK, 1996. CBO9780511524424. [CrossRef]
23. Moniz, P.V. *Quantum Cosmology-The Supersymmetric Perspective-Vol. 1. Fundamentals*; Springer: Berlin/Heidelberg, Germany, 2010; Volume 803. [CrossRef]
24. Moniz, P.V. *Quantum Cosmology-The Supersymmetric Perspective-Vol. 2: Advanced Topics*; Springer: Berlin/Heidelberg, Germany, 2010; Volume 804. [CrossRef]
25. Ramírez, C. The realizations of local supersymmetry. *Ann. Phys. N. Y.* **1988**, *186*, 43. [CrossRef]
26. Ryan, M. *Hamiltonian Cosmology*; Springer: Berlin/Heidelberg, Germany, 1972.
27. Graham, R. Supersymmetric Bianchi Type IX Cosmology. *Phys. Rev. Lett.* **1991**, *67*, 1381. [CrossRef]
28. Graham, R. Supersymmetric general Bianchi type IX cosmology with a cosmological term. *Phys. Lett. B* **1992**, *277*, 393. [CrossRef]
29. Bene, J.; Graham, R. Supersymmetric Homogeneous Quantum Cosmologies Coupled to a Scalar Field. *Phys. Rev. D* **1994**, *49*, 799. [CrossRef] [PubMed]
30. Obregón, O.; Rosales, J.J.; Socorro; Tkach, V.I. Supersymmetry breaking and a normalizable wavefunction for the FRW ($k = 0$) cosmological model. *Class. Quantum Grav.* **1999**, *16*, 2861. [CrossRef]
31. García-Jiménez, G.; Ramírez, C.; Vázquez-Báez, V. Tachyon potentials from a supersymmetric FRW model. *Phys. Rev. D* **2014**, *89*, 043501. [CrossRef]
32. Obregón, O.; Ramírez, C. Dirac like formulation of quantum supersymmetric cosmology. *Phys. Rev. D* **1998**, *57*, 1015. [CrossRef]
33. D'Eath, P.D.; Hughes, D.I. Supersymmetric Mini-Superspace. *Phys. Lett. B* **1988**, *214*, 498. [CrossRef]
34. Martínez-Pérez, N.E.; Ramírez, C.; Vázquez-Báez, V. 1D Supergravity FLRW Model of Starobinsky. *Universe* **2021**, *7*, 449. [CrossRef]
35. Claudson, M.; Halpern, M.B. Supersymmetric Wave State Wave Functions. *Nucl. Phys. B* **1985**, *250*, 689 . [CrossRef]
36. Obregón, O.; Rosales, J.J.; Tkach, V.I. Superfield description of the FRW universe. *Phys. Rev. D* **1996**, *53*, R1750. [CrossRef]

37. Halliwell, J.J. Decoherence in quantum cosmology. *Phys. Rev. D* **1989**, *39*, 2912. [CrossRef] [PubMed]
38. Misner, C.W. *Magic without Magic: John Archibald Wheeler: A Collection of Essays in Honor of His Sixtieth Birthday*; Klauder, J.R., Ed.; W. H. Freeman: San Francisco, CA, USA, 1972.
39. Martínez-Pérez, N.E.; Ramírez, C.; Vázquez-Báez, V. Inflationary evolution in quantum cosmology from FRLW supersymmetric models. *arXiv* **2021**, arXiv:2104.12914.
40. Tristram, M.; Banday, A.J.; Górski, K.M.; Keskitalo, R.; Lawrence, C.R.; Andersen, K.J.; Barreiro, R.B.; Borrill, J.; Eriksen, H.K.; Fernandez-Cobos, R.; et al. Planck constraints on the tensor-to-scalar ratio. *Astron. Astrophys.* **2021**, *647*, A128. [CrossRef]
41. Akrami, Y.; et al. [Planck Collaboration]. Planck 2018 results. X. Constraints on inflation. *Astron. Astrophys.* **2020**, *A10*, 641. [CrossRef]
42. Baumann, D. TASI Lectures on Inflation. *arXiv* **2009**, arXiv:0907.5424. Available online: https://arxiv.org/abs/0907.5424 (accessed on 21 february 2021).

MDPI AG
Grosspeteranlage 5
4052 Basel
Switzerland
Tel.: +41 61 683 77 34

Universe Editorial Office
E-mail: universe@mdpi.com
www.mdpi.com/journal/universe

Disclaimer/Publisher's Note: The statements, opinions and data contained in all publications are solely those of the individual author(s) and contributor(s) and not of MDPI and/or the editor(s). MDPI and/or the editor(s) disclaim responsibility for any injury to people or property resulting from any ideas, methods, instructions or products referred to in the content.

www.ingramcontent.com/pod-product-compliance
Lightning Source LLC
LaVergne TN
LVHW072339090526
838202LV00019B/2443